A Journey through the Universe

Providing an in-depth understanding for both general readers and astronomy enthusiasts, this highly comprehensive book provides an up-to-date survey of our knowledge of the Universe.

The book explores our Solar System, its planets and other bodies; examines the Sun and how it and other stars evolve through their lifetimes; discusses the search for planets beyond our Solar System and how we might detect life on them; and highlights interesting objects found within our Galaxy, the Milky Way. It also looks at our current understanding of the origin and evolution of the Universe, as well as many other intriguing topics, such as time, black holes and Einstein's theories, dark matter, dark energy and the Cosmic Microwave Background.

The book is uniquely supported by video lectures given by the author, available online. It also includes the very latest astronomical observations, such as those made by the Planck and Kepler spacecraft.

IAN MORISON spent his professional career as a radio astronomer at the Jodrell Bank Observatory, and he has had an asteroid named in his honour in recognition of his work. In 2007 he was appointed Professor of Astronomy at Gresham College, the oldest chair of astronomy in the world. He writes a monthly online sky guide and audio podcast for the Jodrell Bank Observatory and is the author of numerous articles for the astronomical press and of a university astronomy textbook. His most recent book is *An Amateur's Guide to Observing and Imaging the Heavens* (Cambridge University Press, 2014).

A Journey through the Universe

Gresham Lectures on Astronomy

IAN MORISON
University of Manchester and Gresham College

University Printing House, Cambridge CB2 8BS, United Kingdom

Cambridge University Press is part of the University of Cambridge.

It furthers the University's mission by disseminating knowledge in the pursuit of education, learning and research at the highest international levels of excellence.

www.cambridge.org
Information on this title: www.cambridge.org/9781107073463

© I. Morison 2015

This publication is in copyright. Subject to statutory exception and to the provisions of relevant collective licensing agreements, no reproduction of any part may take place without the written permission of Cambridge University Press.

First published 2015

Printed in the United Kingdom by TJ International Ltd. Padstow, Cornwall

A catalogue record for this publication is available from the British Library

Library of Congress Cataloguing in Publication data
Morison, Ian, 1943– author.
[Lectures. Selections]
A journey through the universe : Gresham lectures on astronomy / Ian Morison, University of Manchester and Gresham College.
    pages   cm
ISBN 978-1-107-07346-3 (hardback)
1. Astronomy.   I. Title.
QB51.M77   2015
520–dc23
                                                                                    2014016830

ISBN 978-1-107-07346-3 Hardback

Additional resources for this publication at www.cambridge.org/9781107073463

Cambridge University Press has no responsibility for the persistence or accuracy of URLs for external or third-party internet websites referred to in this publication, and does not guarantee that any content on such websites is, or will remain, accurate or appropriate.

This book is dedicated to my friends and colleagues at Gresham College, the Mercers' Company and the City of London Corporation, who have made my years associated with the College the most rewarding of my life.

Contents

Preface   page ix
Acknowledgements   xi

1 Watchers of the skies   1

2 Our Sun   12

3 Aspects of our Solar System   26

4 The rocky planets   38

5 The hunt for Planet X   57

6 Voyages to the outer planets   70

7 Harbingers of doom   91

8 Impact!   100

9 Four hundred years of the telescope   118

10 The family of stars   133

11 Aging stars   145

12 The search for other worlds   164

13 Are we alone? The search for life beyond the Earth   179

14 Our island Universe   192

15 Wonders of the southern sky   209

16 Proving Einstein right   226

17 Black holes: no need to be afraid   239

18 It's about time   255

19 Hubble's heritage: the astronomer and the telescope that honours his name   267

20 The violent Universe   285

21 The invisible Universe: dark matter and dark energy   301

22 The afterglow of creation   320

23 To infinity and beyond: a view of the cosmos   336

Index   347

The colour plates can be found between pages 212 and 213.

Preface

Although I have been a radio astronomer all my working life I have also greatly enjoyed observing the heavens. At the age of 12, I first observed the craters on the Moon and the moons of Jupiter with a simple telescope made from cardboard tubes and lenses given to me by my optician. As I write, I have my father's thin, red bound, copy of Fred Hoyle's book *The Nature of the Universe* on the desk beside me. It was this book that inspired me to become an astronomer.

I was able to study a little astronomy as an undergraduate at Oxford University and was also in the 'signals' section of the University Officers' Training Corps. As I was revising for my finals I spotted an advertisement for a new course in radio astronomy at the Jodrell Bank Observatory. Being interested in both astronomy and radios this seemed a good idea and I began to study there in 1965, initially studying the surface of the Moon by radar.

My supervisor as a PhD student had been giving evening classes in astronomy at a local college and due to illness asked me if I would take them over from him. Giving such evening classes over the majority of my career gave me much experience in trying to explain astronomical concepts to members of the general public, as did the giving of a general course on astronomy to the first year physics students at the University of Manchester.

It was, perhaps, this experience that helped me to be appointed to the post of Gresham Professor of Astronomy in 2007. This is the oldest chair of astronomy in the world, dating from 1597 and once held by Christopher Wren. In this role, over four years, I gave over 25 lectures on astronomy in the City of London. For each of these lectures I was required to write a transcript, and almost all chapters of this book are based on these transcripts but have been, of course, brought totally up to date. In fact, one reason why I did not put this book together sooner was that I wanted to include the results of two major space telescopes that were only released in 2013: Planck, studying the Cosmic Microwave Background, and Kepler, searching for extra-solar planets in the Galaxy.

Each of my lectures was videoed and can be found on the Gresham College website (just search for 'Gresham College lectures by Ian Morison'). These have the same titles as the chapters of this book so, should you wish, you could watch and listen to the lectures to complement the book's text. In this, this book might well be unique. Chapters 3 and 10, on aspects of the Solar System and the properties of stars, though not based on specific lectures, have been added to provide some additional information to make the book complete.

Should you like to explore some of the calculations that lie behind the results given in the book, then you might like to read my textbook *Introduction to Astronomy and Cosmology*, published by John Wiley, and if, perhaps, this book inspires you to observe the wonder of the Universe yourself, then you might like to read my book *An Amateur's Guide to Observing and Imaging the Heavens*, published Cambridge University Press.

I most sincerely hope that this book will help you to increase your knowledge and understanding of the Universe in which we are privileged to live.

Acknowledgements

As the dedication of this book implies, my first acknowledgement goes to Gresham College and its sponsors, the Mercers' Company and the City of London Corporation, who gave me the opportunity to become Gresham Professor of Astronomy.

I would like to thank Peter Shah, Damian Peach, Koen van Gorp and Greg Piepol who have allowed me to use their beautiful images to help illustrate the book.

I would also like to thank Vince Higgs, Lindsay Stewart and Jonathan Ratcliffe at Cambridge University Press who have steered this book through to publication. I must thank too Margaret Patterson, who copy-edited my text, and Kanimozhi Ramamurthy and her team who have prepared the book for printing.

Finally, but not least, I must thank my wife for supporting me as I spent far too many hours at the computer and too few in carrying out domestic chores.

1

Watchers of the skies

Astronomy is probably the oldest of all the sciences. It differs from virtually all other science disciplines in that it is not possible to carry out experimental tests in the laboratory. Instead, astronomers can only observe what they see in the Universe and see whether observations fit the theories that have been put forward. Before we start our journey through the Universe in Chapter 2, I would like to share with you a little of the history of astronomy, looking at some of the astronomers who have, in the past, made great contributions to our knowledge.

Galileo Galilei's proof of the Copernican theory of the Solar System

One of the first triumphs of observational astronomy was Galileo's series of observations of Venus, which showed that the Sun, not the Earth, was the centre of the Solar System so proving that the Copernican rather than the Ptolemaic model was correct. He had made observations of Jupiter that showed four moons – now called the Galilean moons – weaving their way around it. This showed him that not all objects orbited the Earth.

In the Ptolemaic model of the Solar System (which is far more subtle than is often acknowledged) the planets move round circular 'epicycles' whose centres move around the Earth in larger circles called deferents. This can account for the 'retrograde' motion of planets such as Mars and Jupiter when they appear to move backwards in the sky. It also models the motion of Mercury and Venus. In their case, the deferents, and hence the centres of their epicycles, move round the Earth at the same rate as the Sun. The two planets thus move round in circular orbits whose centres lie on the line joining the Earth and the Sun, being

A Journey through the Universe

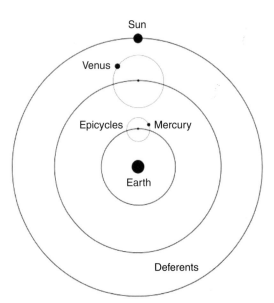

Figure 1.1 The centre points of the epicycles for Mercury and Venus move round the Earth with the same angular speed as the Sun.

seen either before dawn or after sunset. Note that, as Mercury stays closer to the Sun than Venus, its deferent and epicycle are closer than that of Venus – in the Ptolemaic model Mercury is the closest planet to the Earth!

As seen in Figure 1.1, in the Ptolemaic model Venus lies between the Earth and the Sun, hence it must always be lit from behind so could only show crescent phases whilst its angular size would not greatly alter. In contrast, in the Copernican model Venus orbits the Sun. When on the near side of the Sun it would show crescent phases, whilst on its far side, but still visible, it would show almost full phases. As its distance from us would change significantly, its angular size (the angle subtended by the planet as seen from the Earth) would likewise show a large change.

Figure 1.2 shows a set of drawings of Venus made by Galileo when using his simple refracting telescope. They are shown in parallel with a set of modern photographs which illustrate that not only did Galileo show the phases he also correctly drew the changing angular size. These drawings showed precisely what the Copernican model predicts – almost full phases when Venus is on the far side of the Sun and hence has a small angular size coupled with thin crescents having a significantly larger angular size when it is closest to the Earth.

So Galileo's observations, made with the simplest possible astronomical instrument, were able to show which of the two competing models of the

Figure 1.2 Galileo's drawings of Venus (top) compared to photographs taken from Earth (below).

Solar System was correct. In just the same way, but using vastly more sophisticated instruments, astronomers have been able to choose between competing theories of the Universe – a story that will be told later.

The celestial sphere

Looking up at the heavens on a clear night we can imagine that the stars are located on the inside of a sphere, called the celestial sphere, whose centre is the centre of the Earth.

As an aid to remembering the stars in the night sky, the ancient astronomers grouped them into constellations representing men such as Orion, the Hunter, women such as Cassiopeia, mother of Andromeda, animals and birds such as Taurus, the Bull, and Cygnus, the Swan, and inanimate objects such as Lyra, the Lyre. There is no real significance in these stellar groupings – stars are essentially seen in random locations in the sky – though some patterns of bright stars, such as the stars of the 'Plough' (or 'Big Dipper') in Ursa Major, the Great Bear, result from their birth together in a single cloud of dust and gas.

A second major observational triumph

We are now in a position to describe the observations that led to a further major improvement in our understanding of the Solar System.

In 1572, Tycho Brahe, a young Danish nobleman whose passion was astronomy, observed a supernova (a very bright new star) in the constellation of Cassiopeia. His published observations of the 'new star' shattered the widely held belief that the heavens were immutable, and he became a highly respected astronomer.

He realised that in order to show when further changes in the heavens might take place it was vital to have a first-class catalogue of the visible stars. Four years later, Tycho was given the Island of Hven by the King of Denmark and money to build a castle, which he called Uraniborg, named after Urania, the Greek goddess of the heavens. Within its grounds he built a semi-underground observatory called Stjerneborg. For a period of 20 years his team of observers made positional measurements of the stars and, of critical importance, the planets.

Figure 1.3 shows his observatory and indicates how his measurements were made. An observer sighted a star (or planet) through a small window on a south-facing wall. Two things were measured. Firstly an assistant noted the time of transit as the star crossed the meridian. The meridian is the half-circle that runs across the sky through the zenith between the north and south poles and intersects the horizon due south. Secondly, by using a giant quadrant equipped with vernier scales the observer was able to measure the elevation (angular height above the horizon) of the star at the moment of transit. The assistant is standing beside the clock at the lower right of the figure to measure the time at which the star transits and the scribe seated at a table at the lower left would then note the elevation of the star and time of transit in the log book. He could thus determine the position of the star on the celestial sphere.

Not only had Tycho produced a star catalogue 10 times more precise than any previous astronomer – the errors of the 777 star positions never exceeded 4 arcminutes (one arcminute is one sixtieth of a degree) – he had charted the movement of the planets during the 20-year period of his observations. It was these planetary observations that led to the second major triumph of observational astronomy in the sixteenth and seventeenth centuries: Kepler's three laws of planetary motion.

When King Frederik II died in 1588, Tycho lost his patron. The final observation at Hven was made in 1596 before Tycho left Denmark. After a year travelling around Europe he was offered the post of Imperial Mathematician to Rudolf II, the Holy Roman Emperor, and was installed in the castle of Benátky. It was

Figure 1.3 A quadrant used by Tycho Brahe to measure the elevation of a star or planet as it crosses the meridian (due south).

here that a young mathematician, Johannes Kepler, came to work with him. Tycho gave him the task of solving the orbit of the planet Mars. Kepler thought that it would take him a few months. In fact it took him several years!

There was a fundamental problem. The observations of Mars had been made from the Earth – which was itself in orbit around the Sun. Unless one knew the precise orbit of the Earth one could not find the parameters of the Martian orbit. In what has been described as a stroke of genius, Kepler realised that every

687 days (the orbital period of Mars) Mars would return to exactly the same location in the Solar System, so observations of the Earth from Mars made on a set of dates separated by 687 days could, in principle, be used to solve for the orbit of the Earth. (As could, of course, observations made on those days of Mars from the Earth.) Having solved for the Earth's orbit, Kepler was then able to deduce the orbital parameters of Mars.

The laws of planetary motion

From the invaluable database of planetary positions provided by Tycho, Kepler was able to draw up his three empirical laws of planetary motion. The word 'empirical' indicates that these laws were not based on any deeper theory but accurately described the observed motion of the planets. The first two were published in 1609 and the third in 1618.

The first law states:

> **Planets move in elliptical orbits around the Sun, with the Sun positioned at one focus of the ellipse.**

The second law states:

> **The radius vector – that is, the imaginary line joining the centre of the planet to the centre of the Sun – sweeps out equal areas in equal times.**

This implies that the planets, in an elliptical orbit, move faster when closest to the Sun – as they near the Sun they lose potential energy and, as the total energy must be constant, increase their kinetic energy, and so move faster.

The third law relates the period of the planet's orbit, T, with a, the semi-major axis of its orbit, and states:

> **The square of the planet's period, T, is proportional to the cube of the semi-major axis of its orbit, a. (For a circular orbit, the semi-major axis is the radius.)**

A highly significant result of the third law is that it enabled astronomers to make a very accurate map of the Solar System. The relative positions of the planets could be plotted precisely *but the map had no scale*. It was like having a very good map of a country but not knowing, for example, how many centimetres on the map related to kilometres on the ground. A way to solve this problem would be to make an accurate measurement of *one* reasonably large distance across the area covered by the map. This would then give the scale, and thus the distance between any other two points on the map could be found.

Isaac Newton and his Universal Law of Gravitation

Isaac Newton was born in the manor house of Woolsthorpe, near Grantham, in 1642, the same year in which Galileo died. His father, also called Isaac Newton, had died before his birth and his mother, Hannah, married the minister of a nearby church when Isaac was two years old. Isaac was left in the care of his grandmother and effectively treated as an orphan. He attended the Free Grammar School in Grantham and took up lodgings there. He showed little promise in academic work and his school reports described him as 'idle' and 'inattentive'. His mother later took Isaac away from school to manage her property and land, but he soon showed that he had little talent and no interest in managing an estate.

An uncle persuaded his mother that Isaac should prepare for entering university and so, in 1660, he was allowed to return to the Free Grammar School in Grantham to complete his school education. He lodged with the school's headmaster, who gave Isaac private tuition, and he was able to enter Trinity College, Cambridge, in 1661 as a somewhat more mature student than most of his contemporaries. He received his bachelor's degree in April 1665 but then had to return home when the university was closed as a result of the plague. It was there, in a period of two years and whilst he was still under 25, that his genius became apparent.

There is a story (which is probably apocryphal) that Newton was sitting under the apple tree in the garden of Woolsthorpe Manor. He might well have been able to see the first or last quarter Moon in the sky. It is said that an apple dropped on his head (or thudded to the ground beside him) and this made him wonder why the Moon did not fall towards the Earth as well.

Newton's moment of genius was to realise that it *was* falling towards the Earth! He was aware of Galileo's work relating to the trajectories of projectiles, and in 1686 he considered what would happen if one fired a cannon ball horizontally from the top of a high mountain where air resistance could be ignored. The cannon ball would follow a parabolic path to the ground. As the cannon ball was fired with greater and greater velocity it would land further and further away from the mountain. As the landing point becomes further away the curvature of the Earth must be considered. In a popular work published in the 1680s called *A Treatise of the System of the World* he included the diagram shown in Figure 1.4. The mountain is impossibly high in order for it to reach above the Earth's atmosphere. But this is a thought experiment, not a real one! One can see from this that if the velocity is sufficiently high the cannon ball would never land – it would be in an orbit around the Earth! (To be brutally accurate it would hit the back of the cannon after one orbit, but we will ignore that.)

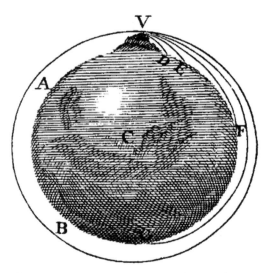

Figure 1.4 Newton's thought experiment using a cannon ball.

Newton applied the same logic to the motion of the Moon, realising that, if the gravitational attraction between the Earth and the Moon caused it to fall by just the right amount, it too would remain in orbit around the Earth. He knew enough about the Moon to be able to calculate the value of the acceleration of gravity at the distance of the Moon. For this he needed to know the radius of the Moon's orbit, assumed to be circular, about the centre of the Earth and also the period of its orbit around the Earth. Newton used a value of the radius of the Moon's orbit of 384,000 kilometres and a period of 27.32 days or 2.36×10^6 seconds and showed that the value of g (the acceleration due to gravity) at the distance of the Moon was 1/3,606 that at the surface of the Earth.

Newton also knew that the radius of the Earth was 6,400 km so that the Moon, at a distance of 384,000 km, was precisely 60 times further away from the centre of the Earth than the Earth's surface. So the value of g at the distance of the Moon had fallen almost precisely by the ratio of the distance from the centre of the Earth squared!

This led Newton to his famous inverse square law: the force of gravitational attraction between two bodies decreases with increasing distance between them as *the inverse of the square of that distance*.

But Newton had a problem. He felt that he could not publish his law until he could prove that the gravitational pull exerted by a spherical body was precisely the same as if all the mass were concentrated at its centre. This can only be proved by calculus, and it took Newton a while to develop the ideas of calculus, which he called 'fluxions'. It was only then that he felt confident to publish his law.

Newton realised that the force of gravity must also be directly proportional to the object's mass. Also, based on his third law, he knew that when the Earth exerts its gravitational force on an object, such as the Moon, that object must exert an equal and opposite force on the Earth. He thus reasoned that, due to this symmetry, the magnitude of the force of gravity must be proportional to both the masses.

His law thus stated that the force, F, between two bodies is directly proportional to the product of their masses and inversely proportional to the distance between their centres. The force is then equal to the product of these two factors multiplied by a constant called G, the constant of gravitation.

Newton derived a value of G by estimating the mass of the Earth, assuming it had an average density of 5,400 kg/m^3. He suspected that the Earth got denser with depth and simply doubled the value of 2,700 kg/m^3 that is measured at the surface of the Earth. (This was a pretty good – and very lucky – estimate as it is actually 5,520 kg/m^3!) The value that he obtained for G was 6.76×10^{-11} N m^2 kg^{-2}. Due to his lucky estimate of the mean density of the Earth, this was a very good result – the now accepted value of G being 6.67×10^{-11} N m^2 kg^{-2}.

His law was called the 'Universal Law of Gravitation', but why Universal? Using his second law (force = mass × acceleration) and his Law of Gravitation, Newton was able to deduce Kepler's third law of planetary motion. This deduction showed him that his law was valid throughout the whole of the then known Solar System. To him that was Universal!

Charles Messier: a great observer

Messier, a French astronomer, who made his observations from the Hôtel de Cluny in Paris, observed all types of astronomical phenomena such as occultations, transits and eclipses, but his great love was discovering and observing comets. His 13 comet discoveries brought him great fame and he was made a fellow of the Royal Society and elected to the French Academy of Sciences.

Whilst scanning the heavens in August 1758 when searching for new comets, Messier came across a faint nebulosity in the constellation of Taurus, the Bull. Thinking first that this might be a comet, he observed it on following nights and soon realised that it was not a comet as it did not move across the sky. To prevent both himself, and others, wasting observing time in the future he decided to produce a catalogue of nebulous objects that might be first thought to be comets. This object in Taurus thus became the first object in his catalogue: M1. It is the remnant of a supernova that was observed in 1054 and is now known as the Crab Nebula, as a nineteenth-century astronomer, the Third Earl of Rosse, having observed it with his great 72-inch Newtonian, thought that it resembled a horseshoe crab. The task of compiling his catalogue did not begin in earnest until 1764, and within seven months a further 38 entries has been added, such as the globular cluster in Hercules, M13, the Dumbbell Planetary Nebula, M27, and the Andromeda Galaxy, M31.

The first of Messier's catalogues, listing 45 objects, was published in 1774 and his final list of 103 objects was published in 1781. Since then the list has grown to 110 objects as astronomers and historians found evidence of another seven deep-sky objects that had been observed either by Messier or his assistant Pierre Méchain not long after the final list had been published. His list has bequeathed a wonderful resource for amateur astronomers as it contains many of the most beautiful celestial objects that can be seen in a small telescope: covering every type of object from open and globular clusters, diffuse and planetary nebulae to many of the brightest galaxies that can be observed from northern mid-latitudes. Because Messier had discovered them with relatively small refractors (albeit without the light pollution that many of us now suffer) these objects can be observed visually with virtually all amateur telescopes and are among the most attractive deep-sky objects in the heavens.

Edmond Halley and the measurement of the astronomical unit

Edmond Halley is best known for the comet that bears his name and whose story is covered in Chapter 7, but he also had an insight that enabled the distance of the Earth from the Sun (one astronomical unit – 1 AU) to be determined.

In the case of the Solar System, the obvious measurement to make is the distance between the Earth and either of the two planets nearest to it, Venus or Mars. Once this is known, Kepler's third law can be used to calculate the distance of the Earth from the Sun. In 1678 Halley realised that if the transit of a planet across the surface of the Sun could be observed from two widely spaced locations on the Earth, the slight difference observed in the passage across the Sun from the two locations would, in principle, enable its distance to be found by the method of parallax. The problem is that the widest separation possible on the Earth, its diameter of 12,756 km, is small compared to the distances of the planets and thus the angular difference that has to be measured is very small and prone to errors.

There was a major effort to observe the two transits of Venus in the 1700s. In 1761, the most successful observations were made by two surveyors, Jeremiah Dixon and Charles Mason, at the Cape of Good Hope. (They are best known for surveying the Mason–Dixon Line, which they surveyed between 1763 and 1767 to resolve a border dispute between British colonies in Colonial America.) Famously, Captain James Cook travelled to Tahiti with his astronomer Charles Green and naturalist Joseph Banks to observe the transit of 1769 and built a fort at Point Venus. Banks became very friendly with the Tahitians (apparently he was 'entertaining' three young ladies during the transit) but proved invaluable to Cooke when the sextant, vital for measuring their precise location, was stolen. He was able to get the local chieftain to help him recover the parts of the disassembled sextant. It was then rebuilt by the clockmaker who was charged with looking after the clock that was used to make the timing measurements when Venus first fully entered the Sun's limb and then just before it left. The value of the astronomical unit deduced from both the eighteenth-century transits was 152.4 million km.

Cook then went on to circumnavigate New Zealand and charted what he termed Dusky Bay. He returned to what is now called Dusky Sound on his second world voyage and spent five weeks in Pickersgill Harbour. Having cleared an acre of forest, his astronomer, William Wales, set up an observatory on Astronomer's Point beside the harbour and measured its precise position, which was the most accurately known position in the southern hemisphere for many years.

Having observed the transit of 2004 in England under clear skies, I was invited to take some astronomers to Pickersgill Harbour to observe the 2012 transit. With luck, the skies were clear and I was able to time both the ingress and egress of Venus. Combining my timing observations with those of an observer in Anchorage, Alaska (giving a baseline of nearly 10,000 km – almost certainly the greatest baseline that has ever been used) I obtained a value for the astronomical unit of 146,000,000 km, which was within striking distance of the established value and with about the same accuracy as that obtained in the 1700s.

A really accurate measurement of the astronomical unit had to wait until 1962, when powerful radars using large radio telescopes at Jodrell Bank in the UK, in the USA and the USSR were able to obtain echoes from the surface of Venus. The accepted value now is 149,597,870.691 km, just less than 150 million km or 93 million miles.

Suggestions for further reading on the history of astronomy:

> *The History of Astronomy: A Very Short Introduction* by Michael Hoskin (Oxford University Press).
> *History of Astronomy* [Free Kindle Edition] by George Forbes.

2

Our Sun

Any journey (and you must admit ours is a pretty epic journey) must start somewhere. Perhaps following the ideas of the Copernicus rather than Ptolemy, it seems that a logical place to start would be at the centre of the Sun. Here, the temperature is ~15 million degrees kelvin (K), which has no real meaning to us except that it is very hot indeed. This is simply telling us how fast the particles within the central core of the Sun are moving and this is the key to how the Sun creates its energy. The core largely comprises protons (the nuclei of hydrogen), alpha particles (the nuclei of helium, consisting of two protons and two neutrons) and electrons.

Up to the late 1800s, scientists could not understand how the Sun could create so much light and heat. Had the Sun been entirely made of something like coal (along with the oxygen it would need to burn) it would have burnt itself up in about a thousand years. Since the Sun had been providing heat and light for at least several thousand years, a chemical source for the Sun's energy was clearly impossible. Around 1870, Hermann von Helmholtz realised that, if the Sun were contracting in size, energy, derived from potential energy, could be released. Knowing the mass and size of the Sun and how much energy the Sun is continuously creating and sending out into space, he calculated how much the Sun would have to reduce in size to provide its observed output, and deduced that the Sun would be able to sustain its energy output for around 20 million years. In the late 1800s people were quite happy to assume that the Solar System was less than 20 million years old, so his idea was almost universally accepted as the likely way that the Sun creates its energy.

However, during the late 1800s, geologists established that many Earth rocks and the fossils within them are definitely hundreds of millions of years old, so Helmholtz must have been wrong. In 1905, Einstein published the famous

$E = mc^2$ equation as part of his Special Theory of Relativity. One could thus surmise that, as c (the speed of light) is very large, a tiny amount of mass (m) might be converted into an enormous amount of energy (E). By around 1925, physicists had determined the mass of a proton (a hydrogen nucleus) and had also determined that an alpha particle (a helium nucleus) has a mass slightly less than that of four protons. They realised that four hydrogen nuclei might be able to 'fuse' together into a helium nucleus (a process called fusion) and the mass that apparently disappears could be converted into energy.

This is difficult! Since the hydrogen nuclei are all positively charged protons, they repel each other. In order to overcome this mutual repulsion, the protons must be moving towards each other at nearly the speed of light. They can only do this if it is very hot – of the order 10 million K. Due to the great mass of the Sun, the pressure at its centre (called its core) must be very high to oppose the mass of the overlying layers that form the greater part of the Sun, and calculations show that the core would reach and exceed the required temperature. Thus the source of this energy is a nuclear fusion reactor within the core of the Sun. As the dust and gas that made up our Sun collapsed under gravity, the temperature at the centre increased and the protons began to move faster. When the temperature exceeded ~10 million K, the proton's kinetic energy became sufficient for two protons on a collision course to get sufficiently close to allow an effect called 'quantum-mechanical tunnelling' to come into play. This allows one of them to overcome the potential barrier due to the electrostatic force between them.

In quantum theory, particles, such as the two protons approaching each other, can be described by wave functions, which represent the probability of finding a particle in a certain location. If a particle is adjacent to a potential barrier, its wave function decays exponentially through the barrier, but will have still have a very small amplitude on the far side of the barrier. There is thus a very small probability that the particle can 'appear' on the other side of the barrier, in which case it is then said to have tunnelled through it. Quantum tunnelling thus allows a particle, in this case a proton, to violate the principles of classical mechanics by passing through a potential barrier higher than the kinetic energy of the particle.

This allows the two protons to come sufficiently close for the strong nuclear force to momentarily bind them together before one of the protons decays into a neutron, a positron and an electron neutrino. This leaves a deuteron, the combination of a proton and a neutron, which is the nucleus of deuterium. The positron then annihilates with an electron, and their mass energy is carried off by two (sometimes more) gamma-ray photons. This is the first step in what is called the proton–proton cycle outlined below and illustrated in Figure 2.1.

14 A Journey through the Universe

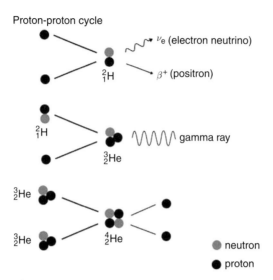

Figure 2.1 The three steps in the proton–proton cycle. Steps 1 and 2 are carried out twice to provide the two $^3He_2$ nuclei required for the third step.

This first step is extremely slow, with a proton typically waiting a billion years before carrying out this reaction. It is thus the limiting step in the chain of nuclear reactions and so determines the overall reaction rate. It might just be worth pointing out that the slowness of this reaction is vital to our presence here on Earth. If the reaction rate was just 10 times faster, the Sun would burn up its energy supply in 1 billion years rather than the ~10 billion years that we will calculate below, and there would not have been sufficient time for our human race to evolve.

The proton–proton cycle

(1) Two protons react to give rise to a deuteron comprising one proton and one neutron. The charge of one of the protons is carried away by a positron (the antiparticle of the electron). An electron neutrino is also created.

(2) A further proton then reacts with the deuteron to give a nucleus comprising two protons and one neutron. It is thus an isotope of helium called helium-3. A gamma ray (very high energy photon) is emitted.

(3) Two helium-3 nuclei react to give one helium nucleus (also called an alpha particle) and two protons are emitted to take part in further reactions.

This is not quite the end of the story. The positrons given off in step 1 of the reaction will quickly meet electrons and annihilate to give further gamma-ray photons. The pressure generated within the core as a result of these nuclear reactions prevents the Sun's collapse.

The photons, initially gamma rays, work their way towards the Sun's surface, continuously interacting with matter. Their direction of motion following each interaction is random, so they carry out what is called a 'random walk' and, as a result, the energy takes of order 100,000 years to pass through the 'radiative zone' that surrounds the core and extends for about two-thirds of the Sun's radius. As they work their way outwards, the temperature drops, and as the photons are in thermal equilibrium with the gas their wavelengths increase.

The region, one-third of the Sun's radius in width, between the outer edge of the radiative zone and the surface is called the 'convective zone'. Here, the energy is carried outwards by first large and then small convective cells, as shown in Figure 2.2. As a result, the Sun's surface, called the photosphere, shows granulations – honeycomb-like variations in brightness. The granulations are brighter in the centre where the convection currents bring the energy to the surface and darker around the edges where material, cooled as it radiates energy into space, returns towards the centre.

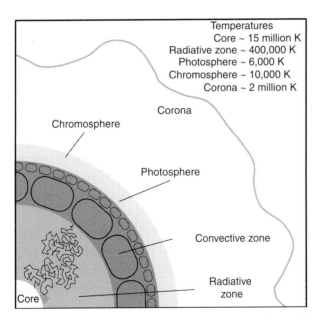

Figure 2.2 A cross section of the Sun showing the regions referred to in the text and their approximate temperatures.

The helium atom is made up of two protons, two neutrons and two electrons. Their individual masses add up to 4.0331 atomic mass units (amu). But helium-4 has a lower mass, 4.0026 amu, and is thus 0.0305 amu lighter than its components. This is the mass that has been converted into energy during the nuclear fusion process. The fractional mass loss is 0.0305/4.0026, which corresponds to 0.76%. To provide the energy output of the Sun, 5.7×10^{11} kg of hydrogen has to be converted to helium per second.

We know that the Sun has a mass of 2×10^{30} kg so, dividing by the mass loss per second, one gets a solar lifetime of ~100 billion years. However, not all of the Sun's mass can be converted; partly because only ~75% is composed of hydrogen, but also because temperatures are only high enough (~15 million K) in the central core for nuclear fusion to take place. The models predict that only ~10% of the Sun's mass can be converted. We thus believe that our Sun will shine burning hydrogen to helium for ~10 billion years.

The Sun has been shining for about 4.5 billion years, so we need not worry about any immediate loss of our heat and light, but there is a long-term worry for beings on Earth. Perhaps surprisingly, as the percentage of hydrogen reduces in the core and that of helium increases, the rate at which hydrogen is fused into helium actually increases and the Sun is getting hotter with time. It is now ~20% hotter than when the Solar System was formed and in a billion years or so its energy output will be such that the surface of the Earth will be too hot for life.

The solar neutrino problem

From the known reaction rate of the proton–proton cycle, about 2×10^{38} electron neutrinos are produced per second, and it is worth noting that neutrinos carry ~2% of the Sun's energy output, so only ~98% of the energy is carried away from the Sun by electromagnetic radiation. At this point we need to know that in the standard model of particle physics there are three types of neutrino: electron, muon and tau. Those emitted by the nuclear reactions in the Sun are electron neutrinos.

The proton–proton cycle (known as the ppI chain) provides about 86% of the Sun's energy. The neutrinos given off in the ppI chain have an energy of only 0.26 MeV and we have only been able to detect them recently. However, there are two further pp chains: the ppII chain produces ~14% of the solar energy, whilst the ppIII chain produces only ~0.11%. The importance of the ppIII chain is that the neutrinos emitted have higher energies (from 7 up to 14 MeV), which makes them easier to detect.

In an experiment to detect solar neutrinos in the 1970s, Ray Davis set up a tank filled with 100,000 gallons (380 m^3) of tetrachloroethylene (C_2Cl_4), located 4,900 feet (1,493 m) underground in the Homestake Mine in South Dakota. It was deep underground to prevent cosmic rays (which would be absorbed by the rock strata above) giving rise to false detections. Very rarely, a neutrino from the ppIII chain would react with a chlorine nucleus to give a radioactive isotope of argon. Having left the tank for a month, Davis flushed the tank to extract and detect the argon atoms. On average, about 10 argon atoms were detected per month. The problem was that only about one-third of the expected number of neutrinos were detected. Remembering that these neutrinos come from relatively rare nuclear reactions in the ppIII chain, it could have been due to our lack of understanding of these reactions, but modern experiments have confirmed Davis's results and he was awarded the Nobel Prize in Physics for this work in 2002. The lack of observed neutrinos was called the solar neutrino problem.

At the bottom of a very deep mine at Sudbury, in Canada, there is a 12-metre sphere that holds 1,000 tonnes of heavy water. This is water that has a deuteron rather than a proton as its nucleus. Every day of operation about 10 solar neutrinos react with a deuteron, the reaction giving rise to two protons and a very high energy electron. This electron travels at a speed faster than the speed of light in the liquid and produces a shock wave akin to that produced by a supersonic plane. The shock wave produces blue light in the form of 'Cherenkov' radiation, which spreads out in a cone. The tank is surrounded by 916 photon detectors, which detect the Cherenkov radiation (around 50 photons per event) so it is possible to measure the number of solar neutrinos detected. The Sudbury detector confirmed that only ~33% of the expected number of electron neutrinos were arriving from the Sun.

In a follow-up experiment, 2 tonnes of high-purity table salt (NaCl) were added to the heavy water in order to provide three times better sensitivity to the muon and tau neutrinos. It appears that the *total* number of neutrinos detected with a solar origin does agree well with that predicted for the total number of *electron neutrinos* that the Sun should produce. The only way that this could be the case is if the electron neutrinos can change (the word 'oscillate' is used) into one of the other two types en route from the Sun.

It was thought that electron neutrinos, like photons, were massless. As Einstein showed, anything travelling at the speed of light would experience no passage of time so there would be no way that an electron neutrino could ever change into one of the other two types. However, if the neutrino *does* have mass it will not travel at the speed of light so it *will* then experience the passage of time and thus could 'oscillate' into muon or tau neutrinos. It is thought that on their way from the core of the Sun to the Earth the neutrinos will evenly distribute themselves amongst the three types – thus only one-third of the electron neutrinos emitted by the Sun will remain to be detected – exactly as observed. The neutrino problem is solved!

The solar atmosphere: photosphere, chromosphere and corona

When the Sun is observed (taking very great care using appropriate filters!) it appears to have a sharp edge but there is, of course, no actual 'surface'. We are, in fact, just seeing down through the solar atmosphere to a depth where the gas becomes what is called 'optically thick'. This deepest visible layer of the atmosphere is called the photosphere (as this is where the photons that we see originate) and is ~500 km thick. The temperature falls from ~6,500 K at its base to ~4,400 K at its upper region with the effective temperature of the photosphere being ~5,800 K. The transport of energy from below gives rise to a mottling of the surface – solar granulations that are each ~1,000 km across. Each granulation cell lasts about five to ten minutes as hot gas, having risen from below the surface, radiates energy away, cools and sinks down again.

The region, about 2,000 km thick, above the photosphere is called the 'chromosphere'. The gas density in this region falls by a factor of about 10,000 and the temperature increases from ~4,400 K at the top of the photosphere to ~25,000 K. Above this is the 'transition region', in which the temperature rises very rapidly over a distance of a few hundred kilometres to a temperature of ~1 million K. The transition region leads into the outer region of the Sun called the 'solar corona', where temperatures reach in excess of 2 million K. Its form and extent depends strongly on the solar activity, which varies through what is called the 'sunspot cycle', but typically extends for several solar radii. At the time of solar minima when activity is low, it usually extends further from the Sun at its equator and the pattern of the Sun's magnetic field is often well delineated near its poles. At solar maxima, the overall shape is more uniform and has a complex structure.

The density of the corona is very low, ~10^{14} times less than that of the atmosphere at the Earth's surface, and its brightness is about a million times less than that of the photosphere. It can thus only be observed during a total eclipse of the Sun, or by using a special type of telescope, called a coronagraph, that can block out the light from the solar disc. How the corona can reach such high temperatures is still somewhat of a mystery, but it is thought that energy might be transported into it by magnetic fields.

The Sun's magnetic field and the sunspot cycle

A photograph of Sun's surface will usually show some darker regions on the surface. These are called sunspots and often appear in pairs or groups – a sunspot group. Each spot has a central 'umbra' surrounded by a lighter 'penumbra'. They appear dark because they are cooler than the surface in general. The umbra has a typical temperature 1,000 K less than its surroundings. Around the outside of a sunspot group may be seen an area that is brighter than the normal surface. This is called a 'plage' – the French word for a beach. (French beaches often have white sand!)

The passage of sunspots across the Sun's surface has been used to measure its rotation period. It has been found that at the equator the period is ~28 days, but this increases with increasing latitude and is ~35 days near the Sun's poles, an effect called differential rotation. Sunspots are intimately linked to this differential rotation and how it affects the Sun's magnetic field.

The Sun's field must be created by the movement of charged particles and its cause almost certainly lies in the convective zone, but no really good model yet exists. Imagine that at one particular moment the Sun has a uniform bipolar field just like that shown by iron filings under the influence of a bar magnet. Looking at the Sun's face we could imagine a field line directly down the centre of the Sun's disc that lies just below the photosphere. The field and surrounding material remain locked together in what follows. Now move forwards in time by 35 days. The field close to the poles will have made one complete rotation and be in the same position as seen from Earth, but close to the equator it will have rotated an additional amount ~(35 − 28)/28 of one rotation or 1/4 × 360 degrees, which is 90 degrees, as seen in Figure 2.3. After three further rotations (as measured near the poles) the field near the equator will have one additional

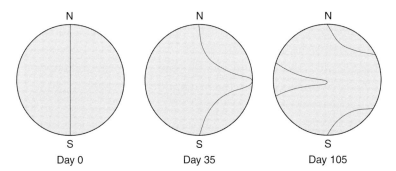

Figure 2.3 The winding of the Sun's magnetic field due to the differential rotation of the Sun.

rotation. You can see that the field is being 'wound up' and becoming more intense. It gains 'buoyancy' and, in places, rises in a loop above the surface. Where it passes through the surface, the magnetic field inhibits the convective flow of energy to the surface so the localised region will be cooler than the surface in general – a sunspot appears. The energy that is inhibited from reaching the surface here will tend to reach the surface in the area surrounding a sunspot group, making this region hotter and thus brighter – a plage.

It is possible to measure the polarity of the field across the Sun's surfaces and we can examine the polarity of the field associated with sunspot pairs. Consider a sunspot pair in the upper hemisphere. Assuming that the north pole is at the top then, as the field leaves the surface, it will be towards us and have positive polarity. As the field re-enters the surface it will be away from us and have negative polarity. So the spots in a sunspot pair will show opposite polarity. In the upper hemisphere the left-hand spot would have positive polarity and the right-hand negative. However, as the field reverses direction at the equator, the sunspots in a pair observed in the lower hemisphere will have the opposite sense.

The twisting of the magnetic field from its initial state gradually produces an increase in the number of sunspots observed across the Sun's surface. This determines what is termed the sunspot number, which reaches a peak after three to four years – called a sunspot maximum. The field then begins to reduce in strength and the sunspot number reduces for a further seven to eight years to a point when the Sun's face can be totally devoid of spots – called a sunspot minimum. (This was the case for 2007–9 during a very long sunspot minimum.)

The whole process then starts over again giving what is called the 'sunspot cycle'. It is often said to be an 11-year cycle, though it can vary in length somewhat and the average length of the cycle over recent decades is 10.5 years. The following cycle has, however, one important difference: the field has the opposite polarity. So perhaps we should call it the 21-year solar cycle, not 11.

Figure 2.4 shows a plot of the sunspot numbers since 1600. The cyclical variation is very apparent. The plot shows two interesting features. A complete lack of sunspots during the late 1600s, called the Maunder minimum, and increasing solar activity during the 50 years up to the millennium but with reduced activity particularly in the maximum that peaked in 2013.

Prominences, solar wind and solar flares

When the Sun is observed in the light of the H-alpha transition at a time of solar eclipse or by use of a special H-alpha telescope, bright columns of gas are often seen stretching up from the chromosphere into the corona. These are

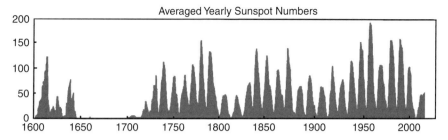

Figure 2.4 The averaged sunspot numbers over the last 400 years.

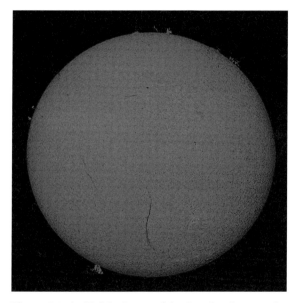

Figure 2.5 An H-alpha image of the Sun showing prominences at the limb of the solar surface. Also shown in colour in Plate 2.5 along with an image in the light of calcium K. Image: Ian Morison.

called 'prominences', as seen in Figure 2.5. They are caused by ionised gas that has been carried upwards by the Sun's magnetic field which then 'rains' back from the corona into the chromosphere again. The Sun is continuously 'boiling off' particles, which are accelerated away from the Sun by its magnetic field forming what is called the 'solar wind'. The interaction of the solar wind with the Earth's atmosphere gives rise to the Aurora Borealis and Aurora Australis, as will be described later.

Solar flares, originating in the Sun's chromosphere and corona, are violent explosions whose energy is believed to be derived from the 'breaking' and then 'reconnection' of the Sun's magnetic field lines and the resulting 'release' of magnetic energy. A total energy of between 10^{29} and 10^{32} ergs, released over periods of minutes to hours, accelerates electrons, protons and heavier ions to relativistic speeds – that is, close to the speed of light. Flares tend occur above the active regions around sunspots, which is where intense magnetic fields emerge from the Sun's surface into the corona, and thus tend to be more frequent at time of solar maximum.

Flares are related to what are called 'coronal mass ejections', which are the ejection of material from the solar corona consisting largely of electrons and protons with small quantities of heavier elements such as helium, oxygen and iron. The material carries with it part of the coronal magnetic field. The streams of highly energetic particles have been observed to take as little as 15 minutes to reach the Earth (so travelling at about one-third the speed of light). They can pose a threat to astronauts and have, in the past, destroyed satellite sub-systems. Interference with short-wave radio communication can also occur, and the interaction of the entrained magnetic field with electricity power transmission cables can cause problems and the possible shutdown of electricity grids – as happened in the Quebec province of Canada in March 1989.

The Aurora

One beautiful manifestation of the interaction of the solar wind with our atmosphere is coloured light displays observed in the night sky (Figure 2.6). They are most often seen within a band centred on the north and south magnetic poles and are known as the Aurora Borealis and Aurora Australis respectively. (Aurora is the Roman goddess of the dawn, Boreas is the Greek name for the north wind, and Australis is the Latin word for south.) The Aurora Borealis is often called the northern lights, as it tends to be seen as a green or reddish glow in the northern sky. It is most commonly seen around the vernal and autumnal equinoxes, though it is not known why this should be so.

Auroras are caused by the collision of charged particles with atoms high in the Earth's upper atmosphere. As the Earth's field lines open out into space above the north and south magnetic poles, charged particles may more easily reach the upper atmosphere in the regions near the magnetic poles. There, they collide with atoms of gases in the atmosphere, and lift atomic electrons into higher energy levels. The electrons then cascade down to their ground states so emitting light. Most auroral light appears to be emissions from atomic oxygen, resulting in a greenish glow at a wavelength of 557.7 nm and a dark-red glow at 630.0 nm. Amongst the many other colours that are sometimes observed, excited atomic nitrogen gives a blue colour whilst molecular nitrogen produces

Figure 2.6 The Aurora Borealis, also shown in colour in Plate 2.6. Image: US Air Force, by Senior Airman Joshua Strang, Wikimedia Commons.

a purple hue. Often the auroral glow is in the form of 'curtains' that tend to be aligned in an east–west direction. Sometimes these curtains change slowly but at other times they seem to be in continuous motion. Their shape is determined by the direction of the Earth's field in the region of the observer and observations have shown that electrons from the solar wind spiral down magnetic field lines towards the Earth. I can testify to the awesome sight that results when field lines guide electrons down to a bright auroral patch directly above the observer. Due to perspective, the converging auroral rays appear as vertical rays of light reaching upwards to what is called a 'corona' directly above.

Lunar and solar eclipses

If the plane of the Moon's orbit was the same as that of the Earth's orbit around the Sun, we would get eclipses of the Sun at new Moon and that of the Moon at full Moon. However, due to the fact that the Moon's orbit is inclined by about 5 degrees, the Moon is usually above or below the Sun–Earth line, and eclipses happen less often. It might be thought that the Moon would totally disappear during a lunar eclipse, but light that is scattered through the Earth's atmosphere falls on the Moon so we can still see it. If we observed the Earth from the Moon during a lunar eclipse we would see that it would have a red limb, as the dust in the Earth's atmosphere scatters blue light more than red, so only the red light reaches around the Earth and it is this light that we see reflected by the Moon. The Moon thus takes on a reddish hue at totality with the brightness and colour very dependent on the amount of dust in the Earth's atmosphere. Following a major volcanic eruption (such as that of Mount St Helens in 1980), so little light reaches the Moon that it can appear a very dull dark grey but conversely, when the atmosphere is relatively free of dust, it can appear a beautiful orange red.

Due to the tidal interactions between the Earth and the Moon, the Moon is gradually moving further away from the Earth. The fact that it now has an angular size that is usually just greater than that of our Sun gives rise to what is probably the most spectacular of celestial events – a total solar eclipse.

To give a total eclipse, the Moon must lie in the plane of the Solar System (the plane containing the Earth and Sun), be at new Moon and have an angular size greater than that of the Sun. The Moon's orbital period around the Earth is 27.3 days so it will cross the plane of the ecliptic (at what are called the nodal points) once every 13.65 days. The Moon is new every 29.5 days, so the two events will coincide at the lowest common multiple of 13.65 and 29.5. In fact, as the Moon does not have to be precisely at the nodal points of its orbit for an eclipse to be total, there is some tolerance, and solar eclipses will recur approximately every 177 days ($6 \times 29.5 = 177$ and $13 \times 13.65 = 177.45$).

But an eclipse is not necessarily total. Both the Earth and the Moon have elliptical orbits. If the Moon is furthest from the Earth (at what is called apogee, so making the Moon's angular size smaller) and the Earth is closest to the Sun (at perihelion, so making the Sun's angular size bigger) then the Moon cannot cover the Sun completely and we get what is called an annular eclipse, when a ring of the Sun is still visible around the edge of the Moon. This is often called a 'ring of fire' and is particularly spectacular when seen close to dawn or dusk when the Sun's light is reddened.

During a total eclipse the Moon's shadow will first touch the Earth's surface at the start of what is termed the 'eclipse track'. The width and length of the track will depend on positions of the Moon and the Earth in their orbits. Close to the Earth's equator, the Moon is nearer than at the poles, so its angular size is greater and the eclipse track is wider, and it is here that the longest period of totality can be observed. The longest possible total eclipse is approximately 7 minutes 30 seconds long. This would be observed at the equator when the Earth is at aphelion (farthest from the Sun) and the Moon is at perigee (closest to the Earth). The longest eclipse that may be observed in the next 3,000 years is on 16 July 2186 and it will have a duration of 7 minutes 29 seconds. The longest in recent times was observed by many from Hawaii on 11 July 1991, when totality lasted 6 minutes and 53 seconds. It is not a coincidence that the dates within the

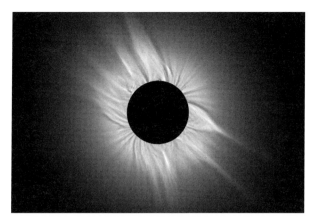

Figure 2.7 The total solar eclipse of March 2006. Image: Koen van Gorp.

year are so close. The Earth is at aphelion on 4 July, so the longest periods of totality will occur close to that date.

The ancients observed that sets of eclipses, both solar and lunar, recurred every 18 years 11 days and 8 hours – a time period called the 'saros'. Because of the one-third of a day in this period, each set of eclipses is observed one-third of the way around the globe at successive saros periods. Solar eclipses will thus be seen in the same region of the Earth after three saros periods: 54 years and 34 days.

It is during a total eclipse that we have a chance to observe the Sun's corona, as shown in the beautiful image in Figure 2.7, taken by Koen van Gorp. Around solar minima, one can usually observe long streamers stretching out into space roughly aligned to the solar equator, whilst near the poles the pattern of the Sun's magnetic field can be seen – with both shown in Figure 2.7. Near solar maxima, the corona appears more chaotic and less structure can be seen.

Suggestions for further reading on the Sun:

Nearest Star: The Surprising Science of our Sun by Leon Golub and Jay M. Pasachoff (Harvard University Press).
The Sun: A Biography by David Whitehouse (John Wiley & Sons).

3

Aspects of our Solar System

How do we know that, given a suitably large bowl, Saturn could float? This chapter (which is not based on a Gresham lecture) will look at our own Solar System as an example of solar systems in general and how astronomers have been able to measure the sizes and masses of the planets – so we can calculate their density and thus discover that Saturn would float in water. It will discuss aspects of planetary orbits, how the Sun's radiation and the properties of the planets determine their surface temperatures, and how atmospheres form and change during the lifetime of a planet.

What is a planet?

There had never been a formal definition of what should, or should not, be a planet and for some time the minor planet Ceres, which is the largest body in the asteroid belt between Mars and Jupiter, had also been classed as a planet.

In 2005, the discovery of a body (initially called 2003 UB_{313}) was announced. It is slightly larger than Pluto and was at a distance of 96.7 AU from the Sun (three times the distance of Pluto). This required a decision as to whether it should become the 10th planet of the Solar System or whether, instead, Pluto should be demoted. Pluto is considerably smaller than it was thought to be when first discovered and has a highly elliptical orbit inclined at a large angle to the plane of the Solar System. Should it have been discovered in recent times, it is highly unlikely that it would have been given the status of a planet.

The Hayden Planetarium

When the prestigious Hayden Planetarium in the American Museum of Natural History in New York came to mount a brand new Solar System

Figure 3.1 The eight planets of our Solar System. Images: NASA: Mercury – Mariner 10; Venus – Magellan; Mars – Mars Global Surveyor; Jupiter – Cassini; Saturn, Uranus and Neptune – Voyager. Wikimedia Commons.

exhibition, its Director, Neil deGrasse Tyson, decided that Pluto would have no place in it – the Solar System would only have eight planets – and that Pluto was just one of countless ice dwarfs that lay beyond Neptune. When the *New York Times* highlighted this fact on 22 January 2001 there was an outcry from both the public and some well-known scientists. In many ways Pluto was a favourite amongst young people, not least because Micky Mouse's dog Pluto had been given that name following its discovery. (Micky's dog had first been called Rover and only became Pluto in *The Moose Hunt*, which premiered on 3 May 1931.) Tyson stuck to his guns and the exhibition went ahead unchanged!

This question as to what should be classed as a planet was not seriously considered until a meeting of the International Astronomical Union (IAU) in August 2006.

The definition that was agreed had three parts:

A planet:
- orbits the Sun,
- has enough mass so that gravity can overcome the strength of the body and so it becomes approximately round and is said to be in hydrostatic equilibrium,
- has 'cleared' its orbit – that is, it is the only body of its size in the region of the Solar System at that distance from the Sun.

The first two parts are fairly obvious, but the third is less so. Essentially it means that there must not be other comparable sized objects orbiting the Sun at similar distances. It is part three that demotes Pluto.

The IAU also produced a definition of what would become known as 'dwarf planets'. These satisfy parts one and two of the definition of a planet, but not part three. In addition they must not be the satellite of another body. As a result, Pluto is now classed as a dwarf planet along with 2003 UB_{313}, which has now been given the name Eris. The minor planet Ceres also satisfies the definition of a dwarf planet, so initially we had eight planets and three dwarf planets in the Solar System. It is likely that, over time, the number of dwarf planets will increase as further large objects are discovered in the region beyond Neptune.

By the end of 2013, the IAU had recognised two others: Haumea and Makemake, though these have not been observed in enough detail to be sure that they fit the definition. If one assumes that the albedo (reflectivity) of a distant object is less than one (as it must be), then, from its brightness and distance one can estimate its minimum diameter. Should this be greater than 838 km, and hence the object will be made spherical by gravity, the IAU is allowing such objects to be named under the assumption that they will be dwarf planets.

In August 2011, Mike Brown, who had led the Palomar Observatory team that discovered Eris, published a list of 390 candidate objects, ranging from 'nearly certain' to 'possible' dwarf planets. He regards eleven of these as 'virtually certain' to be dwarf planets, including Quaoar and Sedna, with another dozen highly likely to be so.

Planetary orbits

Planets orbit the Sun in elliptical orbits with the Sun at one focus. A key parameter is the semi-major axis, a, which is half the major axis of the ellipse. For a circular orbit, a will be the radius of the orbit. Due to the eccentricity, e, of their orbits, their distance from the Sun varies from a minimum distance (at perihelion) of $a(1 - e)$ to a maximum distance (at aphelion) of $a(1 + e)$.

In our Solar System, Venus, whose orbit has an eccentricity of 0.007, is in a virtually circular orbit. Neptune and the Earth have near circular orbits with eccentricities of 0.01 and 0.17 respectively. Mercury and the dwarf planets Pluto and Eris have the most eccentric orbits, with eccentricities of 0.205, 0.249 and 0.441 respectively. It is worth noting that, for part of its orbit, Pluto can come closer to the Sun than Neptune.

An interesting consequence of the eccentricity of the orbit of Mars is that when Mars comes closest to us (every 2 years and 2 months) its distance from us can vary significantly. As a result, its angular size at closest approach will vary by almost a factor of 2 and so determine the amount of surface detail that we can see from Earth. Mars is closest to us within a few days of opposition – that is, when it is on the opposite side of the sky to the Sun – and will thus be seen approximately south at midnight. Mars will be seen with the smallest angular size at opposition when Mars is furthest from the Sun (at aphelion) and the Earth is closest to the Sun (at perihelion) as shown in Figure 3.2(a). Figure 3.2(b) shows the inverse situation when Mars will be seen with the largest angular diameter during opposition.

The Earth is at aphelion, furthest from the Sun, on 4 July each year, so the very closest approaches of Mars will occur in the summer months. The closest approach for nearly 60,000 years occurred on 27 August 2003, when Mars was 55,758,006 km from Earth and had an angular diameter of just over 25 arcseconds. In contrast, if Mars is at aphelion and the Earth at perihelion at the time of closest approach, as shown in Figure 3.2(b), then the angular size is just less than 14 arcseconds – a very significant difference!

The angular sizes observed at opposition are currently increasing until, on 27 July 2018, Mars will have an angular diameter of 24.31 arcseconds – only just less than the absolute maximum.

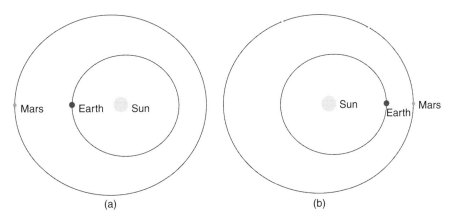

Figure 3.2 The situations when Mars is seen with the smallest (a) and largest (b) angular sizes when at opposition.

Orbital inclination

The orbital inclination of a planet's orbit is the angle at which the orbital plane of a planet is inclined to the plane of the Solar System. The plane of the Solar System is defined to include the Earth, so that the orbital inclination of the Earth's orbit is zero. The inclination angles tend to be small except in the case of Mercury, at 7 degrees, and the dwarf planets, Pluto and Eris, at 17 and 44.2 degrees respectively.

Planetary properties

Planetary masses

It is possible to find the mass of a planet if:

- it has one or more natural satellites in orbit around it, as is the case for the planets Earth, Mars, Jupiter, Saturn, Uranus and Neptune and the dwarf planets Pluto (Charon) and Eris (Dysnomia);
- it has acquired an artificial satellite, as in the case of the Magellan spacecraft in orbit about Venus and the MESSENGER spacecraft orbiting Mercury;
- it has been passed by an artificial satellite, as was the case when Mariner 10 flew by Mercury.

It has thus been possible to calculate the mass of all the Solar System planets and two of its dwarf planets. Let us take Mars as an example.

Calculating the mass of Mars

Mars has a satellite, Phobos, that orbits Mars with a period of 7 hours 39.2 minutes (27,552 seconds) in an almost circular orbit having a semi-major axis of 9,377.2 km or 9.3772×10^6 m. Given these values and knowing the universal constant of gravitation, G, it is a simple calculation to determine the mass, assuming a circular orbit. Having done so, I was gratified to find that my value of 6.43×10^{23} kg was very close to the accepted value for the mass of Mars of 6.42×10^{23} kg. The simple calculation is slightly in error as it does not take into account the slight eccentricity of the orbit of Phobos.

Planetary densities

From the angular size of a planet and its distance one can calculate the diameter of a planet and hence its volume; so, given its mass, one can thus calculate its density. It is interesting to use Saturn as an example. Saturn is not a sphere but an oblate spheroid having a greater equatorial than polar radius;

using an 'average' value of its radius of ~59,000 km, the volume is calculated to be 7.76×10^{23} m^3. Given Saturn's mass of 5.7×10^{26} kg, this gives a density of ~662 kg/m^3, which is a little less than the accepted value of 687 kg/m^3. You will note that this is less than that of water (1,000 kg/m^3 at 4 °C)!

Rotation periods

For some planets, such as Mars, Jupiter and Saturn, one can observe the rotation of a marking on the surface or in the atmosphere – such as the 'red spot' that lies in the atmosphere of Jupiter.

The surface of Mercury is very indistinct as seen from Earth, and Venus is cloud covered! In the case of Mercury, observations of faint markings seemed to imply that it was tidally locked to the Sun (as is the Moon to the Earth) and so its rotation period equalled that of its orbital period of 88 (Earth) days. In this case, one hemisphere would always face the Sun and be hot whilst the other, facing into outer space, would be very cold. A problem arose when the temperature of the surface facing the Earth at inferior conjunction (when Mercury lay between the Sun and the Earth and so we were observing the side away from the Sun) was measured from its infrared and radio emission. It was far hotter than expected, implying that at some point in the recent past it *had* been facing the Sun. The solution to this problem came when its rotation period was measured by radar to be 58.647 days and so Mercury orbits twice in every three of its years. It appears that it is effectively tidally locked when it is closest to the Sun in its elliptical orbit and can rotate about its axis when further away.

Radar observations made in the 1960s showed that Venus had a very slow rotation rate, taking 243.01 days to rotate once round its axis – 18.3 days longer than it takes to orbit the Sun! Even more surprising, it rotates in the opposite direction to that expected. Looking down from above the Solar System all the planets move round the Sun in an anticlockwise direction. The spin of most planets is also in an anticlockwise direction, but Venus (along with Uranus and Pluto) rotates in a clockwise direction, in the opposite sense to its orbital motion, and the rotation is said to be retrograde.

Planetary temperatures

There are three ways that we can measure or estimate the surface temperature of a planet:

- In the case of Mercury, Venus and Mars, spacecraft on the surface have made direct measurements.
- The temperatures of the outer planets can be estimated from their infrared emission.

- We can calculate a nominal temperature on the assumption that a planet acts like a black body and will radiate away the energy that it receives from the Sun. (There must be an equilibrium between the energy absorbed from the Sun and that emitted by a planet. This is not quite true in the case of Jupiter, which is very slowly contracting and so gravitational potential energy is being converted into heat.)

If one carries out this calculation for the Earth one gets a value of 278 K. This is not far off the actual average surface temperature, but should it be? The Earth is, on average, about 50% cloud covered and absorbs only ~77% of the incident solar radiation from the Sun. Taking this into account by reducing the incident energy by 0.77, the Earth's temperature would only be ~260 K. But the 'greenhouse effect' produced by the carbon dioxide, methane and water vapour in the atmosphere prevent the Earth radiating away as much energy as would a perfect black body, so increasing the Earth's temperature. (Greenhouse gases absorb infrared radiation emitted by the Earth and then re-emit it in random directions – so much of the infrared radiation will thus be directed back at the Earth.) The two effects roughly cancel out giving us an average temperature, T_{Earth}, of ~288 K. It is worth pointing out that without the greenhouse gases in our atmosphere our planet would be uninhabitable.

Global warming

The major constituents of the atmosphere, nitrogen, N_2, and oxygen, O_2, are not greenhouse gases. This is because diatomic molecules such as these neither absorb nor emit infrared radiation. Carbon dioxide, CO_2, is the main greenhouse gas in the atmosphere. Over aeons of time its percentage in the atmosphere has remained stable but, unfortunately, the burning of fossil fuels (which have stored carbon within them) is rapidly increasing the amount of carbon dioxide in the atmosphere and this is almost certainly a major contribution to the fact that our Earth's temperature has increased during the last 70 years. This has been termed 'global warming' or 'climate change'. (As the Earth's temperature has not increased during the early years of the twenty-first century, the term 'global warming' has been quietly dropped.)

Water vapour is a naturally occurring greenhouse gas and actually accounts for the largest percentage of the greenhouse effect, somewhere between 36% and 66%. The amount of water vapour in the air from locality to locality is very variable, but overall, human activity does not directly affect water vapour concentrations (except near irrigated fields, for example) and its effects on the Earth's climate are remaining stable.

However, the amounts of two further greenhouse gases are now also increasing:

(1) Methane (CH_4) is 20 times more efficient at retaining heat than carbon dioxide and we are adding up to 500 million tonnes of methane into the atmosphere per year from livestock, coal mining, drilling for oil and natural gas, rice cultivation, and garbage decaying in landfills.
(2) Each year, 7–13 million tonnes of nitrous oxide is added to the atmosphere from the use of nitrogen-based fertilisers, the disposing of human and animal waste in sewage treatment plants, and automobile exhausts.

An increase in the Earth's average temperature of more than 2 degrees could begin to have very harmful consequences for the human race – which explains why the problem is being treated so seriously.

Albedo

So, as the example of the Earth has shown, the actual temperature of a planet is affected by how much of the Sun's incident energy is reflected back into space – called the 'albedo' of a planet – and the effects of greenhouse gases, if any.

The Earth has an albedo of ~0.37, meaning that it reflects ~37% of the Sun's energy and so will absorb 63%. Venus has an albedo of ~0.7 (published values vary from 0.65 to 0.84) so that it only absorbs 30% of the incident solar energy, but its carbon dioxide atmosphere is so thick that its surface temperature is raised significantly. Mars has an albedo of 0.15 so it absorbs much of the incident solar energy, but its thin carbon dioxide atmosphere (about 1/100 that of the Earth) is unable to trap much heat so it is now too cold for carbon/water-based life forms to survive on the surface. However, in the past, when giant volcanoes were emitting vast amounts of gas into the atmosphere (including water vapour, carbon dioxide and methane) its temperature would have been significantly higher and life could, perhaps, have arisen there.

Planetary atmospheres

In every case, a planet's original atmosphere came from the solar nebula out of which the Sun and planets formed. Its composition was thus similar to that of the Sun, so largely composed of the light elements hydrogen and helium. This was the only source of the atmospheres of the outer planets but may not have contributed very much to the terrestrial planets as, by the time they had formed, the solar wind of the young star had ejected much of the solar nebula outwards beyond the inner planets. In addition, as will be seen below, an atmosphere of light gases could not have been kept by these relatively small planets due to their relatively high surface temperatures and low gravity.

Consider an atmosphere made up of a number of different gases, some of light molecules such as hydrogen and helium and some of heavier molecules such as carbon dioxide, ammonia and methane. The law of 'equipartition of energy' states that all species of molecules in the atmosphere will have roughly equal kinetic energies ($\frac{1}{2}mv^2$). This means that, for a given temperature, lighter molecules with smaller masses will have higher velocities than heavier molecules. The average kinetic energies of the gas molecules will depend on the temperature of the atmosphere so, in hotter atmospheres, the molecules will be moving faster.

For a given temperature (and hence kinetic energy), the velocity of a given molecule will be inversely proportional to the square root of its molecular mass, so molecules of hydrogen (molecular mass 2) will move on average four times faster than those of oxygen (molecular mass 32). If a molecule in the upper part of an atmosphere happens to be moving upwards at a sufficiently high velocity, then it could exceed the escape velocity of the planet and so escape into space. The escape velocity depends on the mass of the planet, so it should be apparent that hot, light planets might well lose all the lighter molecules that they might once have had in their atmospheres whilst cooler, more massive planets will be able to hold on to even the lightest molecules within their atmospheres.

For nitrogen (molecular mass 28) and oxygen (molecular mass 32) in the Earth's atmosphere at a temperature of ~300 K, the typical molecular speeds are 0.52 and 0.48 km/s respectively. This is far smaller than the escape velocity of the Earth, which is 11.2 km/s, so we would not expect these gases to escape from our atmosphere. In fact, it is not quite as simple as that. Due to collisions between them, molecules do not all move at the same speed; some are faster and some slower than the average. The relative numbers of molecules at speeds around the average is given by what is called the 'Maxwell–Boltzmann distribution'. A very small fraction of the molecules in a gas have speeds considerably greater than average, with one molecule in two million moving faster than three times the average speed, and one in 10^{16} exceeding the average by more than a factor of 5. So a very few molecules may be moving fast enough to escape, even when the average molecular speed is much less than the escape velocity. Calculations show that if the escape velocity of a planet exceeds the average speed of a given type of molecule by a factor of 6 or more, then these molecules will not have escaped in significant amounts during the lifetime of the Solar System.

In the Earth's atmosphere, the mean molecular speeds of oxygen and nitrogen are well below one-sixth of the escape speed. But consider the Moon: its escape velocity is 2.4 km/s and, assuming that any atmosphere it might have had would have been at the same temperature as our own, the mean molecular speeds of nitrogen and oxygen would be about five times less than the Moon's escape velocity, so it is not surprising that it has no atmosphere. If Mercury had an atmosphere it would have a temperature of ~700 K, giving nitrogen or oxygen an average molecular speed of about 0.8 km/s, significantly more than one-sixth of Mercury's escape velocity of 4.2 km/s. There has thus been ample time for these molecules to escape.

These arguments allow us to see why our own atmosphere contains very little hydrogen. Hydrogen molecules move, on average, at about 2 km/s, which is just more than one-sixth of the Earth's escape velocity. Hydrogen will thus have been able to escape and now makes up only 0.000055% of the atmosphere! In contrast, consider Jupiter: its escape velocity is 60 km/s and it has a surface temperature of only 100 K. In the Jovian atmosphere, the speed of the hydrogen molecules is only about 1 km/s, 60 times less than the escape velocity, and so Jupiter has been able to keep the hydrogen as the largest constituent in its atmosphere.

In summary

Mercury, the Moon and all satellites except for Titan and Triton have effectively no atmospheres, though Mercury has an extremely thin 'transient' atmosphere of hydrogen and helium temporarily captured from the solar wind.

The other terrestrial planets cannot hold on to hydrogen or helium, so will have lost all the initial atmospheres derived from the solar nebula.

The outer planets are both massive and cold and so have been able to keep all of the light gases acquired from the solar nebula. Though similar in mass to the Moon, Titan and Triton are sufficiently cold to have kept atmospheres, largely made up of nitrogen and methane.

The dwarf planets Pluto and Eris are so cold that any nitrogen or other gases would be frozen and form part of the surface.

Secondary atmospheres

Later in their life, the planets Venus, Earth and Mars gained further atmospheres that were the result of out-gassing from volcanoes. It is thought that only 1% of the current Earth's atmosphere remains from its primeval atmosphere. Volcanic eruptions produce varying amounts of gases, which arise from the melting of the planet's crust at great depth. All eruptions differ but, in general, release gases such as water vapour, carbon dioxide, sulphur dioxide, hydrogen sulphide, ammonia, nitrogen and nitrous oxide. It is thought that ultraviolet light falling on water vapour in the upper atmospheres of Venus and Mars would have split the water into hydrogen and hydroxyl (OH). The hydrogen molecules would then have escaped, so removing water vapour from their atmospheres.

The evolution of the Earth's atmosphere

We have seen how the primeval atmosphere of the Earth, largely made up of hydrogen and helium, would have been lost and replaced by a secondary atmosphere that was the result of volcanic out-gassing. This atmosphere was primarily made up of carbon dioxide and water vapour, with some nitrogen but virtually no oxygen, and contained perhaps 100 times as much gas as at present. As the Earth cooled, much of the carbon dioxide dissolved into the oceans and precipitated out as carbonates.

A major change began some 3.3 billion years ago when the first oxygen-producing bacteria arose on Earth and which, in the following billion years, gave us much of the oxygen in our atmosphere. Oxygen and bacteria could then react with the ammonia released by out-gassing to form additional nitrogen. More nitrogen was formed by the action of ultraviolet radiation on ammonia in a process called photolysis.

As the vegetation increased, the level of oxygen in the atmosphere increased significantly and an ozone layer appeared. This gave protection to emerging life forms from ultraviolet light and enabled them to exist on land as well as in the oceans. By about 200 million years ago about 35% of the atmosphere was oxygen. The remainder of the atmosphere was largely nitrogen as, alone of all the gases present in the secondary atmosphere, it does not readily dissolve in water.

Volcanic activity recycles and replenishes the molecules of the atmosphere and has, in particular, recycled the greenhouse gas carbon dioxide – necessary for the Earth's surface temperature to have remained sufficiently warm for the existence of life. The carbonates, formed as carbon dioxide dissolves in the oceans, and shells of calcium carbonate produced by marine life, fall to the ocean beds. Thus, over time, one might expect the amount of carbon dioxide in the atmosphere to reduce, tending to make the Earth colder. However, movements of the oceanic plates of the Earth's crust bring the plates up against the continental plates. As the oceanic plates are denser, they pass under the continental plates – a process known as subduction – and volcanic activity releases the carbon dioxide back into the atmosphere. Over the long term, a process that now gives rise to many problems for mankind, such as earthquakes and tsunamis, has been partly responsible for our presence here on Earth.

More on the Solar System:

> *An Introduction to the Solar System* edited by David A. Rothery, Neil McBride and Iain Gilmour (Cambridge University Press).
> *From Dust to Life: The Origin and Evolution of Our Solar System* by John Chambers and Jacqueline Mitton (Princeton University Press).

4

The rocky planets

Our knowledge of the Solar System has increased dramatically in the last 50 years by the use of spacecraft that have flown by, orbited and even landed on the planets, giving us high resolution images and detailed planetary data. This chapter will cover our Earth and its Moon and the planets Mercury, Venus and Mars, with the following two chapters covering the outer planets, highlighting aspects related to their properties, discovery and satellites. The material included is that which I have found of most interest and I can only hope that what I have found interesting, you will too.

Mercury

Mercury has only ever been visited by two spacecraft, Mariner 10, which made three flybys in the mid 1970s and, early in 2008, by the MESSENGER spacecraft, one of whose images is shown in Figure 4.1. Mercury looks very like the highland regions of the Moon.

In 1991, Mercury was observed by radar using the 70-metre Goldstone antenna with a half-million-watt transmitter. The radar reflections were received by the Very Large Array in New Mexico to provide high resolution radar images. To great surprise, a very strong reflection was received from Mercury's north polar region, which closely resembled the strong radar echoes seen from the ice-rich polar caps of Mars. At very low temperatures water ice is a very effective radar reflector.

One would not expect to find ice on Mercury, but close to the north and south poles the crater floors are permanently in shade, with temperatures as low as 125 K, so any ice existing there could remain for billions of years. It is thought that the ice, also later observed at the south pole, is the remnant of comets that have impacted with Mercury in the past.

Figure 4.1 Mercury as photographed by the MESSENGER spacecraft as it flew by on 14 January 2008. Image: MESSENGER, NASA, JHU APL, CIW.

In 2004, a second NASA spacecraft left the Earth for Mercury. It made its first flyby in January 2008 and entered an elliptical orbit around it in 2011 after two further flybys. One might, perhaps, be surprised that it took seven years to enter an orbit around Mercury, but it is harder to achieve this than one might think. Any spacecraft travelling to Mercury is, in some sense, 'falling' towards the Sun and gains considerable kinetic energy as it does so. So the problem is not getting there, but slowing down sufficiently to be able to orbit it! It actually takes more energy to orbit Mercury than to escape from the Solar System.

Europe and Japan are planning a joint mission to Mercury to be launched in August 2015. Called BepiColombo, it will reach Mercury in January 2020. Two orbiters will be carried to Mercury. The first is the Mercury Planetary Orbiter, which will carry an imaging system consisting of wide-angle and narrow-angle cameras, infrared, ultraviolet, gamma, X-ray and neutron spectrometers along with a telescope to detect near-Earth objects, a laser altimeter and other experiments. The Mercury Magnetospheric Orbiter will carry a set of fluxgate magnetometers, charged particle detectors, a wave receiver, a positive ion emitter and an imaging system.

It had been hoped to include a small lander in the mission, which would have been designed to last for about one week on the surface, but this was cancelled in 2003 due to budgetary restraints.

Venus

As Venus is seen either shining brightly in the east before dawn or, at other times, shining in the west after sunset, it once had two names. The 'evening star' was called Vesperus or Hesperus, derived from the Latin and Greek words for evening, respectively, whilst the 'morning star' was called Phosphorus (the bearer of light) or Eosphorus (the bearer of dawn). It is said that the Greeks first thought that they were two different bodies but later came round to the Babylonian view that they were one and the same. There is a famous sentence in the philosophy of language, 'Hesperus is Phosphorus', that implies an understanding of this fact.

Venus is the brightest object in the night sky after the Moon. As was shown in Chapter 1, Venus's angular size varies by a factor of about 5 as it orbits the Sun. However, when it is further away from the Earth and hence has the smaller angular size, a greater percentage of its surface is seen illuminated. The two effects tend to cancel out, with the brightness staying almost constant for several months at a time. Venus appears bright as it has a very high albedo, reflecting ~70% of the sunlight falling on it, the result of a totally cloud-covered surface.

Twice every 120 years, at an interval of 8 years, Venus is seen to 'transit' across the face of the Sun. (In the twenty-first century: in 2004 and 2012.) Such transits were important historically as they gave a method of calculating the distance of Venus and hence, using Kepler's third law, of measuring the astronomical unit. Captain Cook's exploration of Australia followed an expedition to Tahiti to observe the 1768 transit of Venus.

Much of our knowledge of Venus has come from observations by spacecraft. In December 1962 the Mariner 2 spacecraft flew over the surface at a distance of ~35,000 km and microwave and infrared observations showed that, whilst the cloud tops were very cool, the surface was at a temperature of at least 425 °C. It found no evidence of a magnetic field.

The Russians made many, initially unsuccessful, attempts to land spacecraft on the surface. No one had anticipated an atmospheric pressure of about 100 times that of the Earth. As a result, the descent parachutes were initially too large. This slowed the spacecrafts' descent with the result that their batteries discharged before the craft reached the surface. Other craft were crushed by the great pressures in the lower atmosphere. Finally, in 1970, Venera 7 reached the surface, sending temperature telemetry back for 23 minutes. Veneras 9 and 10 then sent back the first images of the surface revealing scattered boulders and basalt-like rock slabs, as shown in Figure 4.2. On their way to observe Halley's Comet in 1985, two Russian Vega craft sent balloon modules into the

The rocky planets 41

Figure 4.2 Image of a basalt plain by the Venera 14 lander. Image: Soviet Academy of Sciences.

Figure 4.3 The surface of Venus as imaged by radar from the Magellan spacecraft. Image: Magellan Project, JPL, NASA.

atmosphere, which flew at an altitude of ~53 km above the surface and showed that there were high winds within a highly turbulent atmosphere.

Radar observations, initially from Earth, and then from an orbiting NASA spacecraft, Magellan, have given us detailed information about the surface structure. In four and a half years, Magellan mapped 98% of Venus's surface with very high resolution (Figure 4.3). About 80% of the surface comprises smooth volcanic planes with the remainder forming highland regions, one in the northern hemisphere and one just south of the equator. The northern 'continent' is called Ishtar Terra, about the size of Australia and named after the Babylonian goddess of love. It rises to a peak of 11 km above the plains in the Maxwell Montes. The southern 'continent', called Aphrodite Terra, after the Greek goddess of love, is somewhat larger, having a surface area comparable to

South America. The surface has impact craters, mountains and valleys along with a unique type of pancake-like volcanic feature called farra, which are up to 50 km in diameter and 1 km high. The surface is thought to be ~500 million years old and, as a result, we see far more volcanoes present than on Earth, where those older than ~100 million years have been eroded. The amount of sulphur dioxide in the atmosphere appears to be variable, which could indicate ongoing volcanic activity.

The atmosphere has been shown to largely consist of carbon dioxide with a small amount of nitrogen. Its mass is 93 times that of the Earth, resulting in an atmospheric pressure at the surface of about 92 times that on the Earth. It contains thick clouds of sulphur dioxide, and sulphuric acid may even 'rain' in the upper atmosphere! But, due to the surface temperature of over 460 °C, this 'rain' would never reach the surface. This very high surface temperature is the result of the greenhouse gas effect of the very thick carbon dioxide atmosphere.

It is thought that Venus has a very similar internal structure to that of the Earth with a core, mantle and crust. The lack of plate tectonics – the relative movements of parts of the crust – may have prevented its interior cooling to the same extent as has our own planet and it is thought that its interior is partially liquid. Venus is almost a twin to our Earth with a diameter just 650 km less and a mass 81.5% that of our planet but, though once thought to have had an atmosphere much more like ours, evolution has taken it down a very different path!

The Earth

The Earth–Moon system (Figure 4.4) was formed when an object, thought to have had a mass of about 10% that of the Earth, impacted with the Earth soon after the formation of the Solar System. A portion of the combined mass was thrown off into space and formed the Moon, initially far closer to the Earth than now. As the Moon's relative mass and size is comparable to that of the Earth, the two are sometimes thought of as a 'double planet'.

The name 'Earth' derives from the Anglo-Saxon word 'erda' meaning ground or soil. This became 'eorthe' in Old English and 'erthe' in Middle English. Finally, from around 1400, the name 'Earth' was used – the only planetary name not derived from Greek and Roman mythology.

Out-gassing from volcanoes produced Earth's secondary atmosphere and water vapour condensed to form the oceans, with additional water provided by the impact of comets. About 4 billion years ago a self-replicating molecular system – life – arose. Photosynthesis allowed the trapping of solar energy and the by-product, oxygen, accumulated in the atmosphere. A resulting layer of

Figure 4.4 Earthrise as seen by the Apollo 8 crew as they rounded the Moon's far side in December 1968. Image: NASA Apollo 8 Crew.

ozone reduced the flux of ultraviolet radiation on the surface, which allowed life forms to survive on land.

The surface of the Earth has been shaped by plate tectonics – the movement of sections of the crust across the underlying magma – which at times formed vast continents such as Pangaea, allowing species to colonise much of the surface.

In the Cambrian era, which followed a period of extreme cold when much of the Earth was covered in ice, multi-cellular life forms began to flourish. In the ~535 million years since then, there have been five 'mass extinctions' when many species died out. The last of these was 65 million years ago when a large asteroid or comet at least 10 km in diameter formed the Chicxulub crater on (and offshore of) the Yucatan Peninsula in Mexico. The dust produced would have reduced the amount of sunlight reaching the ground and hence the growth of vegetation. This may have been one of the causes of the demise of the dinosaurs at about this time along with ~70% of all species then living on the Earth. Shrew-like small mammals were spared and their evolution finally gave rise to human beings.

Our Earth is a rocky body with a core, mantle and crust. Its rotation causes a bulge at the equator whose diameter is 43 km greater than the polar diameter. As a result, the peak of Mount Chimborazo in Ecuador is the point that lies furthest from the Earth's centre. With an extreme variation from 8.8 km above sea level (Mount Everest) to 10.9 km below (the Mariana Trench), the surface of the Earth is actually smoother than a billiard ball!

The central core has a temperature of ~7,000 K as a result of the decay of radioactive isotopes of potassium, uranium and thorium, all of which have half lives of over 1 billion years. Convection currents within the molten rock brought this heat up towards to the crust giving rise to thermal hotspots and produced the volcanic activity that gave the Earth its secondary atmosphere. The atmosphere is now largely composed of nitrogen (78%) and oxygen (21%), with the remaining 1% made up of water vapour, carbon dioxide, ozone, methane and other trace gases. It is worth pointing out again that without the warming due to the greenhouse gases, carbon dioxide, water vapour and methane, the average surface temperature would be about −18 °C and life would almost certainly not exist.

Ancient peoples put the Earth at the centre of the Universe and regarded the human race as special. As it was found that our Sun is one of billions of stars in our Galaxy, a principle of mediocrity arose. Planets like Earth were thought to be very common and so advanced life like ours was thought to be widespread – we were not special. As we have learnt more about the history of our Earth, how plate tectonics has helped recycle carbon dioxide back into the atmosphere, how Jupiter prevents too many comets from impacting the Earth, and how our large moon has stabilised the Earth's rotation axis, some scientists now believe that the conditions that have allowed intelligent life to arise here may well be very uncommon. It really could be that we *are* special, and it is not impossible that we are the only advanced life form within the Milky Way Galaxy.

The Moon

The Moon is the fifth largest satellite in the Solar System. It has a diameter slightly more than a quarter that of the Earth and its average distance from us is about 30 times the Earth's diameter. The gravitational pull on its surface is about one-sixth that on the Earth. Due to the fact that it has an elliptical orbit, its angular size varies by about ~12% – from 0.5548 degrees at perigee, when it is closest to the Earth, down to 0.4923 degrees at apogee, when it is furthest from the Earth. Partly due to its eccentric orbit, partly due to the inclination of its orbit and partly due to the fact that at moonrise and moonset we see it from different relative positions in space, we can observe a total of 59% of the Moon's surface at one time or another. This effect is called 'libration'.

The well-known 'Moon' illusion makes the Moon appear largest when near the horizon. However, it will of course be closer to us when highest in the sky (about one Earth radius nearer if we were on the equator) and its angular size will actually be about 1.5% larger! Our perception of size is linked with how far away we believe an object is from us. It appears that we 'see' the celestial sphere

Figure 4.5 The Moon. Image: Ian Morison.

above us not as a true hemisphere, but as one that is flattened overhead so that we 'believe' that the objects above us in the sky are 'nearer' to us than those near the horizon. So, observing the Moon above us, we 'believe' it to be closer and mentally reduce its perceived size.

The Moon, seen in Figure 4.5, only reflects about 8% of the light incident upon it and is one of the least reflective objects in the Solar System, reflecting about the same proportion of light as a lump of coal. The side of the Moon that faces Earth is called the near side, and the opposite side the far side (not the 'dark' side). Even with the unaided eye one can clearly see that there are two distinct types of surface on the near side. We see light regions called 'highlands' and darker areas of the surface that we call 'maria', so called because they were thought to be seas and oceans and were given beautiful names such as Oceanus Procellarum, Mare Tranquillitatis and Sinus Iridum – the Ocean of Storms, the Sea of Tranquillity and the Bay of Rainbows. When the far side of the Moon was first photographed by the Soviet probe Luna 3 in 1959, a surprising feature was its almost complete lack of maria.

We now know that the maria regions are covered by basaltic lava. Many are circular, such as Mare Crisium, which implies a number of giant impact events. These would have shattered the lunar crust, allowing lava to well up and fill the impact depression. The fact that the maria are relatively lightly cratered compared to the highland regions shows that they were formed towards the end of the period in the early history of the Solar System when the surfaces of the planets and moons were being bombarded with the debris left over from its formation.

On the near side, maria cover about 32% of the surface, but the far side has only a few patches making up just 2% in area. This requires an explanation, and it appears that several factors may have contributed. Elements required to produce heat by radioactive decay seem to be concentrated in the near-side hemisphere so that volcanic activity – producing the lava that filled the impact basins – would have been more prolific. The crust may also be thinner, making it easier for the lava to have broken through to the surface. Perhaps less likely is the fact that the Earth may have gravitationally 'captured' objects that might have impacted the near side and so destroyed the smooth maria regions.

The lighter-coloured regions of the Moon are commonly called highlands, since they are higher than most maria, though their formal name is 'terrae'. We should expect them to be higher. The basalt rocks that make up the maria are denser than the rocks making up the highland regions. For the Moon to be in hydrostatic equilibrium when at earlier times its interior was partially molten, at some constant depth below the surface the pressure must be equal for all regions. This implies that the column mass above this depth (the mass of, say, a one metre diameter column to the surface) must be equal. This means that the columns of lighter material must be taller, so the highland regions will rise above the maria.

Around the rims of the giant impact craters are seen several prominent mountain ranges – the surviving remnants of the impact basins' outer rims. Interestingly, four regions on the rim of the crater Peary, at the Moon's north pole, are illuminated throughout the lunar day, making this a possible site for a lunar base as solar panels could be used to provide a constant energy supply. This is due to the fact that the Moon has a very small tilt to the plane of the ecliptic. A further consequence of this fact is that regions at the bottom of craters near the pole are in permanent shadow. This would have allowed ice, lodged there from cometary impacts, to have remained and thus be able to provide a source of water, oxygen and hydrogen. However, recent radar observations from Earth suggest that the radar signature indicative of water ice might, instead, be due to ejecta from young impact craters. The presence of significant water ice on the Moon has still to be proven.

Very obvious features of the Moon's surface are the lunar impact craters formed when asteroids and comets collided with the lunar surface. There are about half a million craters with diameters greater than 1 km. The largest is some 2,240 km in diameter and 13 km in depth. Two prominent young craters seen on the near side are Tycho and Copernicus. Near full Moon, light-coloured rays of ejecta can be seen radiating from Tycho, so it is termed a 'rayed crater'.

The surface of the Moon is covered by what is called the 'lunar regolith'. It has a thickness of about 3 to 5 m in the maria regions and 10 to 20 m in the highland regions. Beneath the regolith lies a region of highly fractured bedrock about 10 to 40 km deep. It is believed that, like the Earth, the Moon has a crust (about 50 km thick), mantle and core. When, following its formation, the Moon was molten, the heavier elements fell towards its centre – a process called 'differentiation' – to give a dense core. The core, composed largely of iron, is thought to be small, with a radius of less than 350 km, and is, at least partially, molten.

Tides

Tides in the oceans are the result of the inverse square law of gravity and the size of the Earth. The gravitational force on the side of the Earth nearest to the Moon is thus greater than at the midpoint and even more so than that at the far side. This causes a differential gravitational effect called a 'tidal force', whose effects fall off as the fourth power of the distance. The result is to stretch out the Earth's oceans into an ellipse so that the sea level is higher closest to the Moon (as might well be expected) but also higher on the far side (less obvious) as the force on the oceans there is less than at the midpoint. As the Earth spins on its axis, these two bulges rotate around the Earth, so we normally get two high tides per day. As the Moon is also moving around the Earth in a period not too far off 24 days, the two high tides will appear earlier by about one hour per day.

The Sun causes a tidal force as well, but its effects are less (at about 46%) than that of the Moon. At new and full Moon, the tidal forces reinforce to give what are called 'spring tides', whilst at first and third quarter they partially cancel giving 'neap tides', which have a reduced tidal range. As both the Moon and the Earth are in elliptical orbits, the height of the spring tides can vary significantly. When the Moon is at perigee (nearest the Earth) the tides will be higher, whilst if this is the case when the Earth is at perihelion (nearest the Sun) in winter, the tidal forces are greatest and we get the very highest tides.

The gravitational coupling between the Moon and the oceans affects the orbit of the Moon. Due to the Earth's rotation, the tidal bulges do not point directly towards (and away from) the Moon. The overall effect of this asymmetry is to transfer angular momentum from the Earth's rotation to the Moon. As a result, the Moon is moving to a higher orbit with a longer period and, each year, the separation of the two bodies is increasing by about 3.8 cm.

Lunar exploration

The Moon has been studied more than any other body in the Solar System: it has been imaged from above by Lunar Orbiters, its surface studied by a number of landers – the first being Lunar 9 in 1965, followed by the Russian lunar rovers and the NASA Surveyor craft. Lunar exploration culminated in the NASA Apollo programme when six spacecraft landed men on the Moon. Samples of Moon rocks have been brought back to Earth by three Russian Luna missions (Luna 16, 20 and 24) and the Apollo missions 11, 12, 14, 15, 16 and 17.

As shown in Figure 4.6, the Apollo missions left science packages to measure heat flow, magnetic fields and seismic oscillations along with corner-cube light reflectors. Each reflector assembly contained 100 reflecting elements similar to 'cat's eyes'. Light from laser-equipped telescopes on Earth could be reflected to enable the distance from the telescope to the Moon to be measured to an accuracy of less than a centimetre. These have enabled the Moon's orbit to be measured with very high precision and, one might point out, provide irrefutable evidence that the Apollo missions *did* go to the Moon! There is an interesting

Figure 4.6 Tranquillity Base. The Eagle lander lies behind the seismograph. The data that it acquired were transmitted back to Earth using a cylindrical antenna. Just behind this antenna, at right angles to the Earth, is the Lunar Laser Reflector. Image: Neil Armstrong, Apollo 11 Crew, GRIN, NASA.

aspect of diffraction theory related to these reflectors: if the reflectors were perfect, the light would be reflected directly back to the point where the laser pulse had been transmitted. The problem is that the pulse actually returns to Earth two and a half seconds later, by which time the telescope on the Earth that had transmitted the pulse would have travelled some way around the globe. It would thus no longer be in a position where it could receive the returned laser pulse! The use of multiple small reflectors spreads the returned beam so that the pulse can be detected.

Following a 25-year lull in lunar exploration, in recent years spacecraft from Europe, the USA and China have returned to orbit the Moon and, at the end of 2013, the Chinese soft-landed a lunar rover called Jade Rabbit at the mouth of Sinus Iridum (the Bay of Rainbows). It carried a sophisticated payload, including ground-penetrating radar which could have gathered measurements of the lunar soil and crust. It could reportedly climb slopes of up to 30 degrees and travel at 200 m (660 feet) per hour. Its name derives from an ancient Chinese myth about a rabbit living on the Moon as the pet of the lunar goddess Chang'e. Sadly, its period of operation was short lived.

A Chinese mission to bring samples of lunar soil back to Earth is planned for 2017, which may set the stage for further robotic missions, followed perhaps by a crewed lunar mission in the 2020s.

Mars

Mars is often called the 'red planet' but, to my eyes, appears more of a salmon pink. Shown in Figure 4.7, it is a rocky planet having about half the diameter of the Earth but only one-tenth its mass. The reddish tint is due to oxides of iron on the surface known as hematite or rust, which forms a dust with the consistency of talcum powder. Its atmosphere is very thin, with density about one-hundredth that of the Earth's, and is largely composed of carbon dioxide (95%) along with nitrogen (3%), argon (1.6%) and traces of water vapour and oxygen.

As the Earth and Mars have similar axial tilts, it has comparable seasons, whose lengths are about twice those of the Earth's as the Martian year is approximately two Earth years in length. Martian surface temperatures range from approximately −140 °C during winter up to 20 °C in summer. Mars also suffers from dust storms, which can occasionally cover the entire planet's surface. It has two polar icecaps, which are primarily composed of water ice but covered by a layer of solid carbon dioxide (dry ice). At the south pole the carbon dioxide layer is ~8 m deep overlying ~3 km of water ice within a diameter of ~350 km. The north polar cap has a diameter of ~1,000 km and is ~2 km thick. During the

Figure 4.7 A computer-generated image of Mars. At the bottom right is Valles Marinaris, a canyon four times as deep and three times as long as the Grand Canyon on Earth. On the left are several volcanoes including Olympus Mons, a volcano three times higher than Mount Everest. At the top is the North Polar Cap made of thawing water ice and 'dry ice', solid carbon dioxide. Image: MSSS, JPL, NASA. Also shown in colour in Plate 4.7.

winter, dry ice builds up a ~1 m layer above the water ice and this causes a reduction of carbon dioxide in the atmosphere, so reducing the atmospheric pressure.

A civilisation on Mars?

Mars was first seen through a telescope by Galileo in 1609, but his small telescope showed no surface details. When Mars was at its closest to Earth in 1877, an Italian astronomer, Giovanni Schiaparelli, used a 22-cm telescope to chart its surface and produce the first detailed maps. They contained linear features that Schiaparelli called 'canali', the Italian for channels. However, this was translated into English as 'canals'– which implies a man-made water course – and the feeling arose that Mars might be inhabited by an intelligent race. It should be pointed that a waterway *could not* be detected from Earth, but it was thought that these would have been used for irrigation and so would have irrigated crops growing adjacent to them which *could* be seen from Earth.

Influenced by Schiaparelli's observations, Percival Lowell founded an observatory at Flagstaff, Arizona (later famous for the discovery of Pluto), where he made detailed observations of Mars which showed an intricate grid of canals. However, as telescopes became larger, fewer canali were seen though the surface showed distinct features. They appear to have been an optical illusion, but the myth of advanced life on Mars was not finally dispelled until NASA's Mariner spacecraft reached Mars in the 1960s. (An intriguing image of a huge rock formation, called the 'face on Mars', was obtained by the Viking 1 spacecraft in 1961, leading some to suspect that this was a giant representation of a past civilisation. However, a detailed photograph taken by Mars Global Surveyor in 2001 showed it to be a 'mesa' – a broad flat-topped rock outcrop with steep sides.)

The detailed images taken by the Mariner spacecraft showed giant canyons and vast volcanoes. One of these, Olympus Mons, is the largest known volcano in the Solar System with a caldera of 85 km in width surmounting the volcanic cone whose base is 550 km in diameter. The caldera is nearly 27 km above the Martian surface, three times higher than Everest! It was realised that when these giant volcanoes were active, some 3 billion years ago, they would have given Mars a far thicker atmosphere than it has now, and the effects of greenhouse gases in the atmosphere would have allowed the surface temperature to be sufficiently high for water to exist on the surface. Other visible surface features gave ample evidence of water flow over the surface, leading to speculation that simple life forms might then have existed on Mars.

This possibility led to the sending of two Viking landers to Mars in 1976. As well as imaging the surface and collecting scientific data, their objective was to search for any evidence of life. They conducted three experiments which, though discovering unexpected chemical activity in the Martian soil, provided no clear evidence for the presence of any living organisms. As Mars has a very thin atmosphere (and no ozone layer), far more ultraviolet light reaches the surface than on Earth. This would prevent the existence of life above ground. If life had once arisen on Mars, one could now only expect to find evidence for it beneath the surface.

It was nearly 30 years before the next probe to specifically search for evidence of life was sent to Mars. This was the UK designed and built Beagle II craft, which was due to land on Christmas Day 2003. It had left its mother ship, Mars Express, on a perfect trajectory a few weeks before, but appeared to have crash landed on its arrival at the surface – a sad moment for me as I had been charged with receiving its first signals from the surface using the 76-metre Lovell Telescope at Jodrell Bank.

Figure 4.8 A Martian panorama imaged by the Spirit rover at Gusev Crater. Image: Mars Exploration Rover Mission, Cornell, JPL, NASA.

Orbiters, rovers and the Phoenix lander

Following a period when many Mars probes seemed doomed to failure, recent successes have greatly increased our understanding of its surface, either when viewed with high resolution cameras orbiting the planet, such as those on Mars Reconnaissance Orbiter which began surveying Mars in November 2006, or with the rovers 'Spirit' (Figure 4.8) and 'Opportunity' which landed in 2004. One of these, Opportunity, was still operational at the beginning of 2014.

The rovers' primary scientific mission was to investigate a wide range of rocks and soils that might hold clues to past water activity on Mars. They were targeted to sites on opposite sides of Mars that appear to have been affected by liquid water in the past: Gusev Crater, a possible former lake in a giant impact crater, and Meridiani Planum, where mineral deposits suggested that Mars had a wet past. They have been highly successful, having together traversed over 20 km on the surface of Mars. In 2004, scientists showed pictures revealing a stratified pattern and cross bedding in the rocks inside a crater in Meridiani Planum, suggesting that water once flowed there, whilst an irregular distribution of chlorine and bromine suggested that it was once the shoreline of a, now evaporated, salty sea. To confirm the 'wet past' hypothesis, Opportunity has found hematite, in the form of small spheres nicknamed 'blueberries', which could only have been formed inside rock deposits soaked with groundwater.

Also in 2004, NASA announced that Spirit had found hints of evidence of past water in a rock dubbed 'Humphrey', which appeared to contain crystallised minerals lodged in small crevices. These minerals had been most likely dissolved in water and carried inside the rock before crystallisation. When, in

December 2007, one of Spirit's wheels was not turning properly, it scraped off the upper layer of the Martian soil and uncovered a patch of ground similar to areas on Earth where water or steam from hot springs has come into contact with volcanic rocks. Here on Earth, such locations are often teeming with bacteria as hot water provides an environment in which microbial life can thrive.

The rovers had never been expected to survive so long as it was thought that dust would soon cover their solar panels to such as extent that they could no longer function. But scientists had not realised that dust devils – mini tornadoes that can sweep across the Martian surface – could sweep the panels clean. By March 2005, Spirit's panels had dropped to 60% of their full capacity, but suddenly this increased to 93%! The following day Spirit was able to film a dust devil as it sped across the Martian surface. Sometimes major dust storms can fill the Martian atmosphere; when the orbiter Mariner 9 reached Mars in November 1971, the surface was totally shrouded by dust. As the storm gradually subsided, the first feature to be seen was the caldera of Olympus Mons rising high above the surface. Towards the end of June 2007, a series of dust storms blocked 99% of the direct sunlight to the rovers and they were facing the real possibility of system failure due to lack of power. They were both placed into hibernation to wait out the storms and happily survived to face another Martian year.

Further evidence of water locked up beneath the surface in a permafrost came when the Phoenix lander used its scoop to dig out a trench in the soil. This exposed sub-surface ice which, as would be expected due to the thin atmosphere, began to vaporise over the following days. This confirmed the observations made by the Mars Odyssey and Mars Reconnaissance Orbiter that there is ice beneath the surface of nearly all the northern half of Mars. But until we can drill down into the surface we will not know the depth, and hence the amount, of water ice lying beneath the surface.

Curiosity

In November 2011, Curiosity, a car-sized rover, was launched from Cape Canaveral for the 563,000,000 km journey to Mars. In August 2012 it was successfully lowered onto the surface of Gale Crater from its mother ship, which was hovering above. Curiosity's goals include investigation of the Martian climate and geology, an assessment of whether the landing site ever offered environmental conditions favourable for microbial life. In a set of experiments looking to the future, it also measured the exposure to radiation as it travelled to Mars, and it is continuing to monitor radiation levels as it explores the surface of Mars – providing data that would be important for a future manned mission.

Curiosity is powered by a radioisotope thermoelectric generator (RTG), as used by the successful Viking 1 and Viking 2 Mars landers in 1976, which provides 2.5 kWh per day. The RTG is fuelled by 4.8 kg of (non-fissile) plutonium-238 dioxide. Heat given off by its decay is converted into electricity by thermocouples, providing constant power during all seasons and through the day and night.

The analysis strategy first uses high resolution cameras to look for features of interest. When an interesting rock is found, Curiosity can vaporise a small portion with an infrared laser and examine the resulting spectra to study the rock's composition. If desired, the rover can then use its long arm to swing over a microscope and an X-ray spectrometer to take a closer look, and can even drill into the boulder and deliver a sample to analytical laboratories inside the rover.

Instruments are able to measure the Mars environment: humidity, pressure, temperature, wind speed, and ultraviolet radiation. Hazard avoidance cameras with a 120-degree field of view capture stereoscopic images, which are then processed by internal computers to safeguard against the rover crashing into unexpected obstacles.

In October 2012, the first X-ray diffraction analysis of Martian soil was performed, which revealed the presence of several minerals, including feldspar, pyroxenes and olivine. This suggested that the sample was similar to the 'weathered basaltic soils' of Hawaiian volcanoes.

The names of more than 1.2 million people were etched into silicon now installed on the deck of Curiosity. In addition, a plaque with the signatures of President Barack Obama and Vice President Joe Biden was also installed along with the autograph of Clara Ma, the 12-year-old girl from Kansas who gave Curiosity its name in an essay contest, writing in part that 'curiosity is the passion that drives us through our everyday lives'.

Moons

Mars has two moons, Phobos and Deimos, which may well be captured asteroids. Phobos (Figure 4.9) is a small, irregularly shaped object about 22 km in diameter. It orbits Mars at a height of 9,377 km, closer than any other satellite to its planet, and moves so fast around Mars that it actually rises in the west and sets in the east. Deimos is just 6 km across and in a nearly circular orbit. It takes just over 30 hours to complete one orbit, not that much greater than Mars's rotation period, so it will remain visible for ~2.7 days between rising and setting. It would appear star-like from the Martian surface as it subtends an angle of just 2.5 arcseconds and be about as bright as the planet Venus appears to us.

Figure 4.9 Phobos imaged by the Mars Reconnaissance Orbiter in March 2008. Image: HiRISE, MRO, LPL (University of Arizona), NASA.

Ceres and the minor planets

In 1768, Johann Elert Bode suggested that there might be a planet in orbit between Mars and Jupiter. He based this on the so-called Titius–Bode law, an empirical law proposed by Johann Daniel Titius in 1766, which (roughly) gave the relative orbital distances of the planets from the Sun. The law 'predicted' that a planet should lie at a distance from the Sun of 2.8 AU. With the discovery some years earlier of Uranus, which fitted the law's predictions quite well, a group of 24 astronomers, who became known as the 'Celestial Police', combined their efforts to make a methodical search for the possible planet. Though they did not discover Ceres, they did find several minor planets – also known as asteroids – in what is now called the 'main asteroid belt'.

The body that became known as Ceres was discovered on New Year's Day 1801 by Giuseppe Piazzi, who first thought it to be a comet. With diameter of ~950 km, Ceres has sufficient mass to be in hydrostatic equilibrium (making it round) and so has become the smallest of the dwarf planets. It contains almost one-third of the mass of the main belt asteroids, of which over 170,000 now have computed orbits. In 2006 the International Astronomical Union classed these as 'small Solar System bodies'.

It is thought that Ceres is sufficiently large to have become differentiated, that is, the heavier rocky elements have concentrated in the centre giving Ceres a core over which lies an ice mantle: a mixture of water ice and minerals such as carbonates and clay.

Suggestions for further reading:

> *The Solar System* by Marcus Chown (Faber & Faber).
> *The Planets: A Journey through the Solar System* by Giles Sparrow (Quercus).

5

The hunt for Planet X

This is a story that spanned over 200 years. It began with the discovery that Uranus was not following its predicted orbit and was thus presumably being perturbed by another, as yet undiscovered planet that, once discovered, became known as Neptune. This was followed by the search for what Percival Lowell called 'Planet X' (where X means unknown) that would lie beyond Neptune, and finally the search for a 10th planet beyond Pluto (where X means 10 as well). As we will see, the search for a 10th planet effectively ended in August 2006 when Pluto was demoted from its status as a planet and the number of planets in the Solar System was reduced to eight.

Uranus

Uranus was the first planet to have been discovered in modern times and though it is just visible to the unaided eye without a telescope it would have been impossible to show that it was a planet rather than a star, save for its slow motion across the heavens. Even when telescopes had come into use, their relatively poor optics meant that it was charted as a star many times before it was recognised as a planet by William Herschel in 1781.

William Herschel had come to England from Hanover in Germany where his father, Isaac, was an oboist in the band of the Hanoverian Foot Guards. As well as giving his third child, Friedrich Wilhelm Herschel, a thorough grounding in music, he gave him an interest in the heavens. When 15, William entered the band as an oboist and violinist and first came to England in 1755, when the Foot Guards were sent to help defend England from a feared French invasion. He soon returned to Hanover but in 1757, as the city was overrun by the French, he and his brother Jacob came to England for a second time. Jacob returned to Hanover to take up a position in the court orchestra but his brother, who had anglicised his Christian names to Frederick William, stayed to take up the position of instructor to the band of the Durham Militia.

Following a short period as a concert manager in Leeds, he was offered the post of organist at the Octagon Chapel in Bath, where he set up home. After some years he was able to persuade his family to allow his sister Caroline to come to Bath to act as his housekeeper. She had been acting as a servant in her father's house and William had to pay his father an allowance so that he could pay for a replacement! She repaid William's kindness with great devotion, giving up a career as a singer in order to assist him, and later became a significant astronomer in her own right.

William's interest in astronomy increased and he became very unsatisfied with the telescopes that he could buy, so decided that he would make a reflecting telescope of his own. In those days, the mirrors were cast in speculum metal – an alloy of two parts copper and one of tin – and he made the castings in the kitchen of his house. Occasionally the mould would break and the molten metal would crack the stone floor.

By 1778, he had built an excellent telescope (Figure 5.1) having a mirror of just over 6 inches (150 mm) in diameter and he began to make a survey of the whole sky. On the night of 13 March 1781 he observed an object that did not have the appearance of a star and he first thought that it was a comet.

Observations over the following months showed that it did not have the highly elliptical or parabolic orbit of a typical comet and that it was a planet in a nearly circular orbit, having a semi-major axis of just over 19 times that of the Earth.

It soon became apparent why Herschel had seen that it was a planetary body whilst others had not. Side by side comparisons with telescopes in use by others confirmed his telescope's far higher image quality – Herschel had proven to be a superb telescope maker! Uranus has a maximum angular size of 4.1 arcseconds and, as I have observed with an excellent telescope just a little smaller than Herschel's, it appears as a tiny greenish-blue disc. But unless a telescope has well-figured optics this disc would be very hard to distinguish from a star. The fact that the new planet had been charted, if not recognised as a planet, many times over the previous century allowed an accurate orbit to be computed and, later, this was to be a major factor in the discovery of Neptune.

Herschel quickly received acclaim and was made a Fellow of the Royal Society. The new planet became known as Uranus despite Herschel's wish for it to be named 'Georgium Sidus' (George's star) after King George III – a fellow Hanoverian. In 1782, Herschel demonstrated his telescope to the king and soon received a royal pension to allow him to devote himself exclusively to

Figure 5.1 Replica of the telescope with which William Herschel discovered Uranus.

astronomy. Five years later Caroline, who had become his observing assistant, also received a royal pension. To supplement their royal income Herschel made and sold telescopes, including a 25-foot-long (7.6 m) telescope that he made for the Madrid Observatory.

Neptune

Neptune can be seen in a small telescope and had even been observed by Galileo. Whilst observing Jupiter on 28 December 1612, he recorded Neptune as an 8th magnitude star, and a month later observed it close to a star on two successive nights. He noted that their separation had changed and could easily have reached the conclusion that this was because one was not a star but a planet! It was later observed by John Herschel, William's son, who also believed it to be a star.

The final discovery of the planet Neptune is one of the most interesting stories in astronomy, with its position being predicted independently by two mathematicians, John Couch Adams at Cambridge and Urbain Le Verrier at the Paris Observatory. This resulted from a key fact about Uranus in that it had been observed many times prior to the realisation that it was a planet by William Herschel. John Flamsteed had observed it several times from 1690 and allocated it the name 34 Tauri, and it was recorded several times between then and its eventual discovery as a planet. The French astronomer Pierre Lemonnier observed Uranus at least twelve times between 1750 and 1769. These so-called 'ancient observations', when combined with the more accurate observations made after its discovery as a planet, meant that an accurate orbit for Uranus could be immediately calculated. However, by 1821 it had become obvious that Uranus did not appear to be following its predicted obit accurately and the thought arose that its orbit might be being perturbed by an, as yet undiscovered, planet that lay beyond.

John Couch Adams had studied at St John's College in Cambridge and graduated in 1843 as the 'senior wrangler', the best mathematician in his year. He was elected to a fellowship of his college and decided to devote his study to the resolution of the problem of the orbit of Uranus, believing it to be due to a more distant planet. He derived a solution in September 1845 and, it is thought, gave it to the Professor of Astronomy at Cambridge, James Challis. On 21 September he visited the home of the Astronomer Royal, George Airy, but failed to find him in. He tried again on 21 October, but Airy was having dinner and would not see Adams, who then left a manuscript giving his solution. Airy found this difficult to follow and sent a letter to Adams requesting some clarification of the details, but it appears that Adams failed to reply – possibly because he had been upset at Airy's refusal to see him.

Meanwhile in Paris, Le Verrier, who had been asked to look at the problem by his director, had also derived a position for the planet. Perhaps because he was not well liked, there appears to have been no serious attempt to follow up on his prediction at the Paris Observatory and Le Verrier sent it to Airy. Airy realised the similarity of the predicted positions of the proposed planet and so, in July 1846, asked Challis (at Cambridge) to make a search for it. Unfortunately, Challis did not have a detailed star chart for the region where it was hoped that the new planet might be found so began to make one. If he then re-observed the positions of all the objects that he had charted, he would be able to spot the predicted planet by the fact that it would have moved slightly against the background stars.

On 18 September Le Verrier wrote to Johann Galle at the Berlin Observatory asking them to make a search. Galle received this letter on 23 September and his colleague Heinrich d'Arrest pointed out that they had a newly observed star chart for the appropriate region of the sky and thus they could easily spot the planet if it were present in the field. They made the observation that night and it took only 30 minutes to locate the planet just one degree away from Le Verrier's position!

In fact, Challis had noted on 29 September that one of 300 stars that he charted that night had shown a disc – just as a planet would – but, being a cautious man, had waited for further observations to show its motion through the stars. He had not been able to make them before the announcement of the planet's discovery in *The Times* newspaper of 1 October 1846.

William Lassell, who had made a fortune in the brewery trade in Liverpool, had built a 24-inch telescope and immediately made observations of the new planet in the hope of finding any satellites. He started his observations on the 2nd of October and on the 10th of October discovered Neptune's moon Triton.

The honour of Neptune's discovery is now shared by both Adams and Le Verrier. It is interesting to note that their predictions of Neptune's orbit were not that good and that for only about 10 years, from 1840 to 1850, did their predictions agree reasonably well with Neptune's actual position. More recently, the discovery of some papers that had been 'misappropriated' from the Royal Observatory at Greenwich cast some doubt on Adams' claim, and perhaps he does not deserve equal credit with Le Verrier.

With hindsight, it is actually quite easy to predict where Neptune was to be found. The diagram (Figure 5.2) of Neptune and Uranus has been drawn in a rotating co-ordinate frame so that Neptune appears fixed in space. As Uranus is

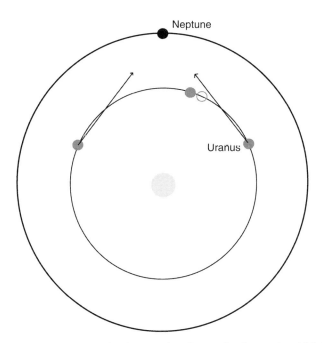

Figure 5.2 The orbit of Uranus in a frame of reference in which Neptune is stationary.

orbiting the Sun more rapidly and its orbit has a shorter circumference, it will pass Neptune 'on the inside track'. You will see that, as Uranus nears Neptune, the gravitational force between them will tend to advance Uranus in its orbit, so it will appear ahead of where it would be expected to be (the solid disc, rather than the open circle). Once Uranus has passed beyond the position of Neptune, their mutual gravitational attraction will slow Uranus down so that, eventually, it will have regained its expected position. One can see that Neptune should lie beyond the position of Uranus when Uranus is furthest ahead of its predicted position.

Pluto

The discovery of Pluto followed on from that of Neptune. Neither the orbit of Uranus nor that of Neptune was well defined, and it was suspected that there might be a more distant planet that was affecting their orbits – called 'Planet X' by Percival Lowell of Mars fame. In 1905 he had predicted its position and he began a photographic search at the Flagstaff Observatory. Nothing was found and, following more refined calculations, a further search was begun in 1914. He and the staff of his observatory continued the search without success until his death in 1916. Though Pluto did actually appear on some of the plates taken on 19 March 1915, it was not recognised as such as the object was far fainter than expected.

Due to problems with Lowell's will following his death in 1916, the observatory virtually ceased to function until 1929 when its then director, Vesto Melvin Slipher, began a new search. A young amateur astronomer, Clyde Tombaugh, had sent Slipher drawings (Figure 5.3) that he had made of Jupiter using his Newtonian telescope in the hope of being offered a job. These impressed Slipher, and he employed Tombaugh to take images with the observatory's 13-inch astrograph – essentially a wide field camera.

Two images taken some time apart were then compared in what is termed a 'blink comparator', in which the images are rapidly viewed in turn. Any object that has moved in the time between the two exposures would appear to jump in position whilst the stars would remain fixed. This allows planetary bodies to be rapidly located. The initial search for Planet X was unsuccessful, so Tombaugh began his own search.

From a pair of plates taken in January 1930, Tombaugh discovered a new planet on the 18th of February. The motion of the planet was confirmed in follow-up observations and on the 13th of March its discovery was announced. Its name was suggested by Venetia Burney, the 11-year-old daughter of an Oxford Professor, when she was told of its discovery the next day. Pluto was

Figure 5.3 Clyde Tombaugh with his home-built telescope along with drawings he made of the planet Jupiter.

the Roman God of the underworld who was able to make himself invisible. As the first two letters of its name were the initials of Percival Lowell, at whose observatory it had been discovered, this suggestion was eagerly accepted.

Initially it was thought that Pluto had a significant size – even though it could only be observed as a point 'star-like' object showing no angular size. It had to have a substantial mass to explain the perturbations in the orbits of Uranus and Neptune. With lobbying from the Lowell Observatory it is not surprising that it was given the status of a planet. There is a way to estimate its size that can be done by assuming its 'albedo' – the amount of sunlight that is reflected from its surface. For example, Venus, which is totally covered by cloud, has a high albedo of 0.65, meaning that it reflects 65% of the light falling on it. The Earth, which is partially cloud covered, has an albedo of 0.37 and Mars has an albedo of just 0.15. At the bottom end of the scale, Mercury and the Moon have albedos of just 0.11 and 0.07, respectively. Objects in the asteroid belt have low albedos, down to about 0.05, whilst a typical comet nucleus has an albedo of 0.04. At the other end of the scale, Enceladus, a moon of Saturn, has one of the highest known albedos of any body in the Solar System, with 99% of the light falling on it being reflected.

The apparent magnitude of a body will thus depend on its distance, its size and its albedo; so as we know Pluto's distance and if we assume its albedo we can estimate its size. Even initially, taking a mid-range albedo, it did not appear that large and, using a typical planetary density, would not have had enough mass to give rise to the assumed perturbations of Neptune's orbit that gave rise to Lowell's prediction of Planet X.

Given enough time, there is a method that will give an accurate measure of the size of a planet: sometimes a planet will pass in front of (occult) a star and the length of the occultation will depend on the distance of the planet from the Sun, its diameter and how closely the centre of the planet passes over the star's position. There will be a maximum length of occultation and from many such observations this maximum will be found and hence the diameter of the planet. Occultations of Pluto are rare, but it soon became apparent that it could be not that large. By 1955, Pluto's estimated mass had reduced to roughly that of the Earth, with further calculations in 1971 bringing it down to that of Mars. (At that rate, Pluto would soon cease to exist!)

A key discovery in 1976 found that Pluto's surface reflection signature matched that for methane ice and would hence have a high albedo of ~0.66, equivalent to that of Venus. Thus Pluto was exceptionally luminous for its size and this, of course, reduced its estimated size. Putting in a reasonable density, Pluto could not be of more than 1 or 2 per cent the mass of the Earth.

Charon

In the 1970s, James Christy had been making planetary observations at the Naval Observatory in Washington in order to refine their orbital parameters. He had been rejecting some images of Pluto as they had appeared somewhat elongated. This effect, as I know to my cost, usually appears when the telescope is not tracking correctly and images 'trail', so becoming elongated. However, he noted firstly that the star images in the photographic plate were perfect – indicating that the telescope *was* tracking correctly – and secondly that the extension of Pluto's image appeared to move around Pluto as time went by. It transpired that Christy was observing the motion of a satellite, now called Charon, orbiting Pluto, and its discovery was announced on 22 June 1978. In 1990, the Hubble Space Telescope was able to image the two separate discs of Pluto and Charon. Charon was the ferryman who carried the dead across the river Styx and had close ties to Pluto, so was an ideal name. Its first four letters were also the same as those of Christy's wife, Charlene.

If the period and semi-major axis of a satellite are known, it is possible to calculate the mass of the planet around which it is orbiting. For the first time, it became possible to accurately calculate the mass of Pluto, which was found to be just 2% that of the Earth. This was far too low to have had the effects on the orbits of Uranus and Neptune by which its existence had been first predicted. Pluto is

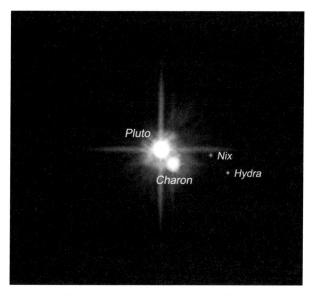

Figure 5.4 Hubble Space Telescope image of Pluto with three moons. Image: NASA, ESA, and STScI.

thus smaller and much less massive than the terrestrial planets. Observations of mutual occultations between Pluto and Charon also allowed an accurate measurement of Pluto's diameter to be made, which is close to 2,300 km. Pluto is thus more than twice the diameter and a dozen times the mass of Ceres, the largest object in the asteroid belt.

In 2005, two further moons of Pluto, shown in Figure 5.4, were discovered; they are now called Nix and Hydra. Appropriately, Nix was the goddess of darkness and night and the mother of Charon. Two further small moons, Kerberos and Styx, were discovered in 2011 and 2012, respectively.

On 18 January 2006 a spacecraft called New Horizons left the Earth for a nine-year voyage to Pluto and beyond. New Horizons passed Jupiter on 28 February 2007, using the planet's gravitational field to increase its speed to around 83,000 km/h (in what is called a slingshot manoeuvre) and, having flown over 4.8 billion km, will fly by Pluto and Charon in July 2015 before hopefully visiting some more distant Kuiper Belt objects in an extended mission.

The search for a 10th planet

Following the impacts of the fragments of the Comet Shoemaker–Levy 9 into Jupiter in July 1994, leaving scars on its surface larger than the size of the Earth (described in detail in Chapter 8), governments became very aware of the threat to the Earth that comets and asteroids might pose. They thus funded observing projects that would identify the so-called 'near-Earth objects' (NEOs) in the hope that we could identify those that were a threat to us in sufficient time for measures to be taken to avert an impact.

The technique is essentially quite simple. Two photographs are taken some time apart (~1 hour) of the same region of the sky. Assuming the telescope is tracking the stars accurately, stellar images will be point-like but, due to their motion round the Sun, asteroid images may appear slightly extended and, of course, they will have moved in position between the two images. In the early days of such searches, the two images were viewed together in a blink comparator – such as that used by Tombaugh when he discovered Pluto – when each image is viewed alternately and an object that has moved will appear to 'hop' back and forth, so making it stand out against the background stars. An alternative approach was to observe the two images as a stereoscopic pair. An asteroid would then appear to 'float' above the stars. This was obviously very time consuming but, using a Schmidt telescope that could image areas of sky several degrees across at a time, a reasonable area of sky could be covered each night of observations.

Ice dwarfs

Technology then moved on to the point that a totally computerised digital solution to the problem could be found by the use of large CCD arrays. Two 'digital images' taken some time apart could be compared by a sophisticated computer program to spot any motion. The further away from the Sun, the less the positional change between the two exposures. This technique has enabled many hundreds of asteroids to be discovered along with comets and, excitingly, what have become known as 'ice dwarfs' lying beyond the orbit of Neptune. This term is used as they are predominantly made up of ice. They are also correctly termed 'trans-Neptunian objects' or even 'Kuiper Belt objects', as this is the name given to a wide region lying beyond Neptune.

The first large ice dwarf was discovered by Bob McMillan using the 90-cm University of Arizona Spacewatch Telescope. It was named after the Hindu god Varuna and initially thought to be about 900 km in diameter. Not long after, Ixion was discovered, thought to have a diameter of 800 km. The only way that their diameters could be estimated was to guess at their albedos, assumed to be about 0.5. It now appears that these objects have higher albedos (~0.7) than initially thought and thus they are smaller – the most recent estimate for Varuna's diameter being 500 km.

In October 2001, following several years' work, a superb instrument saw first light. It comprised the 48-inch Samuel Oschin Telescope on Palomar Mountain (a Schmidt telescope with a very wide field of view) with a new 50-megapixel CCD array called NEAT, the Near Earth Asteroid Tracking camera. Though its prime purpose was to detect NEOs, not long afterwards, in January 2002, Mike Brown and his colleagues at the Palomar Observatory discovered a further ice dwarf with a diameter estimated to be ~800 km, and in the following June discovered an object that may have had an even greater diameter.

It surely could not be long before an object would be found that rivalled Pluto, and it was not long in coming. The team gained success on the very first night of observation when an object ~700 km in diameter and named $2003OP_{32}$ was discovered. As the year went on, the discovery of ice dwarfs became almost routine, and by the February of 2004 more than 30 had been discovered. On 19 February, the team announced the discovery of 2004DW, later given the name Orcus, a Roman god of the underworld. This had an estimated diameter of 1,600 km, even bigger than Charon, Pluto's moon.

Sedna

$2003VB_{12}$, as Sedna was initially called, was discovered from images made in November 2003 and found to have a very elongated orbit. When discovered it was ~11 billion km from the Sun, nearly as distant as Pluto, but when furthest away it would be 146 billion km – 975 AU and 24 times the average distance of Pluto. Its orbital period was 12,000 years. Sedna, with a diameter of ~1,800 km, was the largest object to have been found in the Solar System since Pluto – so surely an object larger than Pluto would be discovered before long. Though Mike Brown did not know it at the time, that object already existed on the hard disks of his computers.

$2003UB_{313}$

If an object was very far out beyond Pluto, its movement across the sky would naturally appear less. To avoid false alarms, the software would not detect objects that moved at less than 1.5 arcseconds per hour across the sky. So Brown's software would reject any objects beyond 12 billion km – Sedna had only just been caught. He therefore reduced the threshold for detection and all the images were reprocessed. Rather embarrassingly, they found an ice dwarf that really should have been detected earlier. It was initially nicknamed 'Santa' as it had been found on 28 December. Santa is egg shaped, rotates rapidly and has a small moon.

Just eight days later, Brown found three more ice dwarfs. Two, K31021A and K31021B, were faint but the third, K31021C, was much brighter. It only moved at 1.4 arcseconds per hour, implying great distance, and was determined to lie 97 AU or 14.5 billion km from the Sun – almost three times further away than

Pluto. To appear so bright at such a distance implied that it must be very large, probably larger than Pluto. Perhaps Planet X had been found! Brown's team already had a name ready for an object that they suspected was the 10th planet and it obviously should begin with an X. They had chosen Xena, the name of a warrior princess who had first appeared in comic books in the 1990s and then in her own television series – of which the team had pleasant memories.

When discovered, Xena was at the furthest point of its orbit and, with a period of 557 years, will not be closest to the Sun for 275 years, when it will close to a distance of 5.6 billon km. Its orbit was inclined at 44 degrees to the plane of the Solar System – not, sadly, a good start for it to be classed as a planet. Brown and his colleagues did not really believe that Pluto should be classed as a planet either. However, it was immediately apparent that it must be larger than Pluto. Given its observed brightness and known distance and assuming it reflected 100% of the light that falls upon it (an albedo of 1), it would have the same diameter as Pluto. To account for its observed brightness, given the typical albedo of an ice dwarf of ~0.7, it would have to be somewhat larger than Pluto. Xena was announced to the press on 29 July 2005. It was later found that Xena had a moon, initially called Gabrielle.

For a major discovery such as this, the IAU's Committee on Small Body Nomenclature could decide on a formal name for Xena within a day. The problem was that if it were just an ice dwarf the committee could name it, but if it were to be classified as a planet they would have no say in the matter. This made the question of what was or was not a planet even more pressing. As described in Chapter 3, in August 2006 the IAU produced a formal definition of what constitutes a planet – and so demoted Pluto – which meant that no more planets could be found but that Xena and other sufficiently large objects that might be found beyond Pluto would be classed as 'dwarf planets'.

A formal name for Xena

Now that the status of Xena as a dwarf planet was established, a formal name for Xena had to be given and Brown proposed the name Eris, the Greek goddess of discord and strife. A better name could hardly have been given considering that it had caused what was one of the bitterest disputes in the history of astronomy! Xena's satellite, which had been originally called Gabrielle, needed a formal name as well and was named Dysnomia, appropriately so as she was Eris's daughter and the demon of lawlessness.

Pluto and Eris were then given their numbers in the list of minor planets, 134,340 for Pluto and 136,199 for Eris. I find this rather demeaning as the minor planet named after Venetia Burney is number 6,235 and even that named after me is number 15,727. The objects that surround them in the list of minor planets are only a few kilometres across – surely this cannot be right. In fact, I proposed a resolution for the IAU to consider that dwarf planets should not be included within the list of minor planets but should, instead, be given their own listings within a new category of 'dwarf planets', in which case Ceres would become DP 01, Pluto DP 02 and Eris DP 03. Sadly, in my view, the IAU decided not to consider this proposal.

How many dwarf planets?

At this point, only Ceres and Pluto have been observed in sufficient detail to prove that they fit the definition of a dwarf planet, but as Eris is more massive than Pluto it is almost certain to be spherical and so is accepted as one by the IAU. The IAU has decided that trans-Neptunian objects with an absolute magnitude less than +1 (and hence, assuming that they are perfectly reflecting, must have a minimum diameter of 838 km) are to be named under the assumption that they are dwarf planets. By the end of 2013 two other bodies had met these criteria and so qualified as dwarf planets and were given the names Makemake and Haumea.

There are many discoveries that have yet to be made – but now, of course, Planet X can never be discovered!

For more on Planet X:

The Hunt for Planet X: New Worlds and the Fate of Pluto by Govert Schilling (Copernicus Books).

6

Voyages to the outer planets

Perhaps one of the greatest achievements of unmanned space flight has been the wealth of information – not to say stunning images – that resulted from NASA's programme to send probes to the outer parts of our Solar System. Here we will chart nearly 40 years of exploration, from the Pioneer and Voyager probes in the seventies and eighties to the Galileo spacecraft's study of Jupiter around the turn of the millennium and, more recently, the Cassini and Huygens probes studying Saturn and Titan.

Pioneer 10

Pioneer 10 was launched from Cape Canaveral on 2 March 1972 and was the first spacecraft to travel through the asteroid belt to reach Jupiter. It entered the asteroid belt on 15 July that year – a region 280 million km wide and 80 million km thick. The material in the belt encompasses sizes from dust particles up to the major asteroids travelling at speeds up to 72,000 km/h, and scientists had feared Pioneer 10 might not be able to negotiate its way through. It was even thought that the debris within the asteroid belt would be so thick that any spacecraft would be destroyed. Happily, these worries proved to be unfounded.

Arriving at Jupiter on 3 December 1973 at an approach speed of 131,000 km/h, Pioneer 10 was the first spacecraft to make direct observations and obtain close-up images of Jupiter. It mapped out the giant gas planet's intense radiation belts, located the planet's magnetic field, and showed that Jupiter was predominantly composed of liquids. Following its encounter with Jupiter, Pioneer 10 continued flying outward to explore the outer regions of the Solar System, where it studied the solar wind – an outflow of energetic particles from the

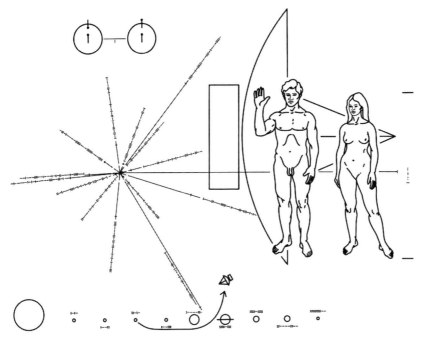

Figure 6.1 The Pioneer plaques. Image: NASA, designed by Carl Sagan and Frank Drake, artwork by Linda Salzman Sagan. Wikimedia Commons.

Sun – along with cosmic rays – highly energetic particles that enter the Solar System from stellar explosions within the Milky Way Galaxy. In 1983, Pioneer 10 became the first human-made object to pass the orbit of Pluto, then the most distant planet from the Sun, and continued to make valuable scientific contributions in the outer regions of the Solar System until its science mission officially ended on 31 March 1997.

Contact was lost with Pioneer 10 in 2003 and it is now heading in the direction of Aldebaran in the constellation Taurus. It was, by some definitions, the first artificial object to leave the Solar System and, as it could theoretically be found by another advanced civilisation, Carl Sagan proposed that it should carry a plaque that would tell them something about ourselves and our Earth. The symbol at the top left in Figure 6.1 symbolises the radio line transition of the hydrogen atom and so gives a length scale – that of the wavelength of the emitted photon of 21 cm. On the extreme right this is used to give the height of the female human being. The 'chart' in the centre left shows the position of Sun relative to the directions of 14 pulsars that lie in the plane of the Galaxy. Thus, in the same way that one could locate the position of a ship from its bearings to known lighthouses (identified by their flash rate), so the position of our Sun could be found. In fact, as the periods of the pulsars, indicated by marks on the bearing lines, change with time, they could even tell when the probe was launched. Neat! The Solar System diagram shows the path of the spacecraft and that we live on the third planet from the Sun. Although the original was slightly censored by NASA, it still received some opprobrium from the American Ladies League of Decency!

Pioneer 11

Launched a year later, in April 1973, Pioneer 11 reached Jupiter in December 1974. It passed within 42,500 km of Jupiter's cloud tops and, despite receiving intense bombardment from Jupiter's radiation belts (which are 40,000 times more intense than Earth's) happily survived. As Pioneer 10 had achieved all of the main mission objectives at Jupiter, Pioneer 11 used Jupiter to provide a 'gravitational slingshot' to increase its speed and send it on a course towards Saturn where, in 1979, it became first spacecraft to fly past Saturn before beginning a long journey out of the Solar System in the direction of the constellation Aquila. It flew within 21,000 km of Saturn and discovered a new ring and two new moons as well as detecting a thick atmosphere on Titan, Saturn's largest moon. Its instruments measured the heat radiation from Saturn's interior, and probed Saturn's magnetosphere, magnetic field and, from the gravitational effects on the spacecraft's trajectory, its interior structure.

The measurements of Jupiter's radiation environment made by the two Pioneers enabled the many later missions to Jupiter and Saturn to have suitably 'hardened' electronic systems that could withstand the intense radiation belts that exist around these planets.

An aid to SETI (the Search for Extra-Terrestrial Intelligence)

The most sensitive and sophisticated SETI program ever undertaken was Project Phoenix which, from 1998 to 2003, observed over 800 Sun-like stars, searching for any signals that might come from ET. It used two of the world's largest radio telescopes, the 300-metre Arecibo dish in Puerto Rico and the 76-metre Lovell Telescope at Jodrell Bank in the UK where I was the project scientist. The use of two very widely spaced telescopes making simultaneous observations meant that local interference would not cause spurious detections. The rotation of the Earth meant that a signal at a specific transmitted frequency would, due to the Doppler effect, be received at different frequencies at the two telescopes. This enabled us to eliminate any signals received from satellites orbiting both the Earth and the Sun. But how could we prove that the system was operating perfectly? Each day, prior to the 12-hour observing period, we detected the very weak signal still being transmitted by the Pioneer 10 space probe, then over 11 billion km from Earth. Its last signal was detected on 23 January 2003 from a distance of 12 billion km – almost twice the mean distance of Pluto!

Voyagers 1 and 2

Voyager 1 was launched by NASA on 5 September 1977, two weeks after its twin spacecraft, Voyager 2, but, as it was sent on a shorter trajectory, it reached Jupiter and Saturn first. Amazingly, it is still in communication with Earth, pursuing an extended mission to locate and study the Kuiper Belt and the outer boundaries of the Solar System. It had been realised that a rare alignment (once every 175 years) of the outer planets would enable the Voyager probes to utilise a technique called 'gravity assist' to gain speed from the gravitational energy of the planets they passed and be able to undertake what was then called 'The Grand Tour'. This was a linked series of gravity assists that would enable a single probe to visit all four of the Solar System's giant planets within a period of just 12 years rather than 30! (If a spacecraft passes close behind the path of a planet, the gravitational pull tries to make the planet fall into its surface and so the probe gains speed along the direction of the planet's orbit. However, if it is moving fast enough, it will not impact the planet but continue onwards with a new trajectory with significantly greater speed. This is also sometimes called a planetary or gravitational 'slingshot'.)

Voyager 1 began imaging Jupiter in January 1979 and its closest approach was on 5 March 1979, just 276,000 km above its cloud tops. Over a 48-hour period, it studied Jupiter's moons, rings, magnetic fields and radiation belt, and made the exciting discovery of volcanic activity on Jupiter's innermost moon, Io.

The gravitational assist trajectories at Jupiter were successfully carried out by both Voyagers, and the two spacecraft went on to visit Saturn. Voyager 1 reached Saturn in November 1980, when the space probe came within 123,000 km of Saturn's cloud tops. A year earlier, Pioneer 11 had detected a dense atmosphere on Titan and it was thought worthwhile for Voyager 1 to investigate this further rather than continue The Grand Tour on to Uranus and Neptune. This close flyby deflected Voyager 1 out of the plane of the ecliptic (within which lie the planets) and so ended its planetary quest. (Had they not done this, Voyager 1 could have flown past Pluto.)

In September 2013, NASA announced that Voyager 1 had crossed the heliopause (the boundary of the Sun's sphere of influence and interstellar space) on 25 August 2012, making it the first human-made object to do so. The probe is expected to continue its mission until 2025, when its generators will no longer supply enough power for its instruments.

Voyager 2 was launched on a slower, more curved trajectory and remained in the plane of the ecliptic, so that having passed Jupiter and Saturn it could continue on to Uranus and Neptune by means of the gravity assists gained during its flybys of Saturn and Uranus in 1981 and 1986. It is probably the most productive single unmanned space voyage carried out so far, having visited all four of the outer planets and their systems of satellites and rings. Voyager 2 carried cameras for imaging along with instruments to make measurements at ultraviolet, infrared and radio wavelengths. It was also able to measure the density of sub-atomic particles, including cosmic rays, in outer space.

On 9 July 1979, Voyager 2 came within 560,000 km of Jupiter's cloud tops and showed that the Great Red Spot was a complex, anticlockwise-rotating storm nestling within Jupiter's complex banded cloud systems along with many other smaller storms and eddies. Perhaps the most exciting discovery made by the two Voyager spacecraft was that of volcanism on Io. Together, the Voyagers observed the eruption of nine volcanoes and found evidence that other eruptions had occurred between the two Voyager flybys.

Voyager 1 had observed a large number of intersecting linear features on the surface of Jupiter's second innermost moon Europa, and scientists had thought that the features might be deep cracks. However, high resolution photos from Voyager 2 showed that they 'might have been painted on with a felt marker'. The idea arose that Europa might have a thin crust of water ice, possibly floating on a deep ocean kept liquid by the tidal heating due to its proximity to Jupiter. Voyager 2 also found three new small satellites: Adrastea, Metis and Thebe.

Just over two years later, Voyager 2 passed behind Saturn. As the radio signals Voyager was transmitting had to pass through Saturn's atmosphere as the craft disappeared and reappeared, scientists were able to gather information on Saturn's atmospheric temperature and density profiles. At the cloud tops the temperature was about −203 °C whilst at the lowest depths measured it increased to −130 °C.

What we know about Uranus

Uranus was the first planet to be discovered in modern times. It is just visible to the unaided eye without a telescope and it would have been impossible to show that it was a star rather than a planet, save for its slow motion across the heavens. Even when telescopes had come into use, their relatively poor optics meant that it was charted as a star many times before it was recognised as a planet by William Herschel in 1781.

Uranus revolves around the Sun once every 84 Earth years at an average distance from the Sun of roughly 19 AU. The surface cloud layers are seen to rotate with a period of as little as 14 hours, but this is due to high winds in the upper atmosphere, and the nominal rotational period of Uranus is 17 hours 14 minutes. Whilst for the majority of planets the rotation axis is roughly at right angles to the plane of the Solar System, Uranus has an axial tilt of 98 degrees, so in effect 'rolls' around the Sun. Each pole gets around 42 years of continuous sunlight followed by 42 years of darkness. In contrast to the Earth, this makes the poles warmer than the equator.

Uranus is the least massive of the giant planets at 14.5 Earth masses and has the second lowest density 1,290 kg/m^3. It probably has a central rocky core of about 2 Earth masses, above which is a mixture of various ices, such as water, ammonia and methane, along with an outer gaseous layer made up of about 1 Earth mass of hydrogen and helium. As the ices make up a far greater proportion of its mass than gas, Uranus is often termed an ice giant rather than a gas giant.

The rings of Uranus

On 10 March 1977, using a telescope mounted in the Kuiper Airborne Observatory, observations were to be made of the occultation by Uranus of a star, SAO 158687. Just before the star's light is lost its light will have passed through the atmosphere of Uranus, and comparing the spectra of the star at this time with that prior to the occultation it is possible to learn about the planet's atmosphere.

The telescope was observing the star well before the expected time of occultation when the astronomers were somewhat perturbed as the star's light suddenly disappeared. The signal did return after a rather tense period but this was then followed by four partial losses of signal. Now reasonably confident that it was not their equipment that was faulty, they continued to observe the star following the occultation when the sequence was seen to repeat in the inverse order. They realised that the light from the star must have been eclipsed by material in five rings about Uranus with the outermost (called the epsilon ring) being the thickest. From the times of the ring's occultations they could calculate the diameters of the rings and found that the outermost was ~44,000 km from the centre of Uranus (Figure 6.2).

Voyager 2 at Uranus

On 24 January 1986, when Voyager 2 came within 81,000 km of the planet's cloud tops, it directly imaged the ring system and more rings were

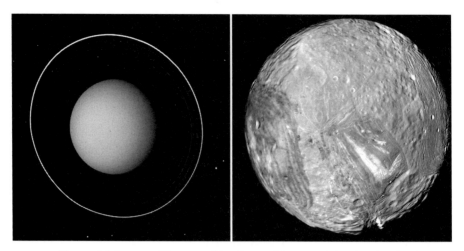

Figure 6.2 Images of Uranus and its rings and its moon Miranda taken by the Voyager 2 spacecraft in 1986. Images: JPL, NASA.

discovered, bringing the number up to 11. Observations showed that the Uranian rings are distinctly different from those at Jupiter and Saturn. The Uranian ring system appears to be relatively young, and it did not form at the same time that Uranus did. It is thought that the particles that make up the rings might be the remnants of a moon that was broken up by a high-velocity impact or torn up by tidal effects.

The radiation belts of Uranus were found to be of similar intensity to those of Saturn. This radiation is such that 'irradiation' would darken any methane that is trapped in the icy surfaces of the inner moons and ring particles and may have contributed to the darkened surfaces of the moons and ring particles, which are almost uniformly dark grey in colour. A high layer of haze was detected around the sunlit pole of Uranus. This area was also found to radiate large amounts of ultraviolet light, a phenomenon that is known as 'dayglow.' The average atmospheric temperature is about −213 °C.

The Uranian moon Miranda, the innermost of the five large moons, was shown to be one of the strangest bodies in the Solar System. Voyager 2's detailed images (Figure 6.2) showed huge canyons made from geological faults as deep as 19 km, terraced layers, and a mixture of old and young surfaces. It is thought that Miranda might consist of a reaggregation of material following an earlier event when it was shattered into pieces by a violent impact. As shown in Figure 6.2, there is a cliff face (a scarp) called Verona Rupes thought to be about 7 km high, making it the tallest cliff in the Solar System. Due to the very low gravity on Miranda, should one fall off, one would have about six minutes to ponder on one's past life.

Voyager 2 at Neptune

Neptune can be seen in even a small telescope and had even been observed by Galileo. Whilst observing Jupiter on 28 December 1612 he recorded Neptune as a faint star and a month later observed it close to another star on two successive nights. He noted that their separation had changed and could easily have reached the conclusion that this was because one was not a star but a planet!

The honour of Neptune's discovery is shared by Adams at the University of Cambridge and Le Verrier at the Paris Observatory, who had both computed its position from the perturbations its gravitational attraction had caused to the orbit of Uranus. Neptune is the fourth largest planet by diameter, and the third largest by mass, slightly more massive than its near-twin Uranus. Neptune's atmosphere is primarily composed of hydrogen and helium along with ~1% of methane, which may help contribute to its vivid blue colour. Winds in its atmosphere can reach 2,000 km/h, the highest of any planet. As Voyager 2 passed Neptune in 1989 it observed a Great Black Spot (Figure 6.3) comparable to Jupiter's Great Red Spot. It measured the cloud-top temperature to be −218 °C. Neptune's sidereal rotation period is roughly 16.11 hours long and it has a similar axial tilt to Earth.

Neptune and Uranus are often considered 'ice giants', given their smaller size and greater percentages of ice in their composition relative to Jupiter and Saturn. Neptune's core is composed of rock and ice, having about 1 Earth mass. The mantle is made up of ~12 Earth masses largely made up of water,

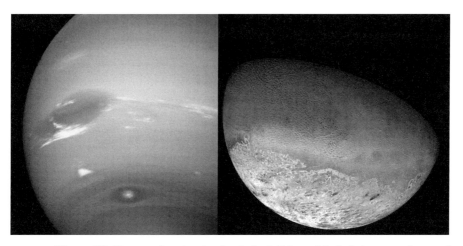

Figure 6.3 Neptune showing the clouds, both light and dark, in its atmosphere and its moon, Triton, as imaged by Voyager 2 in 1989. Images: NASA, ESA, JPL.

ammonia and methane. The atmosphere contains high clouds that cast shadows onto the blue-coloured surface layers.

Like Uranus, Neptune also has a ring system. The rings, which have a reddish hue, may consist of ice particles coated with silicates or carbon-based material. In sequence measuring from the centre of Neptune are the broad, faint Galle Ring at 42,000 km, the Le Verrier Ring at 53,000 km and the narrow Adams Ring, at 63,000 km. The largest of Neptune's 13 moons, and the only one massive enough to be spherical, is Triton which, unlike all other large planetary moons, has a retrograde orbit. This implies that it has been captured from the Kuiper Belt, a region containing many small bodies beyond Neptune's orbit. It keeps one face towards Neptune and is slowly spiralling inwards, where it will eventually be torn apart by gravity, so giving Neptune a more extensive ring system. Triton consists of a crust of frozen nitrogen over an icy mantle believed to cover a substantial core of rock and metal; its surface is relatively young. Part of its crust is dotted with geysers that are believed to be erupting nitrogen (Figure 6.3).

Its voyage continues

Its planetary mission over, Voyager 2 is continuing to travel outwards from the Sun and is now at a distance of over 13.6 billion km – more than twice as far from the Sun as Pluto but not yet beyond the outer limits of the orbit of the dwarf planet Eris. It is not headed towards any particular star but should pass near the star Sirius, currently 8.4 light years from the Sun, in about 296,000 years. It is hoped to be able to receive its (now very weak) radio signals until at least 2025 – a space mission that will then have lasted over 48 years since its launch!

The Voyager message

The Voyager spacecraft became the third and fourth human artefacts to escape from the Solar System. Following the example of the Pioneer plaques, NASA placed a more comprehensive message aboard Voyager 1 and 2 to make a kind of time capsule, intended to communicate something of ourselves and our world. It was in the form of a record whose contents were selected by a committee chaired by Carl Sagan of Cornell University. They put together images, sounds and music along with greetings in 55 languages.

The images include many photographs and diagrams in both black and white and colour. The first images are scientific, showing mathematical and physical quantities, the Solar System and its planets, our genetic code, and human anatomy and (very discretely) reproduction. (NASA, following the criticism of the line drawings of a naked man and woman on the Pioneer Plaque, only allowed a silhouette of the couple to be included.) Images of our human race depict a broad range of cultures going about their lives. There are also images of landscapes and architecture and of animals, insects and plants. The sounds include those made by surf, wind and thunder along with animal sounds such as birdsong and from whales. The musical selection features composers such as Beethoven, Mozart, Bach and Stravinsky. The gilt record cover shows a diagram of the hydrogen atom – to give a length and frequency reference – along with pulsar map – both of which had been on the Pioneer plaques. In the upper left-hand corner is an easily recognised drawing of the phonograph record and the stylus carried with it. Electroplated onto the record's gold-plated copper is a pure sample of the isotope uranium-238, which has a half life of 4.51 billion years. Any civilisation that encounters the record will be able to use the ratio of remaining uranium to its daughter elements to determine its age!

The Galileo mission to Jupiter

The Galileo spacecraft was launched in 1989 and arrived at Jupiter on 7 December 1995, following gravitational assist flybys of Venus and Earth. Amongst its achievements, Galileo conducted the first asteroid flybys (Gaspra and Ida) and discovered the first asteroid moon (Ida's moon Dactyl). It also launched a probe into Jupiter's atmosphere. En route it was able to image the impacts of the fragments of Comet Shoemaker–Levy 9 into the atmosphere of Jupiter – these impacted beyond Jupiter's visible limb as seen from Earth. Due to its distance from the Sun, solar panels would not have been practical, so power was provided by two radioisotope thermoelectric generators, which powered the spacecraft through the radioactive decay of plutonium-238. (Prior to the launch, antinuclear groups, concerned over what they perceived as an unacceptable risk to the public's safety from the plutonium should the spacecraft crash, sought a court injunction to prohibit Galileo's launch.)

On arrival at Jupiter for its two-year prime mission, Galileo orbited Jupiter in extended ellipses, so allowing it to sample different parts of the planet's extensive magnetosphere. The orbits also allowed it to carry out close flybys of Jupiter's largest Galilean moons. Having successfully completed its prime mission, the spacecraft made a number of very close flybys of Jupiter's moons Europa and Io.

In July 1995, five months before reaching Jupiter, Galileo released a probe to enter Jupiter's atmosphere. It collected nearly an hour's data as it descended through 150 km of atmosphere until the pressure reached 23 times that of the

Earth and the temperature rose to 153 °C. The probe found that the atmosphere through which it had passed was rather more turbulent and hotter than expected.

During its mission, Galileo carried out some experiments not directly related to its planetary studies. As it passed Earth on its second gravity assist flyby it tested the feasibility of the communication to and from satellites using pulses from powerful optical lasers. This proved very successful, and I believe that some military satellites are now using this technique to rapidly download their data to Earth in a way that would be very difficult to eavesdrop. The late Carl Sagan devised a series of experiments to see if Galileo could detect signs of simple or advanced life here on Earth as a way of indicating how we might detect life on other planets. It found, for example, very strong absorption of red light over the continents due to the chlorophyll in photosynthesising plants, along with narrow-band radio transmissions that could only come from advanced life.

Finally, when its plutonium-powered generator could no longer supply sufficient power, Galileo was intentionally commanded to crash into Jupiter to eliminate any chance of a future impact with Europa that could contaminate the icy moon. After 14 years in space and 8 years surveying Jupiter and its moons, it finally dived into the Jovian atmosphere at a speed of ~48 km/h on 21 September 2003 – the end of one of NASA's most successful missions.

Damian Peach's image of Jupiter shown in Figure 6.4 and Plate 6.4, which was taken from Barbados when the seeing conditions were near perfect, rivals those from the Hubble Space Telescope.

What we now know about Jupiter and its moons

With Saturn, Uranus and Neptune, Jupiter is one of the gas giants of the Solar System and its mass exceeds that of all the other planets combined by two and a half times. Its interior mass is primarily made up of hydrogen (~71%) and helium (24%) with ~5% of heavier elements. Its composition thus closely follows that of the solar nebula from which it was formed. Interestingly, if Jupiter were to acquire more mass, its diameter would actually decrease, so it is about as large as a planet of its composition could be.

Jupiter is thought to consist of a dense core surrounded by a layer of liquid metallic hydrogen lying under an outer layer, about 1,000 km thick, composed very largely of molecular hydrogen. Jupiter is perpetually covered with a cloud layer about 50 km thick. The clouds are composed of ammonia crystals arranged into bands of different latitudes made up of light-coloured zones between darker belts. The orange and brown colours in the Jovian clouds are caused by

Figure 6.4 Image of Jupiter showing the Great Red Spot and the moons Io (lower left) and Ganymede (upper right). This image is also shown in colour in Plate 6.4. At the time when this image was taken the South Equatorial Belt was missing. Image: Damien Peach.

compounds containing phosphorus and sulphur exposed to ultraviolet light from the Sun. At differing latitudes, the darker clouds so formed deeper within the atmosphere are masked out by higher clouds of crystallizing ammonia producing the pale zones seen between the belts.

The Great Red Spot

Wind speeds of up to 100 m/s are common in the atmosphere, and opposing circulation patterns caused, in part, by Jupiter's rapid rotation rate produce storms and turbulence in the atmosphere. The belts and zones are seen to vary in colour and form from year to year, but the general pattern remains stable. The best known feature in the atmosphere is undoubtedly the Great Red Spot. It is a persistent anticyclonic storm, more than twice the diameter of the Earth, which has been observed since at least 1831. It rotates in an anticlockwise direction with a rotation period of about 6 days and is thought to be stable and so has become a permanent, or at least a very long term, feature of the Jovian atmosphere. It is not, however, fixed in position and, though staying at latitude 22 degrees south, has moved around the planet several times since it was first observed. Similar, but smaller, features are common, with 'white ovals' of cool clouds in the upper atmosphere and warmer brown ovals lower down. These smaller storms can sometimes merge to form larger features, as happened in 2000 when three white ovals, first observed in 1938, combined into one. In the following years its colour has reddened and it has been nicknamed Red Spot Junior.

The rings of Jupiter

Jupiter has a very faint planetary ring system composed of three main segments: an inner halo, a brighter main ring, and an outer 'gossamer' ring having two distinct components. They appear to be made of dust, with the main ring probably made of material ejected from the satellites Adrastea and Metis as a result of meteorite impact. Jupiter's strong gravitational pull prevents the material falling back onto their surfaces and they gradually move towards Jupiter. It is thought that the two gossamer rings are produced in similar fashion by the moons Thebe and Amalthea.

Jupiter's Galilean moons

Even a very small telescope can detect the four major moons of Jupiter (Figure 6.5) as they weave their way around the planet. In order of distance from Jupiter, they are called Io, Europa, Ganymede and Callisto and are comparable in size to our Moon. Discovered by Galileo in 1610, they showed him that Solar System objects did not all have to orbit the Sun, giving further evidence for the Copernican model of the Solar System.

Observations in 1676 made by the Danish astronomer Ole Christensen Rømer of the times of their eclipses as they passed behind Jupiter led to the first determination of the speed of light. An eclipse of Io occurs every 42.5 hours – the period of its orbit – and it thus provides a form of cosmic clock. However, Rømer observed that the 40 orbits of Io during the time that the Earth was moving towards Jupiter took a total of 22 minutes less than when the Earth was moving away from Jupiter about six months later. The change in apparent

Figure 6.5 Jupiter's Moons. Images: NASA, ESA, JPL.

period is due to the Doppler effect, and this enabled him to calculate the ratio of the velocity of light to the orbital speed of the Earth around the Sun. He derived a value for this ratio of ~9,300. As the orbital speed of the Earth is ~30 km/s this gave a value (actually calculated by Christiaan Huygens from Rømer's observations) for the speed of light of ~279,000 km/s – the first measurement of the speed of light.

The two innermost moons, Io and Europa, are of great interest. Io is the fourth largest moon in the Solar System with a diameter of 3,642 km. When high resolution images of Io were received on Earth from the Voyager spacecraft in 1979, astronomers were amazed to find that Io was pockmarked with over 400 volcanoes. It was soon realised that giant tidal forces due to the close proximity of Jupiter would pummel the interior, generating heat, and so give Io a molten interior. As a result, in contrast with most of the other moons in the outer Solar System, which have an icy surface, Io has a rocky silicate crust overlying a molten iron or iron sulphide core. A large part of Io's surface is formed of plains covered by red and orange sulphur compounds and brilliant white sulphur dioxide frost. Above the plains are seen over a hundred mountains, some higher than Mount Everest – a strange world indeed.

In contrast, Europa, the sixth largest moon in the Solar System with a diameter of just over 3,000 km, has an icy crust above an interior of silicate rock overlying a probable iron core. The icy surface (Figure 6.6) is one of the

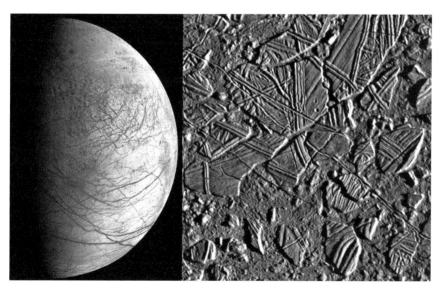

Figure 6.6 The surface of Europa showing cracks caused by tidal flexure and 'icebergs'. Images: NASA, ESA, JPL.

smoothest in the Solar System. Close-up images show breaks in the ice as though parts of the surface are breaking apart and then being filled with fresh ice. This implies that the crust is floating above a liquid ocean, warmed by the tidal heating from its proximity to Jupiter. This could thus conceivably be an abode for life, and some ambitious proposals have been made for a spacecraft to land and burrow beneath the ice to investigate whether any life forms are present!

In December 2013, NASA reported that images taken by the Hubble Space Telescope indicate the presence of hydrogen and oxygen above the moon's southern hemisphere. The observations are consistent with 200-kilometre-high plumes of water vapour. If this can be proved, then it might be possible to detect organic molecules or even evidence of life without having to drill down through the ice. The ESA 'Juice' mission, due to be launched in 2022, will make two close flybys in the 2030s and might even be able to fly through any plumes that may exist near the moon's equator. NASA has made some preliminary plans for an extended mission to Europa called 'Europa Clipper', which would spend a year or more in the vicinity of the enigmatic moon.

The Cassini mission to Saturn

The Cassini–Huygens spacecraft is a joint NASA/ESA/ASI mission, launched in 1997, which continues to study Saturn and its satellites. It was composed of two main elements: the NASA Cassini orbiter and the ESA-developed Huygens probe, which was to descend to the surface of Saturn's largest satellite, Titan. Cassini–Huygens entered into orbit around Saturn on 1 July 2004 and, on 25 December of that year, projected the Huygens probe towards Titan, which it reached on 14 January 2005. Huygens made a descent through Titan's atmosphere to the surface, making the first landing ever accomplished in the outer Solar System.

En route to Saturn, Cassini made a close approach to Jupiter and produced the most detailed global colour portrait of Jupiter yet – showing features just 64 km across. Cassini's observations of the light scattering by particles in Jupiter's rings showed that the particles were irregularly shaped and, most likely, result from ejecta released by micrometeorite impacts on the Jovian moons Metis and Adrastea.

Tests of Einstein's General Theory of Relativity

On its way to Jupiter, Cassini passed behind the Sun, so giving a way of testing Einstein's General Theory of Relativity. Due to the curvature of space cause by the Sun, the radio signals that reached us from Cassini have to travel along a longer path than if the Sun were not present. This causes a delay in their arrival time of about 200 microseconds (called the Shapiro delay), which agreed with Einstein's theory to an accuracy of about one part in 50,000 – one of the very best tests of his theory made to date.

Arriving at Saturn

After travelling for seven years, on 1 July 2004, the spacecraft flew through the gap between the F and G rings and, having passed within 33,600 km of Saturn's cloud tops, went into orbit. Just a day later, it had its first (though distant) flyby of Saturn's largest moon, Titan. Images showed methane clouds above the south pole and many surface features. Radar studies of Titan made in October 2004 showed a relatively smooth surface having a height range of just 54 m. Not surprisingly perhaps, Cassini has discovered several more moons orbiting Saturn – given names such as Methone, Pallene, Polydeuces, Daphnis and Aegaeon. On 11 June 2004, Cassini flew by the moon Phoebe and the very bright, close-up images indicated that a large amount of water ice exists under its immediate surface.

Cassini's primary mission ended on 30 July 2008, but given the excellent condition of the orbiter, the mission was extended to the end of June 2010 and then in February 2010 was extended again until 2017 – the time of summer solstice in Saturn's northern hemisphere.

What we now know about Saturn and its moons

Galileo first observed Saturn with his telescope in 1610 and became somewhat perplexed. He described the planet as having 'ears' and being composed of three bodies that almost touched each other, with the one at the centre about three times the size of the outer two whose orientation was fixed. Galileo became even more perplexed when, two years later, the outer two bodies had gone. 'Has Saturn swallowed his children?' he wondered. He became further confused when they reappeared in 1613. In 1655 Christiaan Huygens observed Saturn with a far superior telescope and suggested that Saturn was surrounded by a ring system. He wrote, 'Saturn is surrounded by a thin, flat ring, nowhere touching, inclined to the ecliptic'.

As telescopes improved, more details could be seen and, in 1675, Giovanni Domenico Cassini observed that Saturn's ring system was composed of a number of smaller rings separated by gaps, the largest of which has become known as the 'Cassini Division'. In the mid 1800s, James Clerk Maxwell showed that a solid ring could not be stable and would break apart, so that they must be made up of myriads of particles individually orbiting Saturn. This would imply that different annuli of the rings would be moving at different speeds around Saturn, and this was proved when James Keeler of the Lick Observatory made spectroscopic studies of the ring system in 1895.

There is no doubt that, due to its ring system, Saturn is the most beautiful object in the Solar System that can be observed with a small telescope. The key to understanding Galileo's confusion lies in Huygens' description that the ring system was inclined to the ecliptic due to Saturn's axial tilt. Assume that Saturn's north pole is, at some point in its orbit, tilted closest to the Sun. Close to the Sun, we see much of the northern hemisphere and the rings at their most open. Just under 15 years later, Saturn is on the opposite side of its orbit and the north pole is tilted away from the Sun. We then see the southern hemisphere best and again the rings are wide open. Halfway between these extremes we see the rings edge-on and, just as Galileo observed, they effectively disappear. So the Earth lies in the ring plane twice every orbit, about once every 15 years.

It is not surprising that the rings effectively disappear as it is thought that they are less than 1 km in thickness. The ring particles range in size from dust particles up to boulders a few metres across and are largely composed of water ice (~93%) along with amorphous carbon (~7%). Three rings can be observed from Earth that extend from 6,630 km to 120,700 km above Saturn's equator. The outer ring, the A Ring, has a significant gap within it called the Encke Division (though it is so thin, he could never have observed it) whilst the Cassini Division separates the A from the middle B, or Bright Ring. Inside the B Ring is the fainter C, or Crepe Ring. Two further rings have been discovered more recently; within the C Ring there is a very faint D Ring, whilst outside the A Ring is a very thin F Ring.

The rings, seen in Figure 6.7, are thought to have been formed when a moon either came within the Roche limit of the planet where tidal forces broke it apart, or was impacted by a large comet or asteroid to give the same result. (As a

Figure 6.7 Saturn and its rings. The little dot just beyond the bright rings on the left is the Earth. This mage is shown in colour along with that of Jupiter in Plate 6.4.
Image: CICLOPS, JPL, ESA, NASA.

small body nears a massive one, the gravitational force on the nearer side of the body exceeds that on the far side. There is thus a differential force across the body which tends to pull it apart. The Roche limit is the distance from a planet at which this force would break up a typical small body.)

The structure we see within the rings is due to the cumulative effect of the gravitational pull of Saturn's many moons. Where a moon has a period that is a simple multiple of that of particles at a certain (nearer) distance from the centre of Saturn, a 'resonance' occurs which clears particles from that part of the ring system. In this way, the moon Mimas clears particles from the Cassini Division.

Titan

Titan is the largest moon of Saturn and the only the moon in the Solar System known to have a dense atmosphere. It is also the only object other than Earth for which there is evidence of surface liquids, in the form of hydrocarbon lakes in the satellite's polar regions. It is about 50% larger than our Moon and 80% more massive, and is second only to Jupiter's moon Ganymede in size and larger (but less massive) than Mercury. Like our Moon, it is tidally locked and always presents the same face to Saturn. Titan has a relatively smooth crust composed of water ice, which overlies a rocky interior. The atmosphere is quite dense, largely made up of nitrogen (~98%), giving a surface pressure of more than one and a half times that of the Earth. Within the atmosphere are clouds of methane and ethane and an orange haze made up of organic molecules that result from the break-up of methane in the atmosphere by ultraviolet light from the Sun. The source of this methane is somewhat of a mystery, as the Sun's ultraviolet light should eliminate methane from the atmosphere in about 50 million years. Observations, first from the Hubble Space Telescope and then from the Cassini spacecraft (in infrared light to observe through the haze), show that Titan's surface is marked by broad swaths of bright and dark terrain. The largest bright feature is Xanadu, about the size of Australia.

The Huygens probe

On 25 December 2004 the Huygens probe separated from the Cassini orbiter that had carried it to Saturn and, having been lowered through its atmosphere by parachute, landed on the surface of Titan on 14 January 2005. Images of the surface (Figure 6.8) taken from a height of ~16 km showed what are considered to be drainage channels in light-coloured higher ground leading down to the shoreline of a darker sea or plain. Some of the photos even seemed to suggest islands and a mist-shrouded coastline. There was no evidence of any liquids at the time of landing, but strong evidence of its presence in the recent past.

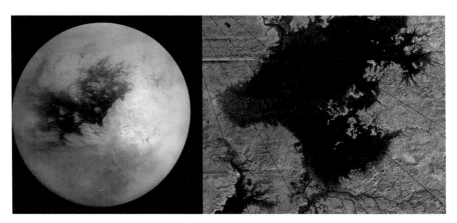

Figure 6.8 A Cassini view of Titan and a methane lake on the surface. Images: (left) VIMS Team, University of Arizona, ESA, NASA; (right) Cassini Radar Mapper, JPL, ESA, NASA.

As the spacecraft landed, a penetrometer studied its deceleration. It was initially thought that the surface had a hard crust overlying a sticky material. One scientist compared the colour and texture of the surface to that of a crème brûlée; another, with stepping on a cowpat! However, it may have been that the craft landed on, and then displaced, a pebble on the surface giving the effect of a surface crust, and the surface may, in fact, consist of 'sand' made up of ice grains forming a flat plain covered with pebbles made of water ice.

Lakes on Titan

Data from Voyager 1 and 2 showed that Titan had a thick atmosphere of approximately the correct temperature and composition to support lakes of liquid hydrocarbons (ethane or methane) on the surface. During a Titan flyby on 22 July 2006, the Cassini spacecraft's radar imaged the northern latitudes and a number of large, smooth regions (Figure 6.8) were seen dotting the surface near the pole. The Cassini–Huygens team concluded that the imaged features are almost certainly hydrocarbon lakes, some of which lie in depressions and appear to have channels leading into them. This was confirmed when, on 8 July 2009, a specular reflection in infrared was seen off the southern shoreline of a lake called Kraken Mare. The Cassini data indicate that Titan hosts within its polar lakes 'hundreds of times more natural gas and other liquid hydrocarbons than all the known oil and natural gas reserves on Earth'. The total volume of these seas is about 15 times the volume of Lake Michigan. The desert sand dunes along the equator, whilst devoid of open liquid, nonetheless hold more organics than all of Earth's coal reserves!

Enceladus

Enceladus, discovered in 1789 by William Herschel, is Saturn's sixth-largest moon. The Voyager spacecraft showed that Enceladus's icy surface reflects almost all of the sunlight that strikes it. Voyager 2 revealed that despite the moon's small size, just 496 km in diameter, it had a wide range of terrains ranging from old, heavily cratered surfaces to regions as young as 100 million years old.

In 2005, as the Cassini spacecraft performed several close flybys of Enceladus, it discovered a water-rich plume venting from the moon's south polar regions. This discovery, along with the fact that there are very few, if any, impact craters in this region, showed that Enceladus is geologically active. Rather as the reason that Jupiter's moon, Io, shows volcanism, it is thought that Enceladus's proximity to Saturn results in tidal heating of the satellite's interior. Analysis of the venting gas suggests that it originates from a body of sub-surface liquid water, which, along with the interesting chemistry found in the gas plume, has fuelled speculation that Enceladus might even be able to support simple life forms.

Cassini is getting very low on fuel and its mission is due to be terminated in 2017 by putting the probe into a destructive dive into Saturn's atmosphere.

The New Horizons mission to Pluto and its moons

As described in Chapter 5, Pluto was discovered by Clyde Tombaugh from a pair of plates taken in January 1930. Its name was suggested by Venetia Burney, the 11-year-old daughter of an Oxford Professor, when she was told of its discovery the next day. Pluto was the Roman God of the underworld who was able to make himself invisible. As the first two letters of its name were the initials of Percival Lowell, at whose observatory it had been discovered, this suggestion was eagerly accepted.

In 1978, James Christy, working at the Naval Observatory in Washington, discovered a satellite of Pluto, now called Charon, and in 2005, the Hubble Space Telescope discovered two further moons, now called Nix and Hydra. Observations of Charon enabled Pluto's mass to be determined – which was far less than originally thought and there is no doubt that, had it been discovered recently, it would never have been afforded the status of a planet, but it has since become part of our culture and I was somewhat saddened when it was demoted to the status of a dwarf planet in 2006.

New Horizons was originally planned as a voyage to what was then the only unexplored planet in the Solar System as, when the spacecraft was launched, Pluto was still classified as a planet. It was launched on 18 January 2006 for a nine-year voyage to Pluto and beyond. Having flown over 4.8 billion km, it will fly past Pluto and Charon around July 2015 and then hopefully visit some Kuiper Belt objects in an extended mission. New Horizons passed Jupiter on 28 February 2007 using the planet's gravitational pull to increase its speed to ~83,000 km/h in a 'slingshot' or 'gravitational assist' manoeuvre and, in passing, made infrared images of Jupiter and its moons. In addition to the scientific equipment, several cultural artefacts are carried by the spacecraft. These include a US flag, a Florida state quarter and some of Clyde Tombaugh's ashes – a nice thought.

More books about planets:

> *The Planets: A Journey Through the Solar System* by Giles Sparrow (Quercus).
> *Space Travel Guides: The Outer Planets* by Giles Sparrow (Franklin Watts).
> *Voyager: Exploring the Outer Planets* by Joan Marie Verba (FLT Publications).

7
Harbingers of doom

Comets can provide some of the most beautiful sights in our heavens, as that of Comet Hale–Bopp seen in Figure 7.1, but in ancient times, before their true nature was known, were often feared. Aristotle proposed that comets were gaseous phenomena in the upper atmosphere that occasionally burst into flames. He depicted comets as 'stars with hair' and used the Greek word 'kometes' to refer to them, from the root 'kome' meaning 'head of hair'. They were regarded as bad omens foretelling catastrophe or the deaths of kings.

Tycho Brahe made careful observations of the comet of 1577 and, by measuring its position from well-separated locations, was able to show that it lay at least four times further away than the Moon. In 1687, Isaac Newton was able to show that the path of a bright comet observed through the winter of 1680/1 could be fitted to a parabolic orbit with the Sun at one focus. He had thus shown that comets were Solar System bodies orbiting the Sun.

Halley's Comet

Edmond Halley calculated the orbits of 24 comets that had been observed between 1337 and 1698. He found that the comets that had been seen in 1531, 1607 and 1682 had very similar orbital elements and so believed that these were three 'apparitions' of the same comet. Halley could even account for the slight orbital differences by taking into account the gravitational effects of Jupiter and Saturn. The comet's period was ~76 years, so he predicted that it would return around the end of 1757. More accurate calculations by three French mathematicians indicated that it would, in fact, pass closest to the Sun in March 1759. It was first spotted by Johann Georg Palitzsch, a German farmer and amateur astronomer, on Christmas Day 1758. Halley, who had died in 1742, did

Figure 7.1 Comet Hale–Bopp imaged by E. Kolmhofer and H. Raab. Image: Wikimedia Commons.

not live to see the comet's return but it was named after him, with the designation 1P/Halley – the '1P' indicating that it was the first known periodic comet.

Halley's Comet was in the skies at the time of the Battle of Hastings in 1066 and depicted in the Bayeux Tapestry, as seen in Figure 7.2. It was certainly not a good omen for King Harold!

Depending on the Earth's orbital position as Halley nears the Sun, it can appear either bright and spectacular as it did in 1066 and 1910 or, as in its last apparition in 1985/6 when far from the Earth, barely visible to the unaided eye. The comet was first photographed in 1910 when it made a very close approach to the Earth – which even passed through its tail causing some alarm! It is thought that the artist Giotto di Bondone observed the comet in 1301, and used it to depict the Star of Bethlehem in his 1305 fresco *The Adoration of the Magi*. As a result, the spacecraft sent to fly past Halley's Comet in 1985 was called 'Giotto'. Halley's Comet will next appear in our skies in 2061.

A comet is now classified as a small Solar System body that orbits the Sun and, when close to the Sun, exhibits a visible coma (an extended atmosphere) and sometimes a tail. The 'nucleus' of a comet is typically of order 10 km in size and is composed of rock and dust bound together by ice. The term 'dirty snowball' that is sometimes used is thus quite apt. Those that are termed 'long-period' comets are debris left over from the condensation of the solar nebula and come

Figure 7.2 Halley's Comet in the Bayeux Tapestry as it had appeared in the sky in 1066.

from the outermost regions of the Solar System, up to a light year distant from the Sun, in what is usually termed the 'Oort Cloud'.

The cloud is thought to contain of order a trillion comets and is named after a Dutch astronomer, Jan Hendrik Oort, who popularised the idea in 1952. However, the concept was first proposed by an Estonian astronomer, Ernst Opik, in 1932 so it is alternatively, and more correctly, called the 'Opik–Oort Cloud'. Such comets will normally only ever be seen once, but occasionally their orbit will be sufficiently perturbed by Jupiter or Saturn to be 'captured' within the inner Solar System with a relatively short period. They will then become known as 'short-period' comets, which have, by definition, a period of less than 200 years. However, the majority of short-period comets are thought to originate in the Kuiper Belt, which lies beyond the orbit of Neptune. Of the ~3,000 comets now known, several hundred have short periods. On average, about one comet per year will reach unaided-eye visibility, but only about one in ten of these will become easily visible.

Cometary nuclei

The nuclei of comets range from about one-half to fifty kilometres in size and have a very rich composition; they are primarily made of rocks, dust and water ice along with frozen gases such as carbon dioxide, carbon monoxide, methane and ammonia. This 'dirty snowball' idea was proposed in 1950 by Fred L. Whipple, who had discovered six comets himself. It is more formally known as the 'icy conglomerate' theory. Comet nuclei also contain many organic compounds such as methanol, formaldehyde, ethanol and ethane and possibly even more complex molecules such as amino acids. They are far too small to become spherical through the force of gravity and are thus irregularly shaped.

Due to their small size and low albedo, comets cannot normally be seen in the outer Solar System. As a comet approaches the inner Solar System, solar radiation causes water, frozen gases and other volatile materials within the comet to vaporise and stream out of the nucleus. This releases the dust that is bound up within the ice and together the dust and gas form a huge, extremely tenuous atmosphere around the comet called the coma. The forces exerted on the coma by the Sun's radiation pressure and outflowing solar wind cause tails to form, which naturally point away from the Sun – as first shown by the German astronomer Peter Apian in 1531, as shown in Figure 7.3.

Both the coma and tail may become visible from Earth as the comet passes through the inner Solar System. The streams of dust and gas each form their own distinct tail, pointing in slightly different directions. The gas or ion tail is heavily influenced by the Sun's radiation pressure and magnetic field and points directly away from the Sun. It is the left-hand tail shown in Figure 7.1. The ion tail often appears bluish in colour, as the most common ion, that of carbon monoxide (CO+), scatters blue light rather than red. The dust tail is yellowish in

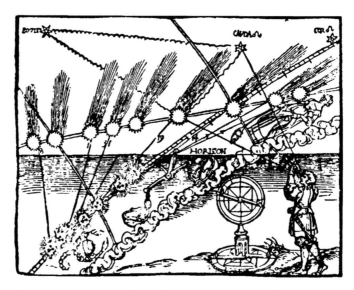

Figure 7.3 The comet of 1532 observed by Peter Apian showing that the tail pointed away from the Sun.

colour and tends to lie along the orbit of the comet; it often appears curved, as the heavier dust particles are less affected by the radiation pressure and solar wind. Sometimes, due to the angle that we observe the comet from, part of the dust tail can actually appear to point from the nucleus of the comet towards the Sun – an antitail. This rare sight typically occurs when the Earth crosses the plane of a comet's orbit and the cometary dust, which lies in a thin sheet, may be seen edge-on.

The majority of comets are too faint to be visible without the aid of binoculars or a telescope, but perhaps once each decade a comet becomes bright enough to be visible with the unaided eye. These are called great comets. Occasionally a comet may experience a huge and sudden outburst of gas and dust and thus the size and brightness of the coma temporarily greatly increase. This happened in late 2007 to Comet Holmes, which was visible to the unaided eye for some time.

The coma may exceed the Sun in size, and ion tails have been known to extend over one astronomical unit in length, and so are the largest objects in the Solar System. The size of the coma and tail increases as the comet nears the Sun, and the comet will tend to be most apparent immediately after it has passed the Sun. We thus tend to observe them in the period before dawn or after dusk. With each passage around the Sun, the comet loses material and will eventually disintegrate into a trail of dust or become an inert, asteroid-like body of fractured rock.

The fact that comets may contain significant amounts of organic compounds is indicated by their very low albedo. The Giotto space probe showed that the nucleus of Halley's Comet reflected only ~4% of the incident sunlight, whilst Deep Space 1 showed that Comet Borrelly reflected only ~2.6%.

The Giotto image (Figure 7.4) showed that the nucleus was not spherical and the outflows were only coming from a number of vents on the sunward side. The fact that the side away from the Sun shows no activity implies that there is little heat conduction through the interior of the comet. The density of the nucleus was just $0.5\,g/cm^3$ – half that of water – so, as ice has a density of $1\,g/cm^3$ and dirt $\sim2.5\,g/cm^3$, this implies that there must be quite a lot of empty space within the nucleus. This image highlights one interesting fact. Comets are the only Solar System objects that have a propulsion mechanism in the form of the jets that form on the sunward side. This means that the orbit of a comet after its solar encounter may be slightly different from that before.

Meteor showers

Each time a comet swings by the Sun in its orbit, some of its ice vaporises and fragments of cometary material are shed. The fragments will vary in size from that of a speck of dust to sand-sized grains and even small pebbles.

Figure 7.4 The image of the nucleus of Halley's Comet observed by the Giotto spacecraft on 13 March 1985. Image: Halley Multicolour Camera Team, Giotto Project, ESA.

Dust-sized particles are far more common than those the size of sand grains and these, in turn, are far more common than those the size of pebbles. These particles gradually spread out along the entire orbit of the comet to form what is called a meteoroid 'stream'. As the Earth orbits the Sun, its orbit can pass through a meteoroid stream and, if it does so, the particles enter Earth's atmosphere at high speed and burn up to form meteors – streaks of light across the sky that usually last for a few seconds. The particles' energy excites the gas through which the meteor passes and strips electrons from the atoms of the gas. These free electrons can reflect radio waves, giving rise to radar echoes which were studied at Jodrell Bank in the late 1940s.

The comets that give rise to most known meteor showers have been identified – for example, Halley's Comet gives rise to the Orionid shower in October. When the meteoroid stream is particularly dense, we occasionally see a spectacular 'meteor storm'. The larger particles form what is called a 'fireball' and can even explode as they near the ground, causing what is called a 'bolide' and giving rise to a sonic boom. The very largest meteoroids can reach the ground, resulting in a meteorite.

Because the particles are travelling in parallel paths, the particles will all appear to radiate away from a single point in the sky. This point is called the 'radiant' and is caused by the effect of perspective, rather as railway tracks appear to diverge from a distant point. Meteor showers are usually named after the constellation from which the meteors appear to originate. Often showers are seen for several nights and then the radiant moves slightly as the Earth's position in its orbit changes.

One of the most reliable meteor showers is the Perseid meteor shower which peaks on the 12th of August, usually giving rise to about 60 meteors per hour. If the Earth passes through a particularly rich part of the stream, the meteor rate can be higher, with well over a hundred meteors seen per hour (as in 1999). The most spectacular meteor shower is probably the Leonids, which peaks around the 17th of November. Approximately every 33 years the Leonid shower produces a 'meteor storm', producing thousands of meteors per hour. The last two massive Leonid storms were in 1933 and 1966 but that in 1999 was less spectacular.

Notable meteor showers

Perseids	mid August	Comet 109P/Swift–Tuttle
Orionids	late October	Comet 1P/Halley
Leonids	mid November	Comet 55P/Tempel–Tuttle
Geminids	mid December	Minor planet 3200 Phaethon

The Stardust mission

Stardust was an American interplanetary mission whose primary purpose was to investigate the makeup of Comet Wild 2 and its coma by collecting a sample of cometary dust and returning it to Earth. It was launched on 7 February 1999 and flew by Comet Wild 2 on 2 January 2004. En route, the spacecraft passed within 3,300 km of the asteroid 5535 Annefrank and took several photographs. During the flyby of Comet Wild 2 it collected dust samples from the comet's coma and took pictures of its icy nucleus.

When the spacecraft flew past the comet, the impact velocity of the particles in the coma was 6,100 m/s, nine times the speed of a rifle bullet! Although the captured particles were each smaller than a grain of sand, such a high-speed capture could have vaporised them entirely or, at the very least, altered their shape and chemical composition. To avoid this happening, the comet particles were collected in ultra low density aerogel. Aerogel is a silicon-based solid with a porous, sponge-like structure, of which 99.9% by volume is empty space. It is 1,000 times less dense than glass, another silicon-based solid. When the particles hit the aerogel, they buried themselves in the material, creating carrot-shaped tracks up to 200 times their own length as they slowed down and came to a stop.

The exposed aerogel was packed in a sample return capsule (SRC), which was released from the spacecraft as it passed close by the Earth on 15 January 2006.

The parachute-borne capsule landed in Utah's Great Salt Lake desert, having travelled through the atmosphere at 46,500 km/h – the fastest re-entry speed into Earth's atmosphere ever achieved by a human-made object. A large fireball and sonic boom were observed in western Utah and eastern Nevada.

The samples returned by the spacecraft were transferred to the Johnson Space Center in Webster, Texas, and to a clean room facility, which had 'a cleanliness factor 100 times that of a hospital operating room to ensure the star and comet dust is not contaminated by earthly grime'. (The Johnson Space Center is also the home of most of the Moon rock samples brought back by the Apollo missions.)

The tracks in the aerogel enabled scientists to find the tiny particles that had been collected, and the December 2006 issue of *Science* magazine detailed their analysis. Among their findings was the discovery of a wide range of organic compounds, including two that contain biologically usable nitrogen. Hydrocarbon compounds were found with longer chain lengths than those observed in the diffuse interstellar medium. The Stardust samples contain abundant amorphous silicates and their presence in Wild 2 is consistent with the mixing of Solar System and interstellar matter, something that had been deduced spectroscopically from previous astronomical observations.

The Deep Impact mission

Deep Impact was a NASA space probe, launched on 12 January 2005, which was designed to study the composition of the interior of Comet Tempel 1 by colliding a section of the spacecraft into the comet. The Deep Impact mission thus became the first to eject material from a comet's surface. On 29 June, the impactor successfully separated from the flyby spacecraft. At this time both the impactor and the flyby spacecraft were headed directly towards the comet. The flyby spacecraft then made a 14-minute burn to change its orbit and so avoid impact and pass by the comet whilst imaging the comet and relaying data captured by camera and instruments on the impactor itself.

It is against international law to fire projectiles towards a space body, so the impactor made three corrections to its orbit so that it would lie in the path of the comet and so be 'hit' by the comet rather than the other way round! During its approach towards the comet it sent images to the flyby spacecraft, which then forwarded them to Earth. The last was taken just 3 seconds before impact. The energy of the final collision (Figure 7.5) was equivalent to that of 5 tonnes of TNT and, briefly, the comet shone six times more brightly than usual.

A great plume of material was ejected from the comet. One slightly sad result of this was that the flyby spacecraft was unable to image the crater that must

Figure 7.5 This spectacular image of Comet Tempel 1 was taken 67 seconds after it obliterated Deep Impact's impactor spacecraft. Image: University of Maryland, JPL-Caltech, NASA.

have been formed. Earth-based observations showed that the comet continued outgassing from the impact site for 13 days and it is thought that a total of 250 million kg of water and between 10 and 25 million kg of dust were lost from the comet due to the impact.

Somewhat surprisingly, the material excavated by the impact contained more dust and less ice than had been expected. In addition, the material was finer than expected and was likened more to talcum powder than sand. From spectroscopic observations, other materials found included clays, carbonates, sodium and crystalline silicates. Measurements of its size and mass revealed that the comet was about 75% empty space, agreeing with the Giotto observations of Halley's Comet.

Two final thoughts

It is thought that cometary impacts on Earth may have brought much organic material to our surface. It is also believed that such impacts may have brought much of the water found on the planet and they most certainly have had an effect on the direction of evolution here on Earth. We have a lot to thank them for – apart from providing us with some of the most beautiful sights in the heavens!

Another book on comets:

> *Comets!: Visitors from Deep Space* by David J. Eicher and David H. Levy (Cambridge University Press).

8

Impact!

An image of the Moon is a salutary reminder that bodies in the Solar System, including the Earth, have suffered millions of impacts in the past – and will continue to do so, but happily at a far reduced rate. If anything, due to its greater mass, the Earth will have suffered more impacts than the Moon but erosion has removed the evidence of all but a few from its surface. The impacts of Solar System debris, such as asteroids and comets, give rise to what are termed 'impact craters', a name that can be applied to any depression resulting from the impact at very high velocity of an object into a larger body. Impact craters are approximately circular depressions that usually have raised rims, and range from small, smooth, bowl-shaped depressions to large craters having terraced rims often with a central peak or peaks. Massive impacts give rise to giant impact basins such as the 'mare' on the Moon, where the lunar crust was breached and so lava was able to well up from below and fill the depression caused by the impact. These 'mare' depressions, also seen on Mercury, were the result of a period of intense bombardment in the inner Solar System that ended about 3.8 billion years ago. Since then, though still appreciable, the rate of crater production within the inner Solar System has been considerably lower, but about every million years the Earth experiences a few impacts large enough to produce 20-km diameter craters. Erosion on the Earth quickly destroys these, but about 170 terrestrial impact craters have been identified. Their size ranges from a few tens of metres up to about 300 km in diameter. Most are less than 200 million years old.

Until the 1930s it was widely believed that the craters found on the Earth were volcanic in origin rather than the result of impacts, and it was not until the 1960s that researchers, notably Eugene M. Shoemaker, found clear evidence that they had been created by impacts, identifying, for example, shocked quartz that could only be formed in an impact event. By 1970, more than 50 impact craters had been found on the Earth.

Crater formation

The speed at which an object hits the Earth can range from ~11 km/s up to 70 km/s with a typical impact speed of ~25 km/s – these speeds are derived from the orbital speeds of objects within the Solar System. It is now possible to simulate such events and it appears that the impacting object will normally penetrate the ground and then cause a below-ground explosion that melts and vaporises the material above. In this case, it does not normally matter at what angle the object has hit the Earth and the resulting craters are nearly always circular, with only very low angle impacts giving rise to elliptical craters. In large impacts, much material will be ejected and mostly fall within a few crater radii, but some may travel significant distances and may form 'rays', as seen centred on the lunar crater Tycho. Some ejected material may even exceed the planet or satellite's escape velocity and then travel within the Solar System to perhaps fall as meteorites on the Earth. This is how we have some Martian rock samples to investigate for signs of life.

Craters on the Earth

Perhaps the best known impact crater on the Earth is Meteor Crater, some 60 km east of Flagstaff in northern Arizona, USA, which was formed about 50,000 years ago when the area was open grassland inhabited by woolly mammoths! It is often referred to as the 'Barringer Crater' in honour of Daniel Barringer who first suggested that it was produced by meteorite impact. Shown in Figure 8.1, it is ~1.2 km in diameter and 170 m deep, surrounded by a rim that rises 45 m above the surrounding plains. At its centre is a ~220 m pile of rubble. The remains of the meteorite are believed to be embedded under the rim at the south side of the crater. It was a nickel–iron meteorite about 50 m across, which, it is thought, impacted the plain at a speed of 12.8 km/s (28,600 mph). The meteorite would have initially weighed ~600,000 tonnes but it is suspected that half may have vaporised as it passed through the Earth's atmosphere. The meteorite that formed the crater is officially called the Canyon Diablo Meteorite – it is named after the town of Canyon Diablo, Arizona, which, now a ghost town, was 19 km to the north.

In 1960, research by Eugene M. Shoemaker confirmed Barringer's hypothesis with the discovery of the presence in the crater of the mineral stishovite. This is a rare form of silica found only where quartz-bearing rocks have been severely

Figure 8.1 Meteor or Barringer Crater in Arizona, USA. Image: Shane Torgerson, Wikimedia Commons.

shocked through an impact event or nuclear explosion. It cannot be created by volcanic action. Since Shoemaker produced the first definitive proof of an extra-terrestrial impact on the Earth's surface, many impact craters have been identified around the world but Meteor Crater is still the most visually impressive. It was used by Shoemaker to train the astronauts in crater geology prior to the Apollo missions in the 1960s.

The Nördlinger Ries and Steinheim Craters in Germany

Two large impact craters in the south of Germany are the most significant impact craters in Europe and are thought to result from the impact of a binary asteroid (two co-rotating asteroids) about 14.4 million years ago. The larger, the Ries Crater, is 24 km across and its floor is about 120 m below the eroded remains of the rim. On a very clear day it is possible to see from one side across to the far rim, but the size is such that the crater does not appear too obvious. As with the Barringer Crater, Eugene Shoemaker showed that it was caused by meteorite impact as, with Edward Chao, he found shocked quartz (coesite) in the stone that had been used to build Nördlingen town church, which lies within the crater. Interestingly, stone buildings in Nördlingen contain millions of tiny diamonds, all less than 0.2 mm across – formed when part of the asteroid impacted a local graphite deposit from which stone has been quarried over the centuries.

The smaller, the Steinheim Crater, is 42 km west-southwest from the centre of the Ries Crater and is 3.8 km across. Viewing from the crater wall at one side, its crater form is quite obvious and there is a low outcrop of material on the crater floor. It is thought that the Ries Crater was formed by an asteroid of 1.5 km diameter and the Steinheim Crater one of 150 m diameter. The pair impacted the region at an angle of around 40 degrees from the surface from a west-southwesterly direction. They impacted at about 20 km/s and had an explosive power of 29 million kilotonnes of TNT. Ejecta from the Ries Crater have been found up to 450 km to the northeast and are believed to be the source of moldavite tektites found in Bohemia and Moravia.

The Chicxulub event

The Chicxulub Crater is an ancient impact crater buried underneath the Yucatán Peninsula in Mexico, with its centre located near the town of Chicxulub. The crater is more than 180 km in diameter, making it the third largest confirmed impact structure in the world – formed by the impact of an asteroid at least 10 km across. The crater was discovered by a geophysicist, Glen Penfield, who had been prospecting for oil during the late 1970s. Within his data, Penfield found a huge underwater arc 70 km in length, which agreed with a gravitational anomaly shown on map of the Yucatán made in the 1960s. This also suggested a giant crater but was not publicised due to commercial considerations. Penfield found another arc on land that, together with offshore arc, formed a circle, 180 km wide. As at the sites of other impact craters, shocked quartz was found, confirming the structural and gravitational evidence of a massive impact.

The age of the rocks and isotope analysis show that this impact structure dates from the end of the Cretaceous Period (called the K–Pg boundary), roughly 65 million years ago and so it is thus implicated in causing the extinction of the dinosaurs. However, this may not have been the sole reason for their demise. The impact released the equivalent of 100 million megatonnes of TNT – 2 million times greater than the most powerful human-made explosion! In 2007, a paper published in *Nature* proposed that the 'Chicxulub asteroid' resulted from a collision in the asteroid belt 160 million years ago that gave rise to the creation of the Baptistina family of asteroids. There is evidence that the impactor was a member of a rare class of asteroids called carbonaceous chondrites, like the Baptistina family of which the largest surviving member is 298 Baptistina. The impact would have caused one of the largest tsunamis in the Earth's history, reaching several thousand metres high, whilst a cloud of superheated dust, ash and steam would have spread from the crater. Debris, from both the impactor and the impact area, would have been thrown out of the atmosphere by the blast, heated to incandescence upon re-entry, so igniting global wildfires. Shock waves from the blast may well have triggered earthquakes and volcanic eruptions across the globe. The emission of dust and particles could have covered the entire surface of the Earth for several years, greatly reducing solar radiation and so interrupting plant photosynthesis and thus affecting the entire food chain.

The main evidence of a giant impact, besides the crater itself, is contained in a thin layer of clay present in the geological record from across the world dating from the K–Pg boundary. It contains an abnormally high concentration of iridium, reaching 6 parts per billion by weight or more as compared to just to 0.4 for the Earth's crust as a whole. In contrast to the Earth, meteorites can contain around 470 parts per billion of iridium. It seemed a reasonable hypothesis that the iridium was spread into the atmosphere when the impactor was vaporised, mixed with other material thrown up by the impact and settled across the Earth's surface, so producing the layer of iridium-enriched clay.

The discovery of the Chicxulub Crater supported the theory that the extinction of numerous animal and plant groups, including the dinosaurs, may have resulted from a giant impact. The time of this 'extinction' event approximately agrees with the Chicxulub event, but recent core samples indicate that the impact occurred about 300,000 years before the mass extinction, and thus it could not have been the direct cause of the dinosaurs' demise. It may have been just one of a number of large impacts spaced over several hundred thousand years. In addition, at around this time the Earth's temperature will have risen due to carbon dioxide released by a massive eruption of lava that formed the Deccan traps of India and this would have been a real problem for the dinosaurs. A further impact, 300,000 years after Chicxulub, possibly in the sea bed beneath the Indian Ocean, may have been the final straw.

Tunguska: 30 June 1908

The Tunguska event was a powerful explosion that occurred near the Tunguska River in Siberia on 30 June 1908. The explosion is believed to have been caused by the air burst of a large meteoroid or comet fragment a few tens of metres across at an altitude of 5–10 km above the Earth's surface. Though the meteor or comet burst in the air rather than directly hitting the surface, this event is still referred to as an impact, which released around 12 megatonnes of TNT – about 1,000 times as powerful as the atomic bomb dropped on Hiroshima. The Tunguska event (Figure 8.2) is the largest impact event over land in Earth's recent history. The blast from the explosion knocked over an estimated 80 million trees spread over 2,150 square kilometres. Had the explosion happened a few hours later it would have destroyed St Petersburg.

At around 7:14 a.m. local time, Tungus natives and Russian settlers in the hills northwest of Lake Baikal observed a column of bluish light, nearly as bright

Figure 8.2 Devastation caused by the Tunguska impact.

as the Sun, moving across the sky followed by a flash and a sound similar to artillery fire. The sounds were accompanied by a shock wave that knocked people off their feet and broke windows hundreds of kilometres away and an explosion equivalent to 5.0 on the Richter scale was registered on seismic stations across Eurasia. Following the impact, ice particles formed at extremely cold temperatures in the high atmosphere increased light levels around dawn and dusk, allowing Londoners to read newspapers by the light!

An eyewitness report

Testimony of S. Semenov, as recorded by Leonid Kulik's expedition in 1930.

At breakfast time I was sitting by the house at Vanavara Trading Post (65 kilometres south of the explosion), facing north. I suddenly saw that directly to the north, over Onkoul's Tunguska Road, the sky split in two and fire appeared high and wide over the forest. The split in the sky grew larger, and the entire northern side was covered with fire. At that moment I became so hot that I couldn't bear it, as if my shirt was on fire; from the northern side, where the fire was, came strong heat. I wanted to tear off my shirt and throw it down, but then the sky shut closed, and a strong thump sounded, and I was thrown a few yards. I lost my senses for a moment, but then my wife ran out and led me to the house. After that such noise came, as if rocks were falling or cannons were firing, the earth shook, and when I was on the ground, I pressed my head down, fearing rocks would smash it. When the sky opened up, hot wind raced between the houses, like from cannons, which left traces in the ground like pathways, and it damaged some crops. Later we saw that many windows were shattered, and in the barn a part of the iron lock snapped.

There was little scientific curiosity about the impact at the time, possibly due to the isolation of the Tunguska region, and any records of early expeditions to the site were likely to have been lost during World War I, the Russian Revolution of 1917 and the Russian Civil War. In 1921, the Russian mineralogist Leonid Kulik visited the Tunguska River basin as part of a survey for the Soviet Academy of Sciences and deduced from local accounts that the explosion had been caused by a giant meteorite impact. The survey returned in 1927 and finally reached the impact site where, to their surprise, they could not find a crater but instead a region 8 km across of scorched tree trunks. At greater distances the trees were knocked down pointing away from the centre. In the 1960s it was found that a butterfly-shaped area 70 km across and 55 km in length had been levelled, but still no crater was found. Tiny spheres of silicate and magnetite containing high proportions of nickel relative to iron were found, leading to the conclusion they were of extra-terrestrial origin, these being found in meteorites. In addition, the bogs that cover much of the region contain an unusually high proportion of iridium, similar to the iridium layer found in the K–Pg boundary. This is believed to result from debris from the impacting body that had been deposited in the bogs.

Using model forests (made of matches on wire stakes) and small explosive charges slid downward on wires, experiments produced butterfly-shaped blast patterns strikingly similar to the pattern found at the Tunguska site and suggested that the object had approached at an angle of about 30 degrees from the ground and had exploded in mid-air.

Controversy remains as to whether the Tunguska body was a meteorite or small cometary body – which would leave no obvious traces. A cometary origin could explain the glowing skies observed across Europe for several evenings after the impact, the result of dust and ice that had been dispersed from the comet's tail across the upper atmosphere. It might have been a fragment of the short-period Comet Encke, which is responsible for the Beta Taurid meteor shower; its timing and direction of approach would have been consistent with this hypothesis. Some argue that a cometary body could not reach so close to the ground, but others suggest that the object was an extinct comet covered with a stony mantle that allowed it to penetrate the atmosphere. To support the opposing view, resin extracted from the core of the trees in the area of impact showed high levels of material commonly found in rocky asteroids and rarely found in comets. In June 2007 it was announced that scientists from the University of Bologna had identified a lake in the Tunguska region as a possible impact crater from the event. They do not dispute that the Tunguska body exploded mid-air, but believe that a one-metre fragment survived the explosion and impacted the ground. The arguments continue!

The Earth is continuously being bombarded by meteoroids usually travelling at a speed of more than 10 km/s. The majority are small but occasionally a larger one enters. Due to the heat generated as they travel through the atmosphere, most burn up or explode before they reach the ground. It appears that stony meteoroids of about 10 m in diameter produce an explosion of around 20 kilotonnes of TNT in the upper atmosphere more than once a year. However, megatonne-range events like Tunguska are much rarer, only occurring about once every 300 years.

Some recent events

On 10 August 1972, a meteor that became known as 'The 1972 Great Daylight Fireball' was witnessed moving north over the Rocky Mountains from the USA into Canada. It was an Earth-grazing meteoroid that passed within 57 km of the Earth's surface. Many saw its passage through the atmosphere and a tourist at the Grand Teton National Park in Wyoming was able to film it with an 8-millimeter colour movie camera.

On 23 March 1989, the 300-metre diameter Apollo asteroid, 4581 Asclepius, missed the Earth by 700,000 km (400,000 miles), passing through the exact position where the Earth was only six hours before. Had the asteroid impacted the Earth, it would have created the largest explosion in recorded history, thousands of times more powerful than the most powerful nuclear bomb ever exploded by man.

On 6 June 2002, an object with an estimated diameter of 10 m entered the Earth's atmosphere over the Mediterranean Sea, between Greece and Libya, and exploded in mid-air. The energy released was estimated to be equivalent to 26 kilotonnes of TNT, comparable to a small nuclear weapon.

On 18 March 2004, a 30-metre asteroid, 2004 FH, passed Earth at a distance of only 42,600 km, about one-tenth the distance to the Moon, and the closest miss ever observed. Similar sized asteroids are thought to come this close about every two years.

On 5 October 2008, scientists calculated that a just-discovered small near-Earth asteroid, 2008 TC3, would impact the Earth on 6 October over Sudan, at 5:46 a.m. local time. The asteroid entered the atmosphere just as predicted – the first time that an asteroid impact on Earth has been accurately predicted. The object was confirmed to have entered the Earth's atmosphere at a speed 12.8 km/s above northern Sudan, its path observed over a wide area and from aircraft. To search for fragments of the asteroid 2008 TC3, students and staff from the University of Khartoum lined up to comb the desert and have so far found some 280 meteorites having a combined weight of 5 kg.

The Chelyabinsk meteor

On 15 February 2013, a near-Earth asteroid entered the Earth's atmosphere over Russia with an estimated speed of 18.6 km/s (over 41,000 mph), almost 60 times the speed of sound. The light from the meteor was brighter than the Sun and was observed over a wide area of the region. Eyewitnesses also felt intense heat from the fireball.

The object exploded in an air burst over Chelyabinsk Oblast, at a height of around 23.3 km (14.5 miles). It produced a powerful shock wave which, happily, was largely absorbed in the atmosphere, though some 7,200 buildings in six cities were damaged by the blast. About 1,500 people were injured seriously enough to seek medical help – mainly from broken glass from windows that were blown in when the shock wave arrived some two minutes after the initial flash. It is thought that the energy in the air blast was approximately 500 kilotonnes of TNT. The sound from the explosion reverberated around the world several times, taking over a day to dissipate. It is interesting to note that the passage of the meteor across the sky was captured by quite a number of car 'dash cameras' that many Russians now use in case of accident. These showed that the meteor approached from the southeast and exploded about 40 km south of central Chelyabinsk above Korkino, at a height of 23.3 km (14.5 miles), with fragments continuing in the direction of Lake Chebarkul.

It is thought that the meteor was between 17 and 20 m in size and weighed about 12,000 tonnes (heavier than the Eiffel Tower) and was the largest known natural object to have entered Earth's atmosphere since the 1908 Tunguska event.

In the following days, fragments of the object were discovered in over 250 locations and, significantly, a 6-metre-wide hole was discovered in the ice of Lake Chebarkul's frozen surface. Magnetic imaging later identified a 60-centimetre-diameter meteorite buried in the mud at the bottom of the lake. This was raised from the lake on 16 October 2013. Weighing 654 kg, it tipped over and broke the scales being used to weigh it before splitting into three pieces.

The comet that crashed into Jupiter

Eugene Shoemaker was a brilliant geologist who had hoped to become one of the astronauts who explored the Moon in the early 1970s, but was rejected because of a medical condition. He was involved in several US space missions, including the Apollo missions when he taught the astronauts about the geology of lunar craters. Shoemaker was perhaps the first person to bring to other scientists' and the public's attention the danger of the impacts of comets and asteroids on the Earth.

He and his wife Carolyn – a great team later joined by David Levy – discovered about 800 asteroids and 20 comets. On the night of 24 March 1993, they took an image that showed what appeared to be a comet having multiple nuclei, whose proximity and motions suggested that it was associated with the planet Jupiter. The existence of this object was soon confirmed by James V. Scotti of the Spacewatch programme at the University of Arizona and the comet was named Shoemaker–Levy 9 (SL9) as it was the ninth short-period comet that the team had discovered during their NEO search.

Further observations revealed that it was orbiting Jupiter rather than the Sun in a highly eccentric orbit with a period of about 2 years. It appeared that in the late sixties or early seventies Jupiter had captured it from its earlier orbit around the Sun and the comet had become, in effect, a temporary satellite of Jupiter. Orbital calculations showed that on 7 July 1992 it had come within 40,000 km of Jupiter's cloud tops. This is within what is called the planet's Roche limit, within which a body can be torn apart by its tidal forces. As a result, the comet's nucleus had been broken up into 23 fragments, which were labelled 'A' to 'W'. Each of these fragments had a slightly different orbit and formed a 'train' that was gradually becoming more spread out. The visible fragments of SL9 were estimated to range in size from a few hundred metres up to a couple of kilometres across, suggesting that the original comet may have had a nucleus up to 5 km across.

As the orbits of the fragments were refined it became apparent that they would collide with Jupiter in July the following year – the train of nuclei ploughing into Jupiter's atmosphere over a period of about five days. This naturally caused great excitement as astronomers had never before seen Solar System bodies collide and provided a unique opportunity for scientists to look inside Jupiter's atmosphere, as the collisions were expected to cause eruptions of material from the layers normally hidden beneath the clouds.

It became apparent that the impact site would lie just beyond the Jovian limb, and so beyond direct view from Earth, but that within a short time Jupiter's rotation would make the effects of the impacts appear on its visible face. However, the Galileo spacecraft, on its way to investigate Jupiter, would be able to observe the impacts as they occurred. The Hubble Space Telescope (HST) was trained on Jupiter, hoping to observe the plume of material that was expected to rise well above the cloud tops and perhaps become visible beyond the limb.

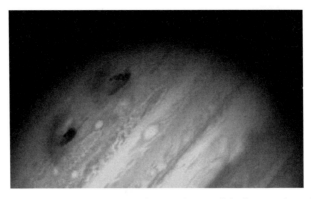

Figure 8.3 Hubble Space Telescope image of the impact sites of fragments F and G – appearing like a pair of eyes on the surface of Jupiter. Image: WFPC2, HST, STScI, SNASA.

The first impact occurred at 20:13 UTC on 16 July 1994, when fragment A of the nucleus slammed into Jupiter's southern hemisphere at a speed of about 60 km/s. Instruments on Galileo detected a fireball that reached a peak temperature of about 24,000 K, compared to the typical Jovian cloud-top temperature of about 130 K. It then expanded and cooled rapidly to about 1,500 K. The plume from the fireball quickly reached a height of over 3,000 km and was observed by the HST.

Astronomers had expected to see the fireballs from the impacts, but did not have any idea in advance how visible the atmospheric effects of the impacts would be from Earth. Observers soon saw a huge dark spot after the first impact. The spot was visible even in very small telescopes, and was about 6,000 km (one Earth radius) across. Over the next six days, 21 distinct impacts were observed with the largest, resulting in a giant dark spot over 12,000 km across, forming from the impact of fragment G on 18 July (Figure 8.3). This impact was estimated to have released an energy equivalent to 6,000,000 megatonnes of TNT (600 times the world's nuclear arsenal).

Hopes that the impact would give further information about the atmosphere were realised as spectroscopic studies revealed absorption lines in the Jovian spectrum due to diatomic sulphur (S_2) and carbon disulphide (CS_2), the first detection of either in Jupiter, and only the second detection of S_2 in any astronomical object. Other molecules detected included ammonia (NH_3) and hydrogen sulphide (H_2S) but, to astronomers' surprise, oxygen-bearing molecules such as sulphur dioxide (SO_2) were not detected. The amount of water detected was also less than predicted, indicating that either the water layer thought to exist below the clouds was thinner than predicted, or that the cometary fragments did not penetrate deeply enough.

The impact of SL9 highlighted Jupiter's role as a kind of 'cosmic vacuum cleaner' for the inner Solar System. The planet's strong gravitational influence leads to many small comets and asteroids colliding with the planet; if Jupiter were not present, the probability of impacts with the Solar System's inner planets would be much greater. Without Jupiter, extinction events such as that at the end of the Cretaceous 65 million years ago would be much more frequent and may not have allowed intelligent life to have developed here on Earth!

Recent impacts on Jupiter

The scars from the impacts were easily visible on Jupiter for many months, so a trawl was made of early observations of Jupiter to see if such an event had occurred before, but none were found. However, such events must have happened before in the life of the Solar System and evidence was found in the Voyager spacecraft observations of crater chains on the surfaces of Ganymede (three) and Callisto (thirteen).

With many amateur astronomers now observing Jupiter (and often taking video sequences to produce high-quality images by eliminating much of the effects of the Earth's atmosphere) there have been several recorded impacts in recent years.

On 19 July 2009, Anthony Wesley observed and first reported an impact that caused a black spot on Jupiter's atmosphere. Observations made by the HST suggest that the observed incident was the result of an asteroid impact about 500 m wide.

On 10 September 2012 at 11:35 UT, amateur astronomer Dan Petersen, using a Meade 12-inch LX200 telescope, visually detected a fireball on Jupiter that lasted one or two seconds. George Hall had been recording Jupiter with a webcam on his 12-inch Meade and, by chance, had captured a four-second clip of the impact. This can be found by searching for 'Jupiter impact video YouTube'. It was estimated that the fireball was created by a meteoroid less than 10 m in diameter. Several collisions of this size may happen on Jupiter on a yearly basis. The 2012 impact was the sixth impact observed on Jupiter, and the fourth impact seen on Jupiter between the years of 2009 and 2012.

There is a sad, yet poignant, end to this story. In 1997, at the age of 69, Eugene Shoemaker was killed in a car crash during an annual trip to Australia in search of asteroid craters. A small vial of Shoemaker's ashes was loaded aboard the spacecraft Lunar Prospector, and now rests with the craft on the surface of the Moon. Around the capsule is wrapped a piece of brass foil inscribed with an image of Comet Hale–Bopp, an image of Shoemaker Crater in northern Arizona, and a passage from William Shakespeare's *Romeo and Juliet*:

> And, when he shall die,
> Take him and cut him out in little stars,
> And he will make the face of heaven so fine
> That all the world will be in love with night,
> And pay no worship to the garish Sun.

Near-Earth objects (NEOs): their discovery and potential threats to Earth

These are objects orbiting the Sun whose orbits bring them to within 1.3 AU of the Sun; many of these orbits will cross that of the Earth. The vast majority are asteroids (NEAs), but a few are short-period extinct comets. By the spring of 2013, 9,683 NEOs had been discovered, of which 93 were comets and 9,590 asteroids. They range in size from ~50 m up to ~32 km (1036 Ganymed) in diameter. The second largest is 433 Eros, which was the target for the Near-Earth Asteroid Rendezvous (NEAR) mission in 2000. The number of near-Earth asteroids over one kilometre in diameter is estimated to be 500 to 1,000. The NEAs only survive in their orbits for a few million years until they collide with the inner planets, are ejected from the Solar System by close approaches with the planets, or fall into the Sun. This implies that, to account for the currently observed number of NEAs, asteroids must be being perturbed from their orbits within the asteroid belt between Mars and Jupiter by the gravitational effects of Jupiter – thus providing a continuing supply of NEAs.

Every hundred years or so, rocky or iron asteroids larger than about 50 m are expected to impact the Earth's surface and cause local disasters or produce tidal waves. Every few hundred thousand years or so, we would expect asteroids larger than a kilometre to impact the Earth, giving rise to global disasters such as that caused by the Chicxulub event. The obvious consequence of the impacts of Comet Shoemaker–Levy 9 on Jupiter alerted governments to the possible threat to Earth and so programmes were instituted to discover NEOs, characterise their sizes and predict their future trajectories and so assess any potential threat.

Programmes to detect NEOs

Lincoln Near-Earth Asteroid Research (LINEAR) project

The LINEAR programme began in 1996, initially using a telescope designed for observing satellites orbiting the Earth. It now uses two 1-metre telescopes and one 0.5-metre telescope based at Socorro in New Mexico, and by 2004 was discovering tens of thousands of objects each year and accounting for 65% of all new asteroid detections. The camera uses a large format CCD array to observe the telescopes' 2 degree field of view. A large area of sky can be observed each night, with each target patch of sky being observed five times each night – largely searching along the ecliptic (the plane of the Solar System) where most NEOs would be expected to be found.

Near-Earth Asteroid Tracking (NEAT)

NEAT used a 1.2-metre telescope located on Maui, Hawaii. It began observing in December 1999 and observed for the six nights each month prior to new Moon. Each 10-hour observing session yielded 15 Mbytes of information (compressed down from 26 Gbytes!) which was transmitted to the Jet Propulsion Laboratory for analysis. Detected objects were immediately reported to the worldwide observing community via the Minor Planet Center (MPC).

Spacewatch

Spacewatch, originally set up in 1980, uses the 0.9-metre telescope sited at the Kitt Peak Observatory in Arizona to hunt for NEOs. The project has recently acquired a 1.8-metre telescope, also at Kitt Peak, which allows a search at 0.7 magnitudes fainter. The 0.9-metre telescope has been upgraded with a mosaic of CCDs, which enables it to cover the sky at least six times faster. This has increased the rate of detection of Earth-approaching asteroids to around 300 per year.

Spaceguard

Spaceguard is the overall name for these affiliated search programmes, designed to detect 90% of NEAs over 1 km diameter by 2008. The name was coined by Arthur C. Clarke in his novel *Rendezvous with Rama*, where SPACEGUARD was the name of an early warning system created following a catastrophic asteroid impact! The number of known NEAs larger than 1 km diameter has now passed the 900 mark. If the population of NEAs larger than 1 km is 1,000 (as predicted from several studies), 900 represent 90% completeness – so the project has reached its target, though taking somewhat more time than the 10-year time period given to NASA by a 1992 US Congressional mandate.

The future

NASA has now been directed to set up a follow-on project from Spaceguard whose objective is to find 90% of NEOs whose diameters are greater than 140 m by the end of 2020. As these are considerably fainter, a new generation of significantly larger telescopes will be required, and these are likely to include the enormous LSST (Large Synoptic Survey Telescope), which will use an 8.4-metre wide field telescope equipped with a 3,200 megapixel camera. Other new telescopes will include the 4.2-metre DCT (Discovery Channel Telescope) and Pan-STARRS (Panoramic Survey Telescope and Rapid Response System).

How is it done?

It is impossible to calculate the orbit of an object from a single observation. With a single observation little can be determined – the asteroid is only known to lie within a cone whose angular dimension is given by the error of the measured position. A second observation will enable a position to be found and three or more observations will enable an approximate orbit to be determined. Once the approximate orbit of an NEO has been found, two further techniques can be used to refine its orbit: firstly, whilst powerful radars cannot provide accurate positions for the NEO, they can give an accurate speed of approach or recession, which is a significant help, and secondly, it is possible to examine archive (earlier) sky images on which the object might be expected to appear to see if the object can be spotted on them. If so, their accurate position at some time in the past is a considerable aid to defining the precise orbit.

Assessment of risk for NEOs

There are two scales that astronomers use to assess the risk posed by individual NEOs. The simpler is called the Torino scale, which uses a range of integers from 0 to 10. A 0 indicates an object has a negligibly small chance of collision with the Earth or is too small to penetrate the Earth's atmosphere intact (and hence poses no threat), whilst 10 indicates that a collision is certain, and the impacting object is large enough to precipitate a global disaster. The value that an NEO is assigned is based on its collision probability and its kinetic energy (expressed in megatonnes of TNT). The Palermo scale is similar but more complex.

The Torino scale

The current record for highest Torino rating was held by the 270-m NEO 99942 Apophis, which in December 2004 was given a rating of 2, which was then upgraded to 4. It is now expected to pass quite close to the Earth on Friday, 13 April 2029 but, as there is no possibility of an impact, its rating has been downgraded to zero. Its orbit may be perturbed as it passes the Earth in 2029, so its rating may increase in the future. This was the first NEO to have been given a Torino scale value higher than 1. In February 2006, the rating for 2004 VD17 was initially given a value of 2 due to a possible encounter in the year 2102, but further observations have again reduced it to zero. Asteroid 2007 VK184, discovered on 12 November 2007 by the Catalina Sky Survey, has a Torino scale value of 1. Observations suggest that 2007 VK184 has a probability of 1 in 31,300 of hitting the Earth during June 2048. The asteroid is estimated to have a diameter of 130 m, and so could cause a significant, but not worldwide, threat to the Earth.

Impact predictions often make the news! Initial observations tend to show an increasing chance of impact, but then further observations rule one out. At first, when there are only a few observations, the error ellipse around the NEO's path is very large and may include the Earth. This leads to a small, but non-zero, impact probability. Further observations may well shrink the error ellipse and will, if it still includes the Earth, raise the impact probability as the Earth now covers a larger fraction of the error region. Finally, as described above, radar observations or the discovery of previous sightings of the object on archival images allow its orbit to be predicted more accurately. This shrinks the error ellipse so that usually the Earth is not included and so the impact probability returns to near zero.

As a result, there have often been problems due to exaggerated press coverage of asteroids that might (on the basis of a limited number of observations) impact the Earth. These are usually shown not to be a threat when further observations are made. Reporting of possible NEO threats is now 'managed' to prevent false alarms and the Torino scale might be abandoned in favour of the Palermo scale (which is more obtuse!). As of November 2013, 2007 VK184 is the only NEO to be listed above zero for potential impacts within 100 years.

What could we do to avert disaster?

As nearly all the NEOs that could present a global threat to the Earth have now been discovered, it is likely that we would know that an object was on a collision course several years in advance. If so, our present technology could be used to deflect the threatening object away from the Earth. The key requirement is to intercept the object when it is furthest from the Sun. At this point, its kinetic energy, speed and (importantly) momentum are at a minimum and hence less energy is required to alter its orbit – the blowing up of an object just before impact as in the film *Armageddon* will simply cause millions of smaller objects to rain down on the Earth rather than one large one – with broadly similar effects.

Both the Earth and the impactor are in orbit around the Sun and an impact can only occur when both reach the same point in space at the same time. The Earth is approximately 12,750 km in diameter and is moving at 30 km/s around its orbit. It thus travels a distance of one planetary diameter in just over seven minutes. So, if we can delay or advance the NEO's passage to the Earth by times of this order (depending on the geometry of the two orbits), it would then miss the Earth. It will thus be obvious that the NEO's time of arrival must be known to this precision in order to forecast the impact and determine what alteration to its orbit is necessary.

Assuming that a rocket can reach the NEO when far from the Sun, there are several ideas for making a small change in its orbit; for example, a nuclear fusion weapon set off above the surface would produce high-speed neutrons that would impact material on the surface of the asteroid facing the explosion. This material would then expand and blow off, thus producing a recoil upon the asteroid that would nudge it towards a fractionally different orbit out of harm's way. It does not have to be a big change – a very modest velocity change in the asteroid's motion would cause the asteroid to miss the Earth entirely. It is important that the explosions are not so energetic as to break up the NEO, so a number of small explosions over a period of time would be likely to be used.

The European Space Agency is already producing the design of a space mission, named Don Quijote, that would attempt to alter the momentum of a NEO by a simple collision. In the case of one possible asteroid threat, 99942 Apophis, a spacecraft weighing less than one tonne could give the required deflection of its orbit. Another idea is to use a 'gravity tractor' (Figure 8.4), a heavy spacecraft that hovers over the NEO which is thus gravitationally attracted towards it. By slowly moving the tractor spacecraft over a number of years, perhaps using an 'ion thruster', the orbit could be sufficiently changed. This has the attraction that it would work with spinning NEOs or those that are 'rubble piles', where nuclear explosions would probably break them apart. A neat idea called a 'mass driver' is an automated system on the asteroid to mine and eject material into space, thus giving the object a slow steady push and, at the same time, decreasing its mass. Finally, the Sun's radiation pressure could perhaps be used, with a large solar sail attached to the NEO so that the pressure of sunlight could eventually redirect the object away from its predicted Earth collision.

The problem with comets

There is no doubt that at some time in the future a comet will impact on Earth – but we have no idea when. The problem is this. In the case of asteroids, it is likely that before long we will know the orbits of all those that are a major threat to Earth and so would be able to predict a possible impact event many years before it is due to happen, so enabling us to take suitable action. The same is true for the short-period comets whose orbits could also be modified. But comets are continuously coming into the inner Solar System that have never been observed before, and it is unlikely that we will have sufficient time to take

Figure 8.4 A 'gravity tractor' to alter the orbit of a possible Earth-colliding asteroid or comet. Image: Dan Durda (FIAAA, B612 Foundation).

appropriate action. Such an event could easily bring about the end of human civilisation, but we should not be too alarmed. The likelihood is that such a cometary impact will only occur once every 300 million years or so, and thus the chance of this happening in our lifetime is exceedingly small.

But there *is* a non-zero chance and, as I suspect that intelligent life is very rare in our Galaxy and thus that the human race is rather special, I feel that something should be done in order to preserve us in the event of such an eventuality. Perhaps two underground retreats could be set up on (not quite) opposite parts of the world, where diverse ethnic groups could go to spend a fortnight's holiday. In the case of a global catastrophe, there would be power and provisions to support them for several years until the dust in the atmosphere had cleared, after which they could emerge into a new future. Is that so stupid an idea?

Finally

A simple rule for life: enjoy every day to the full!

Suggestions for further reading:

Impact!: The Threat of Comets and Asteroids by Gerrit L. Verschuur (Oxford University Press).

Impact Earth: Asteroids, Comets and Meteoroids – The Growing Threat by Austen Atkinson (Virgin Books).

9

Four hundred years of the telescope

In October 1608, a Flemish spectacle-maker by the name of Hans Lippershey applied for a patent for his spyglass which allowed distant objects to be seen as distinctly as if they were nearby. Knowledge of it quickly spread around Europe and soon such telescopes could be bought quite easily. In England, the mathematician Thomas Harriot used one to make what are believed to be the first telescopic observations of the Moon. The first known (rather crude) drawing shown in Figure 9.1 was dated 26 July 1609, followed later with an impressive map of the Moon. Harriot did not publicise his observations and, as a result, it is widely thought that the first astronomical drawings made using a telescope were made by Galileo Galilei.

The spyglass came to the attention of Galileo in July 1609; he quickly worked out the principle of the telescope and built himself an eight-power telescope. Grinding his own lenses and optimising the shape of the objective lens, he gradually improved the power and image quality of the telescope and began to observe the heavens, making his first astronomical observations in the autumn of 1609. In March 1610, he published *The Starry Messenger*, which described his observations of the Moon and planets – particularly those of Jupiter and its moons:

> On the seventh day of January in this present year 1610, at the first hour of night, when I was viewing the heavenly bodies with a telescope, Jupiter presented itself to me; and because I had prepared a very excellent instrument for myself, I perceived (as I had not before, on account of the weakness of my previous instrument) that beside the planet there were three starlets, small indeed, but very bright. Though I believed them to be among the host of fixed stars, they aroused my curiosity somewhat by

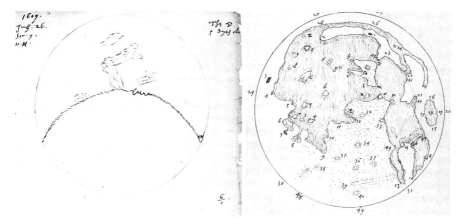

Figure 9.1 Thomas Harriot's Moon map of 26 July 1609 (left) and his later, impressive map of the full Moon.

appearing to lie in an exact straight line parallel to the ecliptic, and by their being more splendid than others of their size ... There were two stars on the eastern side and one to the west. The most easterly star and the western one appeared larger than the other. I paid no attention to the distances between them and Jupiter, for at the outset I thought them to be fixed stars, as I have said. But returning to the same investigation on January eight – led by what, I do not know – I found a very different arrangement. The three starlets were now all to the west of Jupiter, closer together, and at equal intervals from one another ...

On the tenth of January ... there were but two of them, both easterly, the third (as I supposed) being hidden behind Jupiter ... There was no way in which such alterations could be attributed to Jupiter's motion, yet being certain that these were still the same stars I had observed ... my perplexity was now transformed into amazement. I was sure that the apparent changes belonged not to Jupiter but to the observed stars, and I resolved to pursue this investigation with greater care and attention ...

I had now decided beyond all question that there existed in the heavens three stars wandering about Jupiter as do Venus and Mercury about the sun, and this became plainer than daylight from observations on similar occasions which followed. Nor were there just three such stars; four wanderers complete their revolution about Jupiter.

Galileo continued his observations and conveyed some of them in ciphers to Johannes Kepler. These included his observations of Saturn, which showed 'handles' (the ring system), and those of Venus, which when decoded said, 'The mother of lovers (Venus) imitates the shapes of Cynthia (the Moon)' – Galileo had observed that Venus showed phases.

As described in Chapter 1, the most significant observations made by Galileo were those of Venus that proved that the Copernican, rather than the Ptolemaic, model of the Solar System was correct.

In order to produce acceptable images, the length of these simple telescopes had to be very long compared to the diameter of their objective lens. They thus became very unwieldy. The problem is that an objective made of a single lens gives red and blue focuses that are widely separate from the green focus. Using such a lens in a telescope and trying to get the best image – probably where the green light comes to a focus – one would see the sharp green image on top of out of focus blue and red images, which surround the image with a purple glow. This effect is called chromatic aberration.

The achromatic doublet and giant refractors

By combining a biconvex crown glass lens with a plano-concave flint glass lens, it is possible to make what is called an achromatic doublet or achromat – implying that the problem of chromatic aberration is at least largely eliminated. Such a doublet lens was first patented by John Dolland in 1758 but it is believed that the first achromatic lenses were made by Chester Moore Hall in about 1733. This allowed refractors to be made with far larger aperture objective lenses. These not only collected more light and so were able to detect fainter objects, but also improved the image quality or 'resolution' of the telescope and so began the era of the giant refractors, which culminated in the construction of the 40-inch aperture Yerkes Telescope.

Though larger aperture telescopes will theoretically give higher resolution, in practice the resolution is usually limited by what is called the 'seeing' – a function of turbulence in the atmosphere. The atmosphere contains cells of gas with slightly differing refractive indices, which are carried high above the telescope by the wind and act rather like the glass used for screens that blur what is seen beyond. A star is effectively a point source and should theoretically give an image (a central disc surrounded by some faint rings, called the Airy pattern) determined by the telescope aperture. In practice, a stellar image as seen from the UK will probably be of order 2 to 3 arcseconds across and will be highly unsteady. This is one reason why professional telescopes are located on high mountains on islands, such as La Palma in the Atlantic Ocean and Hawaii in the Pacific Ocean, or high in the Chilean Andes. There is far less atmosphere above the telescope and the air tends to be less turbulent as it has been flowing over the sea. Under the best conditions the seeing might limit the resolution to half an arcsecond, so larger aperture telescopes *will* see more detail but not significantly more than a telescope whose aperture is ~400 cm across. The best location for an optical telescope is in space, as in the case of the Hubble Space Telescope, where its full resolution of 1/20 arcsecond at visible wavelengths may be realised. The turbulence of the atmosphere and hence seeing varies from night to night. In bad seeing the image of a star will appear bloated and the Moon can appear to be boiling. On such nights the image quality will be totally determined by the atmosphere. But, rarely, the atmosphere can be still and then the aperture, type of telescope and the quality of the optics will determine what you can see.

When the giant refractors were the prime astronomical instruments, observers tended to concentrate on stellar observations. Precise observations of the star Sirius showed that it was following a sinusoidal path across the sky. In 1844, German astronomer Friedrich Bessel realised that this must be due to the fact that it had a companion star in orbit around it. In this case, it is the barycentre (centre of gravity) of the star system that follows a linear path (called the system's proper motion) with the two stars oscillating on either side. By analysing the path, it was possible to deduce the presence of a star, now called Sirius B and of just 0.06 solar masses, orbiting Sirius A. It had not been directly observed, as with telescopes of the time it would be lost in the glare of Sirius due to light scatter within the telescope. When, in 1862, Alvan Clark was testing the pristine optics of an 18.5-inch refractor by observing Sirius to carry out a 'star test', he spotted the 9th magnitude companion for the first time. It is now known to be a 'white dwarf' star – similar to the final state of our Sun.

One of the most important sets of observations using a refractor was made by Vesto Slipher at the Lowell Observatory, where Percival Lowell had used the observatory's 24-inch refractor to draw the canals on Mars. Slipher made spectroscopic observations of 24 'white nebulae' which showed that they were moving at high speed relative to our Galaxy. Some of the nearby galaxies were coming towards us (gravity is making the members of our Local Group of galaxies collapse down to eventually form one giant amorphous galaxy) but the majority were moving away from us.

By 1915 Slipher had measured the shifts for 15 galaxies, 11 of which were redshifted. Two years later, a further 6 redshifts had been measured and it became obvious that only the nearer galaxies (those within our Local Group) showed blueshifts. From the measured shifts and using the Doppler formula, he was able to calculate the velocities of approach or recession of these galaxies. As we will see, these data were used by Edwin Hubble in what was perhaps the greatest observational discovery of the last century and it is, perhaps, a little unfair that Slipher has not been given more recognition.

Reflecting telescopes

The Newtonian reflecting telescope was invented by Sir Isaac Newton, who did not believe that the problem of chromatic aberration that was suffered by simple refracting telescopes of the time could be overcome. Sadly, it is not thought that he made any significant astronomical observations with it. In the Newtonian telescope, a primary mirror reflects the light to a focus that would lie in the centre of the tube so, to avoid obstructing the light path with one's head, a secondary mirror, often called the flat, reflects the light sideways to form an image just outside the tube where the focuser and eyepiece are placed.

The discovery of Uranus

The first major discovery made using a reflecting telescope was, as described in Chapter 5, that of the planet Uranus by William Herschel, who had come to England from Germany. It soon became apparent why Herschel had seen that Uranus was a planetary body whilst others had not. Side by side comparisons with telescopes in use by others confirmed his telescope's far higher image quality – Herschel had proven to be a superb telescope maker! Uranus has a maximum angular size of 4.1 arcseconds and, as I have observed with an excellent telescope just a little smaller than Herschel's, it appears as a tiny greenish-blue disc. But unless a telescope has well-figured optics this disc would be very hard to distinguish from a star.

The Leviathan of Birr Castle

During the 1840s, William Parsons, the Third Earl of Rosse, built the mirrors, tube and mountings for a 72-inch reflecting telescope that was erected in the grounds of his home at Birr Castle in County Offaly, Ireland. This was, for three-quarters of a century, the largest optical telescope in the world. With this instrument, Lord Rosse made some beautiful drawings of astronomical objects. Perhaps the most notable was that of an object that was the 51st to be listed in Messier's catalogue and known as the 'Whirlpool Galaxy'. It was the first drawing (Figure 9.2) to show the spiral arms of a galaxy and bears excellent comparison to modern day photographic images. (The galaxy is interacting with a second galaxy, NGC 5195, seen at the right of the drawing.)

Though Newtonian telescopes are still widely used by amateur astronomers, the majority of professional telescopes are variants of a type called the

Figure 9.2 M51, the Whirlpool Galaxy, as drawn by the Third Earl of Rosse using the 72-inch telescope at Birr Castle in County Offaly, Ireland. Image: Wikimedia Commons.

Cassegrain telescope. In these telescopes, the secondary mirror is a hyperboloid that reflects the light back down through a central hole through the primary mirror below which lies the focal plane. This a far better place to locate heavy equipment such as a spectrometer. The image quality of a parabolic primary mirror (as used in Newtonian and Cassegrain telescopes) falls off rather rapidly away from the optical axis due to the optical aberration called coma. If, instead, the primary is also a hyperboloid in what is called a Ritchey–Chrétien telescope, coma is eliminated, making it well suited for wide field and photographic observations. It was invented by George Willis Ritchey and Henri Chrétien in the early 1910s. The vast majority of all professional telescopes are Ritchey–Chrétiens.

The 100-inch Hooker Telescope at Mount Wilson

The Leviathan of Birr Castle was the largest telescope in the world until, in 1917, the 100-inch Hooker Telescope was built at the Mount Wilson Observatory in the mountains above Los Angeles. (The increasing light pollution from Los Angeles is a problem!) Edwin Hubble became a staff astronomer at Mount Wilson and used it to make one of the most important discoveries of the last century. In the 1920s he measured the distances of galaxies in which he could observe a type of very bright variable stars called Cepheid variables, which vary in brightness with very regular periods. Their period is related to their brightness and so, by measuring the period it is possible to derive the absolute brightness of such a star in a distant object, such as a galaxy. From its observed (apparent) brightness it is then possible to calculate its distance and hence the distance of the galaxy in which it resides. He was thus able to show that the then-called 'white nebulae' lay beyond our Milky Way Galaxy and that the Universe was far larger than previously thought. As will be described in detail in Chapter 19, he combined his distance measurements with those of the speed of recession provided by Slipher to a produce a plot of speed against distance, and deduced from its linear nature that we live in an expanding universe.

The 200-inch Hale Telescope at Palomar Observatory

The next, and one of the most significant telescopes ever built, was the Hale Telescope built at the Palomar Observatory. In 1928, George Ellery Hale of Mount Wilson Observatory had secured a grant of 6 million dollars from the Rockefeller Foundation for 'the construction of an observatory, including a 200-inch reflecting telescope ... and all other expenses incurred in making the observatory ready for use'. With the increasing light pollution from Los Angeles, Mount Wilson was no longer regarded as an ideal site for an observatory, so a site at an elevation of 1,712 m on Palomar Mountain, 160 km southeast of Pasadena, California, was selected.

Following unsuccessful efforts at making the primary mirror out of quartz, Corning Glass Works were able to cast a mirror out of Pyrex – a low thermal expansion glass that expands and contracts far less than ordinary glass, hence its use in kitchenware. Engineers then started designing the telescope's structure, which would weigh hundreds of tonnes but be capable of moving smoothly and accurately to follow celestial objects across the sky. Whilst tracking, the mirror must maintain its shape to a few millionths of a centimetre. Several revolutionary and ingenious engineering concepts were implemented into the design to meet these requirements, including a Serrurier truss open telescope structure and oil bearings. Russell Porter produced a wonderful set of drawings of the telescope.

Construction of the dome began in 1936 with telescope foundations anchored 6.7 m into the bedrock. The 41.1-m tall and 41.7-m diameter dome was completed in two years. Also in 1936, the mirror blank was transported by rail to Pasadena (at no more than 25 mph (40 km/h)!) taking 16 days. It then spent 11 years in the Caltech Optical Laboratory, where 10,000 pounds (4,500 kg) of glass were ground and polished away to give the required surface accuracy. The extended time was due to World War II, during which all work on the telescope stopped, finally to restart in September 1945. The completed mirror was transported from Pasadena to Palomar in November 1947 and then a further two years were spent to finish polishing, aligning and adjusting the mirror before final completion and dedication in June 1948. In January 1949, Edwin Hubble took the first photographic exposure with the 200-inch, 21 years after the grant for its construction had been given. It was the world's largest optical telescope for 45 years.

The Hale Telescope played an important role in the study of quasars – quasi-stellar objects – so called because on an optical photograph they appeared like stars. They had been first discovered in a set of radio observations at Jodrell Bank Observatory, which had shown that a small number of the then known radio sources had angular sizes less than 1 arcsecond. As the atmosphere caused even point objects to have an angular size of ~2 arcseconds when observed optically, these could well have been nearby star-like objects. A very accurate position of the brightest, called 3C 273 (as it was the 273rd object in the Third Cambridge Catalogue of Radio Sources), was found when it was occulted by the Moon. As the precise position of the Moon's limb at the time of immersion and emersion was known, the position was found to an accuracy of a few arcseconds. It was then possible to use the 200-inch Hale Telescope to take a photograph of the object. Though its image looked very like a star, a jet was seen extending ~6 arcseconds to one side, so it was certainly not a star. Maarten Schmidt, a Dutch astronomer who had immigrated to the USA and was on the staff of CalTech, used the Hale Telescope to take a spectrum of the object. It was unlike any spectrum that had been observed previously and it took some time for Schmidt to realise that the lines that he could see in the spectrum had been redshifted by 16%. This indicated that its distance was about 2 billion light years – it was then the most distant object known in the Universe. But 3C 273 is one of the closer quasars to us and the most distant currently known lies at a distance of 13 billion light years. So quasars are some of the most distant and most luminous objects that can be observed in the Universe.

The Schmidt camera

The Schmidt camera was invented in 1930 by Bernhard Schmidt. He wanted to design a new type of instrument that would have a very large field of view yet be free of aberrations such as coma and have a short focal ratio, so allowing fainter stars to be observed for a given exposure time. To eliminate chromatic aberration the new design would use a mirror as the primary, but to give a large field of view and eliminate coma a spherical mirror would be needed. But spherical mirrors suffer from spherical aberration. Schmidt realised that he could correct for spherical aberration if a corrector plate were placed at the radius of curvature of the spherical mirror. This has a varying thickness

Figure 9.3 Russell Porter's 1941 drawing of the 48-inch Samuel Oschin Schmidt Telescope at Palomar Observatory. At the top can be seen the corrector plate with, down to its left, the photographic plate mounted halfway along the telescope tube.

across its aperture to compensate for the path length difference between the parabolic and spherical mirrors.

Schmidt cameras have become one of the most useful tools of modern astronomy, ideally suited to photographing large star fields in the Milky Way – showing maybe 10,000 stars on one negative. Fritz Zwicky used an 18-inch Schmidt camera to survey the Coma cluster of galaxies and provide the first evidence of 'dark matter', as will be described in Chapter 21. Surveys of the sky are made using such cameras, notably the 48-inch Samuel Oschin Schmidt Telescope at Palomar Observatory (Figure 9.3), which produced the Palomar Sky Survey completed in 1958. The film plates were 14 inches (35.6 cm) square and covered an area of sky 6 degrees across. The survey initially covered the whole of the northern sky down to a declination of −27 degrees. Plates were made with both blue and red sensitive emulsions which were sensitive to stars and galaxies that are about a million times fainter than the limit of human vision.

The UK built a similar telescope at the Siding Spring Observatory in Australia in 1973, which produced a matching survey of the southern sky. On a personal note, an asteroid numbered 15,727, now named 'ianmorison', was discovered in 1990 by the 2-meter diameter Alfred-Jensch-Telescope at Tautenburg Observatory in Germany, the largest Schmidt camera in the world.

Ways of improving the image quality of ground-based telescopes

Given a perfect telescope used in space, resolution is directly proportional to the inverse of the telescope diameter. A plane wavefront from a distant star will form an image with an angular resolution only limited by the diffraction of light, and the telescope is said to be diffraction limited.

However, on Earth, turbulence in the atmosphere distorts the wavefront, creating phase errors across the mirror. Even at the best sites, ground-based telescopes observing at visible wavelengths cannot achieve an angular resolution better than can be achieved by telescopes of about 20 cm diameter. Wavefront errors are also caused by inaccuracies in the mirror's surface and effects caused by gravitational and thermal changes in the mirror and its support structure.

The wavefront errors are thus of two types:

(1) Slowly changing errors due to gravitational and thermal effects on the mirror. These are corrected by what are called 'active optics' systems.
(2) Rapidly changing errors due to the turbulence in the atmosphere. These are corrected by 'adaptive optics' systems.

Active optics

This is the term used to describe the methods used to correct for slow changes in the mirror and its telescope structure. In a typical telescope the mirror – which is relatively thin – is supported by a large number of actuators (perhaps 150 in number) which can be moved to apply forces to the rear surface of the mirror and so adjust the surface profile. Periodically, the image of a star is analysed, taking about 30 seconds, and a computer calculates the errors in the surface that would give rise to the observed image. The computer then calculates the force correction that each actuator has to perform to achieve optimal image quality.

Adaptive optics

This is the term used to describe the correction of phase errors caused by the atmosphere when, across a large mirror, rapidly changing phase errors of a few micrometres equivalent path length result. If there is a reasonably bright star in the field of view (which should give an image which is that of an Airy disc whose angular size is determined by the aperture of the telescope) its actual, distorted, image can be analysed in a computer and corrections applied to correct for the wavefront errors and so produce a diffraction-limited image.

It would be quite impossible to correct the primary mirror every few milliseconds to the required precision of ~0.02 micrometre so, instead, a small (8 to 20 cm) deformable mirror is used in the light path whose surface profile is controlled by hundreds of piezoelectric actuators to compensate for the atmospheric effects. It is easier to fully correct the wavefront and so give a diffraction-limited performance for observations in the near infrared, but such systems can still provide an improvement of perhaps 10 times at visible wavelengths.

Often a suitable reference star will not be found in the field of view that includes the target object, so artificial reference stars are now being created by firing a laser that excites sodium atoms in the upper atmosphere at an altitude of ~90 km. Such an artificial reference star can be created as close to the astronomical target as desired, but a lookout has to be kept for high-flying aircraft.

Spun-cast telescope mirrors in alt-azimuth mounts

A major development in the manufacture of large telescope mirrors was the practice of spinning the mold containing molten glass whilst in the furnace. By spinning the furnace at the appropriate speed whilst the glass is molten, the surface takes on a paraboloidal shape and, when the cooling process is complete, this surface is accurate to a small fraction of a centimetre. This avoids the removal of large amount of glass (cf. the Hale 200-inch mirror) from the centre of the mirror and so greatly reduces both the cost and time required to produce a parabolic mirror. Given the approximately correct shape, the precise paraboloidal shape is ground using a spinning tool impregnated with diamond particles in a numerically controlled milling machine. The process gives a surface with accuracy of about 50 micrometres (0.002 inch), which is then polished to the final surface shape using a very fine polishing compound to give an accuracy of better than 25 nanometres (1 millionth of an inch).

This has enabled a large number of 8-metre class telescopes to be built over the last 20 years. Due to advances in drive systems and their computer control, these telescopes use alt-azimuth (alt/az) mounts. This is a simple two-axis mount for supporting and rotating the telescope about two mutually perpendicular axes: a vertical (*altitude*) axis, and a horizontal (*azimuth*) axis. The biggest advantage of an alt-azimuth mount is the simplicity of its mechanical design resulting in a far reduced cost. To follow an object across the sky, an alt-azimuth mount needs to be rotated at variable rates about both axes and it is also necessary to rotate the photographic plate (or, now, CCD camera) to compensate for the rotation to the field of view that is a consequence of this type of mounting. An alt-azimuth mount also reduces the cost of the dome structure covering the telescope, since the telescope structure is more compact. All current and proposed large optical telescopes use this type of mounting.

Multiple and segmented mirrors

A diameter of 8.4 m seems to be about the maximum size that single mirrors can be made, so other techniques are now being employed to give large effective apertures using two or more smaller mirrors whose light is combined to form one image. An early example was the Multiple Mirror Telescope (MMT), which used six 1.8-m diameter military-surplus mirrors originally manufactured to be incorporated into US spy satellites. Completed in 1979, it had an effective aperture of 4.5 m, almost as big as the Hale Telescope. However, it was always a challenge to keep the individual mirrors perfectly aligned, so in 1999 they were replaced by a single 6.5-m mirror made using the spin technique described above.

In 2008, the Large Binocular Telescope was completed. This uses two 8.4-m mirrors operating in tandem to give an effective aperture of 11.8 m, so is currently the world's largest optical telescope within a single dome. However, the overall size of the pair is 22.8 m, and for some applications it will have an effective resolving power of a telescope of this larger size – about 1/200 arcsecond.

Gemini North and South

The Gemini Observatory comprises two 8.1-metre telescopes: Gemini North is located on Mauna Kea, Hawaii, at a height of 4,214 m, whilst Gemini South is at a height of 2,737 m on Cerro Pachón, Chile. Together, the twin telescopes can give full sky coverage with both sites giving a high percentage of clear weather and excellent atmospheric conditions. They have been designed to operate especially well at infrared wavelengths and, to this end, their mirrors are coated with silver, which reflects significantly more infrared radiation than the aluminium used to coat most other telescope mirrors. As atmospheric water vapour absorbs infrared radiation, both telescopes are located on high mountain tops where the air has a very low water vapour content.

The Keck Telescopes

The twin Keck Telescopes, located at a height of 4,200 m at the top of Mauna Kea, Hawaii, are the world's largest optical and infrared telescopes. They have primary mirrors of 10 m diameter each composed of 36 hexagonal segments whose positions are adjusted (using active optics) to act as a single mirror. Telescopes can take time to thermally stabilise to the nighttime temperatures when the domes are opened. To minimise these effects, during the day the interiors of the Keck domes are chilled close to freezing point.

When observing, the active optics system controls the position of each mirror segment twice a second to a precision of 4 nm to compensate for thermal and gravitational deformations. The Keck Telescopes also use an adaptive optics system using 15 cm diameter deformable mirrors that change their shape up to 670 times per second to cancel out atmospheric distortion, improving the image quality by a factor of 10.

The Keck Telescopes have made a notable contribution to the detection of extra-solar planets by the 'radial velocity' method, as described in Chapter 12.

The Very Large Telescope

The Very Large Telescope (VLT) is operated by the European Southern Observatory (ESO) and consists of four 8.2-metre telescopes that can either work independently or in combined mode, when they are equivalent to a single 16-metre telescope – making it the largest optical telescope in the world. It can observe over a wavelength range from the near ultraviolet up to 25 micrometres in the infrared.

The VLT is located at the Paranal Observatory on Cerro Paranal in the Atacama Desert, northern Chile, at a height of 2,600 m, one of the best observing sites in the world. The four main telescopes have been given the names of objects in the sky in the Mapuche language: Antu – the Sun, Kueyen – the Moon, Melipal – the Southern Cross, and Yepun – the star Sirius.

One notable achievement was the first visual image of an extra-solar planet, albeit around a brown dwarf rather than a normal star, which was observed in the infrared when using an adaptive optics system. By tracking stars orbiting around the super-massive black hole at the centre of our Galaxy, as will be described in Chapter 17, the black hole's mass has been calculated. Using the VLT, astronomers have been able to measure the age, 13.2 billion years, of the oldest star known in our Galaxy and have analysed the atmosphere of an exoplanet.

Optical interferometers

In radio astronomy, many observations have been made by combining the signals from two separated telescopes in what is called an interferometer. This has been extended to use an array of antennas to synthesise a telescope that has a resolution equivalent to a single telescope whose diameter is the size of the array. This has enabled radio astronomers to achieve the same, or greater, resolution as optical telescopes. It is possible to combine the light from two or more of the 8-metre telescopes of the VLT with up to six 1.8-metre telescopes located close by to form an interferometer. In 2008, the VLT was able to image the inner structure of the star Eta Carinae, which could at any time explode in a hypernova event, whilst in 2012, the four VLT telescopes were combined to make a mirror with an effective size of 130 m.

A second major interferometer links the two 10-metre Keck telescopes on Mauna Kea, Hawaii. It had been hoped to build smaller telescopes to develop a similar array to that at the VLT, but currently there is a ban on building any further telescopes on the mountaintop site.

Robotic telescopes

With the use of sophisticated computer systems and high-speed connections over the Internet, it is possible to operate telescopes remotely. Such 'robotic' telescopes, often with mirrors of about 2 m in diameter, such as the two Faulkes Telescopes in Hawaii and Australia, are now spreading around the world. A particularly interesting use is that of observing the effect of gravitational lensing by a foreground star of a more distant one. Observations are usually performed using networks of robotic telescopes, which continuously monitor millions of stars towards the centre of the Galaxy in order to provide a large number of background stars. In the same way that a convex lens can concentrate the light from a distant object into the eye and so make it appear brighter, if a distant star passes behind one of intermediate distance, the brightness of the distant star will undergo a temporary increase, which can last for many days. The peak brightness can be up to 10 times that normally observed. Several thousand such events have now been observed. As will be described in Chapter 12, this technique has been use to detect extra-solar planets which give rise to a secondary brightness increase on the flanks of that produced by the star.

Robotic telescopes are also key instruments in observing the afterglow of the transient events known as gamma-ray bursts. As will be described in Chapter 20, these bursts are very short and are initially detected in gamma rays using, for example, the Swift orbiting gamma-ray telescope. The position is immediately sent around the world, using the Internet, to robotic telescopes that can quickly move to the gamma-ray burst's position and make a precise measurement of the position so that the source of the burst can be found and investigated.

On a smaller scale, robotic telescopes up to ~30 cm in diameter are available for use by amateur astronomers to carry out imaging from high sites and dark skies. I have used a, highly corrected, 8-inch aperture telescope located in Spain to image the galaxy M33 in Triangulum, which is shown in Figure 14.12.

The Hubble Space Telescope

This has been one of the most productive scientific instruments ever built and deserves much of Chapter 19 to give it appropriate coverage.

Three more books on telescopes:

> *The Telescope: Its History, Technology, and Future* by Geoff Andersen (Princeton University Press).
> *The Telescope: A Short History* by Richard Dunn (Conway Maritime Press).
> *The History of the Telescope* by Henry C. King (Dover).

10

The family of stars

Virtually all of the light that we can see in the heavens has come from the surface of stars and they are, certainly for our existence, the most important objects in the Universe. The heat and light of the Sun sustain life here on Earth, and the life cycles of stars that existed billions of years ago produced the elements of which we are made and that of the planet on which we live. This chapter will discuss the properties of stars in general, how these are measured, and where our own star, the Sun, is positioned amongst them.

Stellar luminosity

Stars have a very wide range of intrinsic luminosity – their energy output across the whole electromagnetic spectrum. The luminosity (3.86×10^{26} W) of our Sun, L_{sun}, is taken as the reference point. The Sun actually turns out to be quite a good star to take as the reference luminosity, as it lies roughly halfway (on a logarithmic scale) between the faintest and brightest stars.

The apparent brightness of a star as seen in the sky tells us nothing about the star's intrinsic brightness – a very bright star at a great distance might very well appear less bright than an intrinsically faint star that is much closer to us. Thus, in order to be able to relate the intrinsic brightness and hence luminosity of other stars to that of our Sun we have to eliminate the square law effect on brightness due to the star's distance. To do this we thus need to be able to measure the distances of stars.

Stellar distances

To measure stellar distances the method of 'parallax' is used. This requires one to observe an object from two positions some distance apart

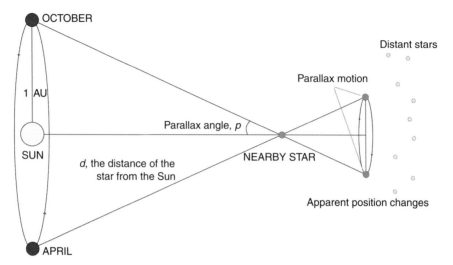

Figure 10.1 The method of parallax.

(forming a baseline) and measure the change in angle as seen against distant objects. The further away the object, the greater the baseline required. In the case of stars, there is no perceptible change in angle from points across the Earth, so a considerably bigger baseline is required. Happily, one is available to us, the distance across the Earth's orbit around the Sun. Many stars (until recently the vast majority) are too far away to show any change in their observed position when observed, say, in the spring and autumn when the Earth is at opposite sides of the Sun (Figure 10.1). These stars can thus be considered as reference points against which the movement in position of nearby stars can be measured. The angular change in their position, coupled with knowledge of the Earth's orbit, can thus be used to find their distance.

The measured angles are very small, an arcsecond or less, and depend on the exact times at which the measurements are made. To determine the parallax, the change in angle of a star's position (which has been measured using a baseline of ~2 AU) is converted into what is termed the 'parallax' of the star, which is the angular shift in position of the star that *would* be observed with a baseline of exactly 1 AU. One can then immediately derive its distance.

The parsec

As the angle and distance are inversely related, it is possible to define a unit of distance such that a star located at this distance would have a parallax of 1 arcsecond. This unit is called the 'parsec', and is the unit of distance normally used by professional astronomers. The simple relationship is

$$d = 1/p \text{ parsecs}$$

(where d is distance in parsecs and p is the parallax of the star in arcseconds). The parsec is usually abbreviated to pc.

A star that subtends an angle of 1 arcsecond will be at a distance of 1 parsec. This is equivalent to a distance of 3.26 light years, so

1 parsec is equal to 3.26 light years.

A star that has a parallax of 1/10 arcsecond will lie at a distance of 10 parsecs or 32.6 light years. The nearest star has a parallax of 0.772 arcseconds, which corresponds to a distance d given by

$$\begin{aligned} d &= 1/0.722 \text{ parsecs} \\ &= 1.295 \text{ parsecs} \\ &= 1.295 \times 3.26 \text{ light years} \\ &= 4.22 \text{ light years}. \end{aligned}$$

It is a star in the Alpha Centauri multiple star system and has, not surprisingly, been given the name Proxima Centauri.

In practice, as is often the case, things are not quite so simple. All stars in the Galaxy are moving around the centre of the Galaxy and, unless they are moving either directly away or towards us, will be slowly moving across the sky. This is called a star's 'proper motion' and is usually expressed in units of arcseconds a year. Thus, if one measures a change in the position of a star when observed from either side of the Earth's orbit, one does not know whether this is caused by parallax, proper motion or a combination of both. To separate the two effects, one needs to observe the star again after a full year when the Earth is back to its original position. Any change in position observed over the full year will only be due to proper motion and so this component of the observed motion can be measured, so enabling the part due to parallax to be found. In practice, observations over several years will give the best results.

Hipparcos and Gaia

Observations from the ground are limited in positional accuracy by turbulence in the atmosphere and, until recently, the number of stars whose parallax was known was limited to the few thousand out to distances of about 40 parsecs (130 light years). In 1989, unencumbered by the atmosphere, a satellite called Hipparcos (High Precision Parallax Collecting Satellite) was able to make three years of observations whose positions were accurate to about a milliarcsecond. It was thus able to accurately measure the distances and proper motions of 118,000 stars out to a distance of ~90 parsecs (300 light years). The data collected by Hipparcos contributed to the prediction of when Comet

Shoemaker–Levy 9 would collide with Jupiter. Even so, Hipparcos was only able to make measurements across ~1% of the size of the Galaxy.

Following on from ESA's Hipparcos mission, the Gaia space telescope was launched in December 2013 to travel to the Sun–Earth Lagrange point L2, located approximately 1.5 million km from the Earth directly away from the Sun. In general, the further away an object is from the Sun the longer its period (or 'year'), but at the L2 point a spacecraft also feels the gravitational pull of the Earth in exactly the same direction – which has just the same effect as if the Sun were somewhat more massive – and so will travel around the Sun more quickly. The clever thing is that at just the right distance from the Earth it will orbit in precisely one Earth year so, relative to the Earth, stays in the same position in space. In position at L2, reached in January 2014, a 10-metre diameter sunshade was deployed which always faces the Sun, keeping the telescope cool whilst powering Gaia using solar panels on its sunlit surface.

The mission aims to compile a three-dimensional space catalogue of approximately 1 billion stars, planets and asteroids representing approximately 1% of the Milky Way population and show, in detail, the structure of our Milky Way. It will measure the positions of the brighter stars to an accuracy of 7 microarcseconds – equivalent to the diameter of a small coin at the distance of the Moon – and by observing each star up to 70 times during the five-year mission will be able to track their movements through space. This very high precision will enable Gaia to spot the telltale 'wobble' in the path of a star that indicates the presence of an orbiting planet and, as their image passes across the billion pixel camera, also detect comets and asteroids. Secondary instruments will measure the physical properties of the brighter stars, such as their luminosity, effective temperature, gravity and composition, which will provide the data to enable the origin, structure and evolutionary history of our Galaxy to be studied.

Meanwhile, a method called 'spectroscopic parallax' can be used to measure distances across the Galaxy, but the discussion of this method will need to follow later in the chapter when stellar spectra have been covered.

Stellar colour and surface temperature

If one spends a little time observing the night sky, it soon becomes apparent that stars can have different colours: Betelgeuse in Orion and Aldebaran in Taurus have an orange tint, Capella in Auriga is somewhat yellow in colour and Rigel, also in Orion, a slightly bluish white.

The colours we perceive are a function of the surface temperature of the star. As the surface temperature increases from ~3,000 K up to ~20,000 K, the colour of the star moves from red, through orange, yellow and white to blue.

Stellar spectra

Dependent on the temperature in the chromosphere of a star, sets of absorption (dark) lines will be observed in the spectrum of a star. As an example, neutral hydrogen produces a distinctive set of spectral lines, called the Balmer series, that range in wavelength from 363.46 nm in the ultraviolet to 656.3 nm in the red. The most prominent is the 635.3 nm red line, which is called the H-alpha line and is seen in clouds of gas where the electrons have been lifted into excited states by incident ultraviolet radiation and then drop back down into lower energy states.

Our eyes are not sensitive to the red light of the H-alpha line and, sadly, we see very little colour in the Universe with our eyes, but our eyes are far more sensitive in the green, and using a telescope of 16 inches aperture, I have observed the central part of the Dumbbell planetary nebula appearing as a vivid green colour – the light from the H-beta line.

Other atoms will produce similar series of lines in either neutral or ionised form depending on the temperature of the stellar atmosphere. The spectrum that we observe is thus a mix of all these lines and depends strongly on temperature. The hydrogen Balmer series, for example, appears strongest when the star's atmosphere is at ~9,000 K. At very high stellar temperatures, virtually all the hydrogen atoms are ionised so the Balmer lines are very weak. However, it takes far more energy to fully ionise helium, so lines of both neutral (He I) and singly ionised (He II) helium are seen.

In the latter part of the nineteenth century, the spectra of thousands of stars were photographed by astronomers at Harvard University and the spectra were used to classify the stars into what are called their spectral types. For example, type A stars were those where the hydrogen Balmer lines were seen to be at their strongest. Stars where the hydrogen lines were weak but helium lines were seen were called type O. In all, the stars were split into seven spectral types: O, B, A, F, G, K and M. Here they have been listed in decreasing order of temperature: O the hottest and M the coolest. Each type is split into 10, so the hottest stars within a spectral type will be classified as, say, G0 and the coolest within that type G9. Our Sun is classified as a G2 star and is thus towards the hotter end of the G-type stars.

- **O-type** stars range from ~60,000 K down to 30,000 K. As we will see when stellar evolution is covered, such stars have a very short lifetime so are relatively rare.
- **B-type** stars are cooler, ranging from 30,000K down to 10,000 K.
- **A-type** stars range in temperature from 10,000 K down to 7,500 K.
- **F-type** stars cover the range from 7,500 K down to 6,000 K.

- **G-type** stars, the type that includes our Sun, cover the temperature range from 6,000 K down to 5,000 K.
- **K-type stars** range from 5,000 K down to 3,500 K.
- **M-type** stars are the coolest with surface temperatures less than 3,500 K.

From the surveys that currently exist, the percentages of stars in the differing spectral classes are as follows.

Type	Colour	Percentage
O	Blue	0.003%
B	Blue-white	0.13%
A	White	0.63%
F	White-yellow	3.1%
G	Yellow	8%
K	Orange	13%
M	Red	75%

You will see that the great majority of stars are cool M-type stars and there are very small percentages of O- and B-type stars.

A point to note from this table is that far from being an 'average' star, as is sometimes stated, the Sun is actually pretty well up in the stellar league table – with ~96% of stars being fainter and only ~4% being brighter!

Spectroscopic parallax

The fact that stars' spectra can be measured, even for stars at great distances, allows the stellar distance scale to be extended across the Galaxy and even out to the nearest galaxies. The method used is called 'spectroscopic parallax' – though it has nothing to do with parallax! It is based on the very simple premise that all stars of the same spectral type, such as a type F0 star, will have the same intrinsic luminosity. Suppose then that we observed an F0 star that appeared 10,000 times less bright than a nearby F0 star that was at a distance of 8 pc as measured by the method of parallax. The inverse square law tells us that the distant star would be $(10,000)^{1/2}$ or 100 times further away, so would lie at a distance of 800 pc.

As a further example, consider two B8 stars, such as Rigel, in Orion, and a star with a similar spectrum observed in the Large Magellanic Cloud (LMC) – a nearby galaxy. Rigel appears 23 million times brighter than the distant star so, using the inverse square law and knowing the distance of Rigel, we can calculate that the LMC lies at a distance of ~48,000 pc or ~156,000 light years.

The method of spectroscopic parallax suffers from two fundamental problems. The first is that stars of the same spectral type do not necessarily have the same luminosity, as the intrinsic luminosity of a star depends to an extent on what is called the star's metallicity – the percentage of elements heavier than helium or hydrogen in its makeup – so reducing the accuracy of the method.

The measurements of the luminosity of distant stars suffer from a further problem: that of 'extinction', which is the name given to the absorption of light by intervening dust. This would reduce the observed brightness of a star and so make it appear to be further away. However, dust absorbs more light in the blue (shorter wavelength) end of the spectrum than it does at the red (longer wavelength) end of the spectrum. Thus, the light from a star whose light has passed through dust clouds will appear redder than it should. From this effect it is possible to estimate the effect on the star's apparent brightness and so make a suitable correction.

The Hertzsprung–Russell diagram

In the early 1900s, Ejnar Hertzsprung in Denmark and Henry Russell at Princeton University in the USA independently plotted a graph of stellar luminosity against temperature which has become known as the Hertzsprung–Russell diagram or H-R diagram (Figure 10.2). Both axes are logarithmic. The x axis represents temperature (NB: it increases to the left of the plot) and, in Figure 10.2, I have also given the spectral type. The vertical axis is a measure of brightness relative to that of the Sun.

The 'main sequence' is an S-shaped region that extends from the top left (very bright, high surface temperature O-type stars) down to the bottom right (faint, low surface temperature M-type stars). It is the region that includes between 80% and 90% of all stars. The stars at the lower right of the main sequence are called 'red dwarfs' as their luminosity is far less than that of out Sun.

Above the main sequence on the right-hand side of the plot is an area of bright stars whose colours range through yellow, orange and red. Because these stars are very bright they are called giant stars, such as the star Aldebaran, in Taurus, which is called a red giant (though they actually appear orange in colour). At the very top of the plot is a region, extending from the blue through to the red, of exceedingly bright stars that are called supergiants. Betelgeuse, a red supergiant in Orion, would lie at the extreme top right of the plot. In contrast, at the extreme top left of the plot one would find the star Rigel, the brightest star in Orion, whose luminosity is ~45,000 times that of the Sun. This is termed a blue supergiant.

140 A Journey through the Universe

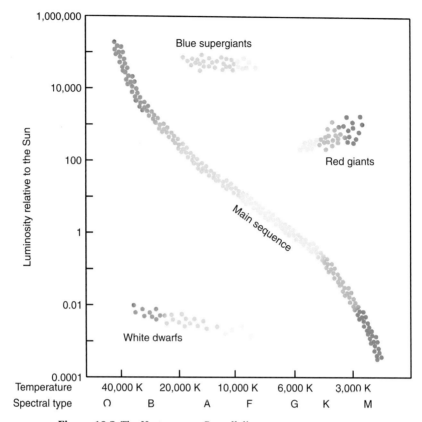

Figure 10.2 The Hertzsprung–Russell diagram.

Below the main sequence lies a region in which white dwarf stars are found. (They encompass a wide surface temperature range and are not necessarily white.) The companion to Sirius, in Canis Major, is a white dwarf. As we will see in Chapter 11, on stellar evolution, white dwarfs are the remnants of stars like our Sun. They are very small, about the size of the Earth, so even those with very high surface temperatures are not very luminous.

The reason why most stars are seen to lie on the main sequence is simply because this is where stars spend the majority of their life as stable objects producing energy by the nuclear fusion of hydrogen to helium. As stars evolve, their position in the H-R diagram changes and a star is said to move along an evolutionary track across the H-R diagram. The giant phase in the life of a star is relatively brief, which is why we see far fewer stars of this type. The white dwarfs are the final states of many stars and gradually cool over billions of years, thus moving down and to the right of the H-R diagram. Over time, as more of the stars in our Galaxy come to the end of their lives, their numbers will increase relative to those on the main sequence.

The size of stars: direct measurement

The angular sizes of relatively nearby stars can be measured directly. The diameter of our Sun, calculated earlier, comes from a knowledge of its angular size and distance. There is only one star, Betelgeuse, a red supergiant in Orion, whose angular size can be directly observed with a normal telescope. In 1995, the Faint Object Camera of the Hubble Space Telescope (HST) was used to capture the first conventional telescope image of Betelgeuse and measured an angular size of ~0.05 arcseconds (Figure 10.3). Given the best estimate of its distance one can then calculate the diameter, which is ~1 billion km, about 700 times that of the Sun.

A method called 'optical interferometry' uses two or more mirrors separated by distances of order tens of metres and combines their light to give the effect of one giant optical telescope, so giving far higher resolution than single telescopes such as the HST, which has a mirror of 2.4 m diameter. The method has been used to measure the angular diameters of nearby stars, so that given their distances as measured by the method of parallax, their diameters can be derived. Figure 10.4 shows an image of the double star Capella where the two stars are separated by just 50 milliarcseconds.

Figure 10.3 Hubble Space Telescope image of Betelgeuse. Image: A. Dupree (CfA), NASA, ESA.

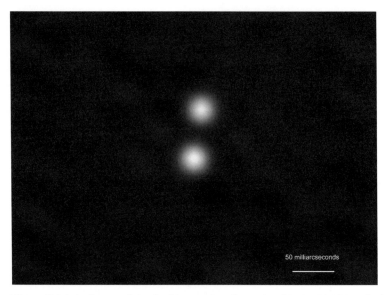

Figure 10.4 An image of the double star Capella made with the COAST optical interferometer at the Mullard Radio Astronomy Observatory of Cambridge University.

In 2002, the light from two 8.2-metre telescopes of the Very Large Telescope (VLT) array in Chile was combined to form an interferometer with a baseline of 102.4 m. Its resolution was thus equivalent to an optical telescope ~100 m in diameter. It measured an angular diameter of Proxima Centauri, the nearest star to the Earth, of 1.02 ± 0.08 milliarcseconds – incidentally about the angular size of an astronaut on the surface of the Moon as seen from the Earth. Proxima Centauri lies at a distance of 1.3 pc giving a diameter of 1.4×10^9 m, about one-seventh that of the Sun.

In general, the radii of stars on the main sequence range from ~20 times that of the Sun at the upper left of the H-R diagram down to 0.1 that of the Sun at the lower right (Figure 10.5). Giant stars lie in the region of ~10 to 100 times the radius of the Sun; an example is Aldebaran, in Taurus, which has a radius 45 times that of the Sun. Supergiant stars such as Betelgeuse often pulsate, its radius varying between around 700 to 1,000 times that of the Sun with a period of ~2,100 days.

Stellar masses

The mass of our Sun provides the starting point for determining that of other stars. We can determine the mass of a second star if we can observe one in a binary system orbiting around a G2 star like our Sun. The method uses the

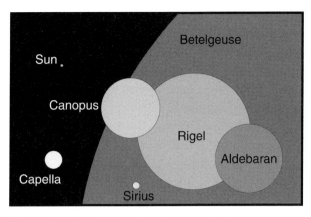

Figure 10.5 The relative sizes of some typical stars.

generalisation of Kepler's third law that was derived by Isaac Newton. Newton's derivation showed that the square of the orbital period is inversely proportional to the sum of the two masses orbiting each other. So, if the period and orbital major axis of a binary system can be measured, then the sum of the masses can be found. Suppose we observe a binary system in which one star was a type G2 (assumed to have the same mass as our Sun) and the second a type A0 star. If the combined mass of the system were calculated to be 4 solar masses, then the mass of the other star would be 3 solar masses. We thus know that an A0-type star would have a mass of 3 solar masses. If then, a binary star system, one of which was a type-A0 star, was observed to have a mass of 4.5 solar masses, we could deduce that the other star in the system would have a mass of 1.5 solar masses. From its temperature of ~7,000 K and spectra we would then know that a star of type F2 would have a mass of 1.5 solar masses.

From such observations it has been found that the most massive stars are about 50 times more massive than the Sun with the least massive being about 1/15 the mass of the Sun.

Stellar lifetimes

More massive stars have higher temperatures in their cores and burn up their hydrogen fuel far more rapidly than less massive stars. Assuming that stars can burn similar percentages of their total mass in the core, then a natural consequence is that more massive stars will have shorter lives on the main sequence.

Let us take Rigel as an example: its mass is 17 times that of our Sun whilst its visual luminosity is 45,000 times that of the Sun. It will thus only stay on the main sequence for 17/45,000 of the lifetime of our Sun. We believe that the Sun will remain on the main sequence for ~10 billion years, which implies that Rigel can only remain there for ~1/2,600 of this period – about 4 million years! This is actually an overestimate as Rigel, being very hot, emits a good percentage of its radiation in the ultraviolet, which is not accounted for in the visual luminosity. There is a second measure of luminosity called the 'bolometric luminosity', which measures the total energy output across the whole of the electromagnetic spectrum, and this is ~66,000 times that of our Sun rather than ~45,000, so reducing the main sequence lifetime to ~2.7 million years.

At the other end of the main sequence, a red dwarf star might have a luminosity of only ~1/10,000 that of our Sun, but a mass of one-fifth that of the Sun. This would give it a lifetime of ~2,000 times that of the Sun and thus their lifetimes are far longer than the present age of the Universe. In fact, whereas in a star like the Sun only ~10% of the star's mass will be converted into helium during its main sequence phase, the less massive red dwarfs mix their interiors by convection, so allowing a greater proportion of their mass to be converted into helium and thus extending their lifetimes even further.

Suggestions for further reading:

Stars: A Very Short Introduction [Kindle Edition] by Andrew King (Oxford University Press).
Introduction to Astrophysics: The Stars by Jean Dufay (Dover).

11

Aging stars

This chapter will look at how stars evolve during the later stages of their lives, and describe the remnants left when they die: white dwarfs, neutron stars and black holes.

For a collapsing mass of gas to become a star, nuclear fusion has to initiate in its core. This requires a temperature of ~10 million K and this can only be reached when the contracting mass is greater than about 10^{29} kg, about 1/20 the mass of the Sun, or 20 times that of Jupiter.

In low-mass stars, less than ~0.5 solar masses, the conversion of hydrogen to helium by nuclear fusion is the same as in our Sun. However, whereas in stars of greater mass nuclear fusion only converts ~10% of the mass of the star (that residing in its core), in the lowest mass stars it is thought that convection currents mix the star's interior and so will allow much of the star's mass to undergo nuclear fusion, so increasing the time during which they can carry out the fusion of hydrogen to helium – a period that is significantly longer than the present age of the Universe. We thus have no direct observational evidence of what happens when fusion ceases in such stars and can only use computer modelling to investigate what might happen.

In order for helium to fuse into heavier elements, temperatures of order 100 million K are needed and this requires sufficient mass in the star's envelope to provide the required pressure to enable such temperatures to be reached. In stars of mass less than 0.5 solar masses there is simply not enough pressure to give the temperatures that would allow helium fusion to begin. So when nuclear fusion, converting hydrogen to helium, finally ceases – and modelling of a 0.1 solar mass red dwarf suggests that this might be after 6 trillion years – the star will slowly collapse over a period of several hundred billion years to form what is called a 'white dwarf'. Over many trillions of years, the white dwarf will cool until its surface temperature is below that at which significant light is emitted and the inert remnant will become a 'black dwarf'. No white dwarfs derived from low-mass stars yet exist, but they will be discussed in detail in the following section on mid-mass stars, as their evolution also produces white dwarfs that *can* now be observed.

All stars in the mass range ~0.5 to ~8 solar masses have a common end state in the form of a white dwarf. There is, however, a difference in the process of nuclear fusion from hydrogen to helium in stars above and below about 2 solar masses.

For stars whose mass is less than ~2 solar masses, like our Sun, the bulk of their energy is produced by what is called the proton–proton cycle, as described in Chapter 2. However, there is a more complex process called the carbon–nitrogen–oxygen (CNO) cycle, which provides 1% to 2% of the Sun's total energy output. In stars greater than about 2 solar masses this process, proposed independently by Carl von Weizsäcker and Hans Bethe in 1938 and 1939 respectively, becomes dominant. It provides a very efficient way of converting hydrogen to helium, so the hydrogen in more massive stars burns more quickly so increasing the energy output of the core. As the greater energy output of the star must be balanced by radiation from its surface, the star becomes bluer and has a greater luminosity.

The net result of the cycle is to fuse four protons into an alpha particle along with two positrons, two electron neutrinos (which carry some energy away from the star) and three gamma-ray photons. The carbon acts as a catalyst and is regenerated. The two positrons annihilate electrons releasing energy in the form of gamma rays – each annihilation giving rise to two gamma-ray photons.

When the CNO process reaches an equilibrium state, the reactions of each stage will proceed at the same rate. The slowest reaction within the cycle is that which converts ^{14}N into ^{15}O so, in order for this reaction to have an equal reaction rate, the number of nitrogen nuclei must be significantly larger than those of carbon or oxygen. Thus, over time, the relative amount of nitrogen increases until equilibrium is established and nitrogen becomes the most numerous of the three. This process produces essentially all of the nitrogen in the Universe and thus has great significance for us as nitrogen is an essential element of all life-forms here on Earth.

The triple alpha process

Eventually, either by the proton–proton chain or the CNO cycle, the core of the star will be converted into ^4He. At this point, nuclear fusion stops so that the pressure in the core that prevents gravitational collapse drops. The core thus reduces in size but, as it does so, its temperature will rise. Finally, when it reaches ~100 million K, a new reaction occurs – the 'triple alpha process (3α)', so called because it involves three helium nuclei, which are known as alpha particles. This is an extremely subtle process. The first obvious nuclear reaction that would happen in a core composed of helium is that two ^{4}He nuclei fuse to form ^{8}Be. But ^{8}Be is very unstable – it has a lifetime of only 10^{-19} s – and virtually instantly decays into two ^{4}He nuclei again. Only when the core temperature has increased to 100 million K does it become likely that a further ^{4}He nucleus can fuse with ^{8}Be before it decays. The result is a ^{12}C nucleus. It is highly significant to our existence here on Earth that there is such a difference in temperature between that (~15 million K) at which the hydrogen fuses to helium and that (~100 million K) at which ^{12}C can be formed. If this were not the case, and the process could happen at the core temperatures close to that at which the proton–proton or CNO cycles operate, there would be no long period of stability whilst the star remains on the main sequence with a relatively constant luminosity. This, of course, has allowed stable temperatures to exist on Earth for billions of years and so enabled intelligent life to evolve.

But there is a further real problem in attempting to form ^{12}C. A temperature of 100 million K is required to give the ^{4}He nuclei a reasonable chance to fuse with a ^{8}Be nucleus before it has a chance to decay. The ^{4}He nuclei are thus moving very fast and so have appreciable kinetic energy. It would be expected that this energy would prevent a stable ^{12}C nucleus arising as it would be sufficient to split the newly formed nucleus apart. (If a white billiard ball (^{4}He) approached a red ball (^{8}Be) very slowly they might just 'kiss' and remain touching, but if it came in at high speed the energy of impact would split them apart.)

So why is ^{12}C so common? This problem was pursued with great vigour by the British astrophysicist Fred Hoyle, in the early 1950s. As he then stated, 'Since we are surrounded by carbon in the natural world and we ourselves are carbon-based life, the stars must have discovered a highly effective way of making it, and I am going to look for it.'

He realised that the excess energy that was present in the reaction (and thus expected to break up the newly formed ^{12}C nucleus) could be contained if there happened to be an excited state (called a 'resonance' by particle physicists) of the carbon nucleus at just the right energy above its ground state. This is because, due to the quantum nature of matter, though atomic nuclei usually exist in their ground state, it is possible for them to absorb energy (such as an interaction with a gamma-ray photon) and jump into an excited state. This will later decay back to the ground state with the emission of a gamma-ray photon of the same energy. This is analogous to an atom absorbing a photon of energy that lifts an electron to a higher energy level. The electron will then, in one or more steps, drop back down the energy levels, emitting photons as it does so.

Hoyle realised that a stable carbon nucleus could only result if it had an excited state that was very close in energy to that of the excess energy of the three ^{4}He that came together in its formation. This would thus lift the resulting ^{12}C nucleus into an excited state from which it could drop back to the ground state by the emission of a gamma-ray photon and so reach a stable state.

Some experiments in the late 1940s had suggested that such an excited state might exist, but Hoyle had been told that these were in error. Hoyle argued that there *must* be an appropriate excited state otherwise we could not exist, and he pestered the particle physicists at the California Institute of Technology (Caltech), led by William Fowler, to repeat the experiments. Fowler did so (it is said, only so that Hoyle would go away) and found that there was indeed an excited state within 5% of the energy predicted by Hoyle! Hoyle was essentially using the 'anthropic principle', which says that our existence as observers puts constraints on the Universe in which we live. William Fowler received the Nobel Prize in part for this work. Many believe that Hoyle should also have won the Nobel Prize for this incisive observation and his following work in showing how the elements are synthesised in stars.

For a given mass of gas, the 3α process only releases ~10% of the energy produced in forming helium nuclei from hydrogen, so the length of the helium burning phase will be ~10% of the star's life on the main sequence.

During its helium burning phase the core will be compressed to perhaps 1/50 of its original size and have a temperature of ~100 million K, with, in addition, a shell of hydrogen burning surrounding the core. The energy so produced causes the outer parts of the star to also undergo significant changes. The radius of the star as a whole increases by a factor of ~10, but at the same time the surface cools to (in the case of a 1 solar mass star) a temperature of ~3,500 K. The star will then have an orange colour and it becomes what is called (perhaps perversely) a 'red giant'.

For mid-mass stars less in mass than our Sun, this is about as far as nuclear fusion can take the formation of elements as there is not enough overlying mass above the core to allow its temperature to rise sufficiently for further nuclear fusion reactions to be carried out. The stars in the upper part of this mass range are able to carry out one further nuclear reaction and combine a further alpha particle with a carbon nucleus to form oxygen. This reaction and the 3α process are thought to be the main source of carbon and oxygen in the Universe today. But we could not exist if these elements stayed within the star – they *must* lose much of their material into space – another example of the anthropic principle – and this is exactly what is observed.

Variable stars

In the later stages of their lives, stars become less stable and may well oscillate in size. As a star's size increases, its surface area will also increase, tending to increase its luminosity but, at the same time, the surface temperature will reduce so reddening its colour. As the emitted energy per unit area decreases as the fourth power of the temperature, a star's luminosity actually falls as the size increases. Conversely, as the size of a star reduces, the colour will shift towards the blue and the luminosity will increase. The periodic changes in colour and luminosity result in what is called a 'variable star'. During this phase of their lives, stars often have intense solar winds and so lose much of their outer envelopes into space.

Planetary nebula

Finally, it appears that a star becomes so unstable that the outer parts of the star are blown off to form what is called a 'planetary nebula' surrounding the core remnant. Planetary nebulae are some of the most beautiful objects (Figure 11.1) that we observe in the Universe and many, such as the Ring and Dumbbell nebulae, may be observed with a small telescope. Planetary nebulae are relatively common with over 1,500 known, but it is expected that many more, perhaps over 50,000, exist in the Galaxy but are hidden by the dust lanes. The name 'planetary nebula' is, of course, a misnomer as they have nothing to do with planets, but many do have a disc-like appearance. They are large tenuous shells of gas which are expanding outwards at velocities of a few tens of kilometres per second. They also contain some dust and have masses of typically one-tenth to a fifth of a solar mass. Of order 10 planetary nebulae are thought to be formed each year, so the interstellar medium is being enriched by around 1 solar mass per year.

White dwarfs

At the centre of a planetary nebula lies a white or blue-white star. They are not very bright so that relatively large telescopes are required to see them visually. (I have once, using a 16-inch telescope under perfect conditions, observed the star at the centre of the Ring Nebula.) This star is approaching the final stage of its life when it will become a white dwarf. Once nuclear reactions have ceased, what is left at the centre of the star will contract under gravity. It is composed mainly carbon and oxygen, and devoid of its outer layers through a combination of the intense stellar winds and the ejection of a

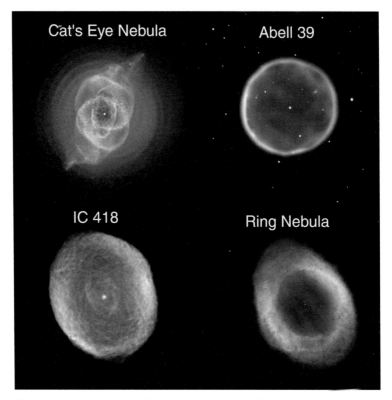

Figure 11.1 Planetary nebulae. In all cases, the stellar remnant can be seen at the centre. This is also shown in colour in Plate 11.1. Images: Hubble Space Telescope, STScI, ESA, NASA.

planetary nebula. The fact that contraction finally ceases is due to a quantum-mechanical effect known as 'degeneracy pressure'. In 1926, R. H. Fowler realised that, as a result of the Pauli exclusion principle, no more than two electrons (of opposite spin) could occupy a given energy state. As the allowed energy levels fill up, the electrons begin to provide a pressure, called 'electron degeneracy pressure', that finally halts the contraction. This pressure only depends on density, not temperature, and this has the interesting result that the *greater* the mass of the white dwarf, the *smaller* its radius.

A further consequence of being supported by electron degeneracy pressure is that there is a limiting mass that cannot be exceeded. This depends on the composition of the star; for a mix of carbon and oxygen, it turns out to be ~1.4 solar masses. This result was published in 1931 by Subrahmanyan Chandrasekhar when he was only 19. In 1983, Chandrasekhar rightly received the Nobel Prize for this and other work. We will see later what happens when the mass of the collapsing stellar remnant exceeds the Chandrasekhar limit.

White dwarfs range in radius from 0.008 up to 0.02 times the radius of the Sun. The largest (and thus *least* massive) is comparable to the size of our Earth, whose radius is 0.009 times that of the Sun. The masses of observed white dwarfs lie in the range 0.17 up to 1.33 solar masses, so it is thus obvious that they must have a very high density. As a mass comparable to our Sun is packed into a volume one million times less, its density must be of order one million times greater – about 1 million g/cm^3. (A tonne of white dwarf material could fit into a matchbox!)

The discovery of white dwarfs

The first known white dwarf was discovered by William Herschel in 1783; it was part of the triple star system 40 Eridani. What appeared surprising was that although its colour was white (which is normally indicative of bright stars) it had a very low luminosity. This is of course due to its small size so, although each square metre is highly luminous, there are far fewer square metres.

The second white dwarf to be discovered is called Sirius B, the companion to Sirius, the brightest star in the northern hemisphere. Friedrich Bessel made very accurate measurements of the position of Sirius as its proper motion carried it across the sky. The motion was not linear and Bessel was able to deduce that Sirius had a companion. Their combined centre of mass *would* have a straight path across the sky but both Sirius and its companion orbit the centre of mass, thus giving Sirius its wiggly path. Due to the close proximity with Sirius, Sirius B is exceedingly difficult to observe as it is usually obscured by light scattered from Sirius within the telescope optics. A very clean refractor has the least light scatter, and it was when Alvan Clark was testing a new 18-inch refracting telescope in 1862 that Sirius B was first observed visually.

The future of white dwarfs

The observed surface temperatures of white dwarf stars range from 4,000 up to 150,000 K, so they can range in colour from orange to blue-white. Their radiation can only come from stored heat unless matter is accreting onto them from companion stars. As their surface area is so small it takes a very long time for them to cool; the surface temperature reduces, the colour reddens and their luminosity decreases. The lower the surface temperature the less the rate of energy loss, so a white dwarf will take a similar time to cool from 20,000 K down to 5,000 K as it will from 5,000 K to 4,000 K. In fact, the Universe is not old enough for any white dwarfs to have cooled much below 4,000 K; the coolest observed so far, WD 0346+246, has a surface temperature of 3,900 K.

Black dwarfs

Eventually a white dwarf will cool sufficiently so that there is no visible radiation and it will then become a 'black dwarf'. It could still, however, be detected in the infra-red, though will be very faint, and the presence of black dwarfs in orbit around normal stars could still be deduced by the effect they have on the motion of their companion stars.

High-mass stars

Stars whose masses exceed about 8 solar masses have sufficient mass overlying the core so that the temperature of the core can increase beyond that in less massive stars. This allows the capture of alpha particles to proceed further. Having made carbon and oxygen, it is then possible to build up the heavier elements having atomic numbers increasing by 4 – produced by the absorption of alpha particles – so that, in turn, the ^{16}O fuses to ^{20}Ne, the ^{20}Ne fuses to ^{24}Mg and then ^{24}Mg fuses to ^{28}Si, producing a core dominated by silicon.

For each successive reaction to take place the temperature has to increase as there is a greater potential barrier for the incoming alpha particle to tunnel through. In the melee, protons can react with these elements to form nuclei of other atoms with intermediate atomic numbers such as ^{19}Fl and ^{23}Na, though these will be less common. A shell-like structure results, with layers of the star containing differing elements, the heaviest nearest to the centre.

When the temperatures reach the order of 3×10^9 K, silicon can be transformed through a series of reactions passing through ^{32}S, ^{36}A and continuing up to ^{56}N. The silicon burning produces a core composed mostly of iron (the majority) and nickel. Iron and its close neighbours in the atomic table have the most stable nuclei, and any further reactions to build up heavier nuclei are endothermic (they would absorb energy rather than provide it) so this is where nuclear fusion has to stop. The star is then said to have an iron core. This core is surrounded by shells in which the lighter elements are still burning, giving an interior like that shown in Figure 11.2.

The energy released by each stage of burning is reduced and, as a result, the time spent carrying out each successive reaction becomes shorter: a star of mass 20 times that of our Sun will spend about 10 million years on the main sequence burning hydrogen to helium, then spend about 1 million years burning helium to carbon and about 300 years burning carbon to oxygen. The oxygen burning takes around 200 days and silicon burning is completed in just 2 days.

Once the core reaches its iron state, things progress very rapidly. At the temperatures that exist in the core (of order 8×10^9 K for a 15 solar mass star) the photons have sufficient energy to break up the heavy nuclei, a process

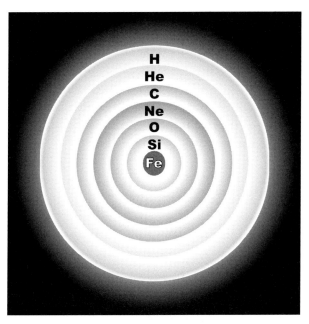

Figure 11.2 The 'onion-like' shells of fusion burning during the later stages in the evolution of a giant star. Image: R. J. Hall, Wikimedia Commons.

known as photodisintegration. An iron nucleus may produce 13 helium nuclei in the reaction, which then break up to give protons and neutrons. As energy is released when the heavy elements are produced, these inverse processes are highly endothermic (requiring energy to progress) and thus the temperature drops catastrophically. There is then not sufficient pressure to support the core of the star, which begins to collapse to form what is called a 'neutron star'.

In the forming neutron star, free electrons combine with the protons produced by the photodisintegration of helium to give neutrons and electron neutrinos. These barely interact with the stellar material, so can immediately leave the star carrying away vast amounts of energy – the neutrino luminosity of a 20 solar mass star exceeds its photon luminosity by seven orders of magnitude for a brief period of time! The outer part of the core collapses at speeds up to 70,000 km/s and, within about a second, the core, whose initial size was similar to that of the Earth, is compressed to a radius of about 40 km. This is so fast that the outer parts of the star, including the oxygen, carbon and helium burning shells, are essentially left suspended in space and begin to infall towards the core.

The core collapse continues until the density of the inner core reaches about three times that of an atomic nucleus, $\sim 8 \times 10^{14}$ g/cm^3. At this density, the strong nuclear force, which in nuclei is attractive, becomes repulsive – an effect caused by the operation of the Pauli exclusion principle on neutrons and termed 'neutron degeneracy pressure'. As a result of this pressure, the core rebounds and a shock wave is propagated outwards into the infalling outer core of the star. As the material above is now so dense, not all the neutrinos escape immediately and give the shock front further energy, which then continues to work its way out to the surface of the star, there producing a peak luminosity of roughly 10^9 times that of our Sun. This is comparable to the total luminosity of the galaxy in which the star resides!

Type II supernovae

This sequence of events is called a 'Type II supernova'. The peak brightness drops by a factor of around 100 each year so that it gradually fades from view. We believe that such supernovae will occur in our Galaxy on average about once every 44 years. Sadly, the dust in the plane of the Galaxy only allows us to see about 10% to 20% of these and so they are not often seen.

The Crab Nebula

On 4 July 1054, a court astrologer during the Sung dynasty, Yang Wei-T'e, observed a supernova in the constellation Taurus. The gas shell thrown out in the supernova explosion was first discovered in more modern times in 1731 by John Bevis, who included it in his sky atlas, *Uranographia Britannica*. Later, in 1758, it was independently discovered by Charles Messier whilst he was searching for the return of Halley's Comet. It became the first object in the Messier catalogue, with the name M1. The Third Earl of Rosse, who drew its form using his 72-inch telescope in Ireland, thought that it appeared similar to a horseshoe crab and so he called it the 'Crab Nebula', the name by which it is usually known.

The Crab Nebula, shown in Figure 11.3, is still, nearly 1,000 years after it was first observed, expanding at rate of 1,500 km/s and its luminosity is about 10,000 times greater than our Sun. Much of this radiation appears to be the result of electrons, moving at close to the speed of light (called relativistic electrons), spiralling around magnetic field lines in the nebula. The fact that the nebula still appears so energetic remained a puzzle until a neutron star (which is the remnant of the stellar core) was discovered in 1969 at the centre of the nebula. This will be described in detail below. The gas shell, now of order 6×4 arcminutes in size, can still be observed with a small telescope.

The Crab Nebula is thought to be the remains of a Type II supernova. In 1987, a supernova (1987A) was observed in a nearby galaxy, the Large Magellanic Cloud, and is also thought to have been a Type II supernova, but those observed

Figure 11.3 The Crab Nebula. The neutron star is the lower right of the pair of stars at the centre of nebula. Image: Hubble Space Telescope, STScI, ESA, NASA.

by Tycho Brahe in 1572 and Johannes Kepler in 1604 are thought to have been caused by a different mechanism and are termed 'Type I supernovae'.

Supernova 1987A

In February 1987, a supernova was observed in the Large Magellanic Cloud, a galaxy close to our own Milky Way Galaxy. Visible for a while to the unaided eye (Figure 11.4), it became the closest observable supernova since that of 1604.

There is an aspect of its explosion that merits mention which was the result of a wonderful piece of serendipitous timing. In the late 1970s a particle physics model called the 'Grand Unified Theory' (GUT) suggested that protons would decay with a half life of 10^{31} years. This means that if one observed a number of protons for 10^{31} years half would have decayed. This is obviously not an experiment that can be mounted, but the possible proton decay could be detected if one observed a very large number of protons for a relatively short period. It was thought that the proton would decay into a positron and a neutral pion, which would then immediately decay into two gamma-ray photons. The positron would then annihilate with an electron to form two more gamma-ray photons.

To this end, a number of detectors were built in the 1980s, including that at the Kamioka Underground Observatory located 1,000 m below ground in Japan.

Figure 11.4 Supernova 1987A in the Large Magellanic Cloud. Image: ESO.

To provide the protons, 3,000 tonnes of pure water was contained in a cylinder 16 m tall and 15.6 m in diameter. The cylinder was surrounded by 1,000 photomultiplier tubes attached to its inner surface, which would be able to detect the gamma rays produced in the proton decay. It came into operation in 1983 and was given greater sensitivity in 1985. To date, even with a new detector containing 50,000 tonnes of water, no convincing proton decays have been detected and later versions of GUT suggest that the decay half life might be nearer to 10^{35} years. But what is critically important was that the detector, which came into full operation at the end of 1986 after its upgrade in 1985, could also detect neutrinos.

Einstein's Special Theory of Relativity states that no particle can travel at the speed of light in a vacuum. However, in a dense medium, like water, light travels at lower speeds. It is thus possible for a particle to travel through water faster than the speed of light. If the particle is charged, it will emit light radiation called Cherenkov radiation. The process is analogous to the formation of a sonic boom when an airplane exceeds the speed of sound. Neutrino interactions with the electrons in the water can transfer almost all the neutrino momentum to an electron, which then moves at relativistic speeds in the same direction.

The relativistic electron produces Cherenkov radiation which can be detected by the photomultiplier tubes surrounding the tank. The expanding light cone will trigger a ring of photomultiplier tubes whose position gives an indication of the direction from which the neutrino has travelled. This makes it more than just a detector – it forms a very crude telescope.

When SN1987A was seen to explode just a few months later (this being the serendipitous timing referred to earlier), the Kamiokande experiment detected 11 neutrinos within the space of 15 seconds. A similar facility in Ohio detected a further 8 neutrinos within just 6 seconds and a detector in Russia recorded a burst of 5 neutrinos within 5 seconds. These 24 neutrinos are the only ones ever to have been detected from a supernova explosion. Perhaps surprisingly, the neutrinos were detected some three hours before the supernova was detected optically. This is not because they had travelled faster than light! They had, of course, travelled out directly from the collapsing core of the star, whereas the visible light was not emitted until later when the shock wave reached the surface of the star. The detection of those 24 neutrinos was a perfect confirmation of the theoretical models that had been developed for the core collapse of a massive star and consistent with theoretical prediction that ~10^{58} neutrinos would be produced in such an event. These observations also allowed an upper limit to be placed on the neutrino mass. If one assumes that the neutrinos began their trip somewhat ahead of the light from the supernova, and given the fact that they arrived before the light having travelled through space for ~169,000 years, this means that they must have been travelling very close (within one part in 10^8) to the speed of light. This, together with the fact that the higher and lower energy neutrinos arrived at the same time, allows an upper limit to be put on the mass of a neutrino. It cannot be greater than about three-millionths the mass of an electron.

Neutron stars

What remains from this cataclysmic stellar explosion depends on the mass of the collapsing core. When stars whose total mass is greater than ~8 solar masses but less than ~12 solar masses collapse the result is a 'neutron star' – the core being supported by neutron degeneracy pressure as described above. The typical mass of such a neutron star would be ~1.4 solar masses so that it is, in effect, a giant nucleus containing ~10^{57} neutrons. As the theoretical models are not all that precise, neutron stars will have a radius in the range 10 to 15 km. Assuming a radius of 10 km, the average density would be 6.65×10^{14} g/cm^3 – more than that of an atomic nucleus!

Gravity at the surface would be intense; for a 1.4 solar mass star with a radius of 10 km, the acceleration due to gravity at the surface would be 190 billion times that on the surface of the Earth and the speed of an object falling from a height of 1 m onto the surface would be 6.88 million km/h! A simple Newtonian calculation of the escape velocity from the surface gives a value of $0.643c$. This implies that both special and general relativity need to be invoked when considering neutron stars. The structure of a neutron star is very complex; part may even be in the form of a superfluid sea of neutrons, which will thus have no viscosity. This can give rise to an observable consequence, as will be described later.

A neutron star may have an outer crust of heavy nuclei, the majority being of iron and nickel. Within this is an inner crust containing elements such as krypton, superfluid neutrons and relativistic degenerate electrons. The inner crust overlies an interior of superfluid neutrons intermixed with superconducting protons and relativistic degenerate electrons. Finally, there may be a core of pions or other elementary particles.

Like white dwarfs, neutron stars become smaller and denser with increasing mass, but there will come a point when the neutron degeneracy pressure can no longer support the mass of the star. So, in an analogous manner to the Chandrasekhar limit for the maximum mass of a white dwarf, there is a limit, believed to be about three to four solar masses, beyond which the collapse continues to form a black hole.

Stars rotate as, for example, our Sun, which rotates once every ~25 days at its equator. The core of a star will thus have angular momentum. As the core collapses, much of this must be conserved (some is transferred to the surrounding material), so the neutron star that results will be spinning rapidly with rotational periods of perhaps a few milliseconds. The neutron star will also be expected to have a very intense magnetic field. This rotating field has observational consequences that have allowed us to discover neutron stars and investigate their properties.

When the neutron star is first born its surface temperature may approach 10^{11} K but rapidly falls to about 10^9 K. Neutrinos carry away much of the star's energy for about 1,000 years, whilst the surface temperature falls to a few million K. Photons – in the form of X-rays – then carry energy away from the surface, which stays close to 1 million K for the next 1,000 years. Its luminosity will then be comparable to that of our Sun.

This explains why the Crab Nebula is still visible. It was known that a star close to the centre of the nebula had a very strange spectrum. If this were the neutron star associated with the supernova explosion, its energy output would have kept the gas thrown out into the interstellar medium excited, so remaining visible. The way in which this observation was confirmed and how, to date, nearly 2,000 neutron stars have been discovered is one of the most interesting stories of modern astronomy.

The discovery of pulsars

When stars are observed through the Earth's atmosphere they are seen to scintillate ('twinkle' is a rather nice, if not scientific, term that is often used). This is because irregularities in the atmosphere passing between the observer and the star act like alternate convex and concave lenses that sequentially converge the light from the star (so making it appear slightly brighter) and then diverge it (so reducing its brightness).

There is a similar effect related to radio sources caused by irregularities in the solar wind – bubbles of gas that stream out from the Sun expanding as they do so. It was realised that this could give a way of investigating the angular sizes of radio sources by studying the amount of scintillation observed when the source was at different angular distances from the Sun. It would also be a way of discovering radio sources with very small angular sizes – known as 'quasars'. To carry out this experiment a very large antenna was required, and Antony Hewish at the Mullard Radio Astronomy Laboratories at Cambridge recruited a PhD student called Jocelyn Bell to first help build the antenna – which was made up of an array of 2,048 dipoles – and then carry out and analyse the observations. The array observed radio sources as they passed due south, so a given radio source would be observed every day as it appeared on the meridian.

The signals from radio sources appeared on a roll of chart, about 400 feet (122 m) of which was produced each day. Soon Bell (Figure 11.5) was able to distinguish between the scintillating signal of a radio source and interference, often from cars passing the observatory. In July 1967 she observed a 'little bit of scruff' that did not look like a scintillating radio source but did not appear like interference either. A second intriguing feature was that it had been observed at night when a radio source would be seen away from the direction of the Sun and scintillation would not be expected to be seen. Looking through the charts, she discovered that a similar signal had been seen earlier from the same location in the sky. She observed that it reappeared again at always a precise number of sidereal days later, which implied that the radio source, whatever it was, was amongst the stars rather than within the Solar System. Hewish and Bell then equipped the receiver with a high-speed chart recorder to observe the 'scruff' in more detail and discovered to their amazement that it was not random, but a series of precisely spaced radio pulses having a period of 1.33724 s.

Observations using a different telescope at Cambridge confirmed the presence of the signal and also that fact that the pulse arrived at slightly different times as the frequency of observation was changed. This effect is called dispersion, and is exactly similar to the fact that different wavelengths of light travel at different speeds in glass. The interstellar medium is not a perfect vacuum and so

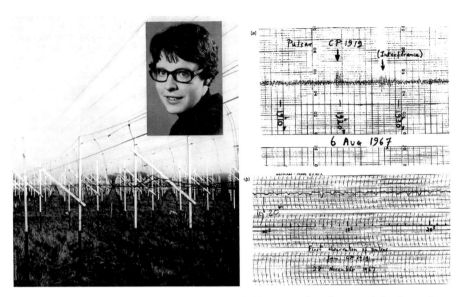

Figure 11.5 Jocelyn Bell with the Cambridge array that discovered the first pulsar, and the discovery record. Image composite: Ian Morison with components from Jocelyn Bell.

can cause this effect – but it would be only observed if the source of the pulses was far beyond the Solar System.

At that time, none of the theoretical astronomers in the Cambridge group could conceive of a natural phenomenon that could give rise to such highly precise periodic signals – it seemed that no star, not even a white dwarf, could pulsate at such a fast rate – and they wondered if it might be a signal from an extra-terrestrial civilisation. Bell, who called the source LGM1 (Little Green Men 1), was somewhat annoyed about this as it was disrupting her real observations. When, later, a second source with similar characteristics but a slight faster period of 1.2 s was discovered she was somewhat relieved as 'it was highly unlikely that two lots of Little Green Men could choose the same unusual frequency and unlikely technique to send a signal to the same inconspicuous planet Earth!'

A few days before the paper presenting these discoveries was published in *Nature* in February 1968, Hewish announced the discovery to a group of astronomers at Cambridge. Fred Hoyle was amongst them, and suggested that the signal might be pulsed emissions coming from an oscillating neutron star – the theoretical remnant of a supernova but never previously observed. After a press conference following the publication of the *Nature* paper announcing the discovery, the science correspondent of the *Daily Telegraph* coined the name 'pulsar' for these enigmatic objects.

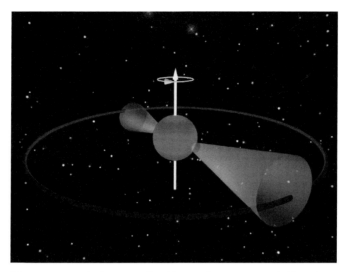

Figure 11.6 Twin beams emitted by a pulsar. Image: Jodrell Bank Centre for Astrophysics, University of Manchester.

Some three months later, in a paper also published in *Nature*, Thomas Gold at Cornell University in Ithaca, USA, gave a satisfying explanation for the pulsed signals. Gold suggested that the radio signals were indeed coming from neutron stars, but that the neutron star was not oscillating but was instead spinning rapidly around its axis. He surmised that the rotation, coupled with the expected intense magnetic field, generates two steady beams of radio waves along the axis of the magnetic field lines, one beam above the north magnetic pole and one above the south magnetic pole. If (as in the case of the Earth) the magnetic field axis is not aligned with the neutron star's rotation axis, these two beams would sweep around the sky rather like the beam from a lighthouse (Figure 11.6). If then, by chance, one of the two beams crossed our location in space, our radio telescopes would detect a sequence of regular pulses – just as Bell had observed – whose period was simply the rotation rate of the neutron star.

Gold, in this paper, pointed out that a neutron star (due to the conservation of angular momentum when it was formed) could easily be spinning at such rates. He expected that most pulsars should be spinning even faster than the first two observed by Jocelyn Bell and suggested a maximum rate of around 100 pulses per second.

Since then, nearly 2,000 pulsars have been discovered. The majority have periods between 0.25 s and 2 s. It is thought that as the pulsar rotation rate slows the emission mechanism breaks down, and the slowest pulsar detected has a period of 4.308 s. There is a class of 'millisecond' pulsars where the proximity of a companion star has enabled the neutron star to 'pull' material from the outer envelope of the adjacent star onto itself. This also transfers angular momentum, so spinning the pulsar up to give periods in the millisecond range – hence their name. The fastest known pulsar is spinning at just over 700 times per second, with a point on its equator moving at 20% of the speed of light and close to the point where it is thought theoretically that the neutron star would break up.

The energy radiated away by pulsars is derived from their angular momentum but, as this is so high, the rate of slowdown is exceptionally slow and so pulsars make highly accurate clocks; some may even be able to challenge the accuracy of the best atomic clocks. The periods of all pulsars slowly increase (except when being spun up to form a millisecond pulsar) and a typical pulsar would have a lifetime of a few tens of millions of years.

The linking of pulsars with supernova neutron star remnants was confirmed when the 'odd' star close to the centre of the Crab Nebula was shown to be a pulsar with a period of 0.0333 s – rotating just over 30 times per second. A second pulsar was discovered within the Vela supernova remnant, and both this and the Crab pulsar emit beams of radiation not just at radio waves but across the whole electromagnetic spectrum including visible light, X-rays and gamma rays.

Most pulsars are seen along the plane of the Galaxy, just as one would suspect as they are the remnants of stars but, perhaps surprisingly, a significant number are observed away from the plane. The 217-km MERLIN array at Jodrell Bank Observatory is capable of making very precise measurements of the position of pulsars and has observed, from positional measurements made over a number of years, that many are moving at speeds comparable to, and even exceeding, the escape velocity of the Galaxy, which is ~500 km/s. The highest pulsar speed so far measured (in this case by the USA's 5,000-km VLBA) is 1,100 km/s, at which speed it would take just five seconds to travel from London to New York!

These pulsars have obviously been ejected from the supernova explosion that gave rise to them with very great energies, enabling them to travel around the Galaxy and, in some cases, to leave the Galaxy for the depths of intergalactic space. It appears that, usually, the supernova explosion will be more intense on one side or the other of the central neutron star, which is then ejected at high speeds rather like a bullet from a gun. In some cases it is even possible to track the course of a pulsar back to the gaseous remnant of the supernova. The situation where the resulting pulsar remains within the supernova gas shell, such as in the Crab Nebula, appears to be very rare.

Black holes

In the case of the most massive stars, neutron degeneracy pressure can no longer support the core against collapse and the result is the formation of a black hole. These fascinating objects deserve a later chapter (Chapter 17) to themselves!

For more on the evolution of stars:

> *Extreme Explosions: Supernovae, Hypernovae, Magnetars, and Other Unusual Cosmic Blasts* by David S. Stevenson (Springer).
> *Stellar Evolution and Nucleosynthesis* by Sean G. Ryan and Andrew J. Norton (Cambridge University Press).
> *Stars and Stellar Evolution* by Klaas De Boer and Wilhelm Seggewiss (EDP Sciences).

12

The search for other worlds

This is one of the most exciting areas of research being undertaken at the present time, with the discovery of new planets being announced on a monthly basis. This chapter will describe the techniques that are being used to discover them and then discuss their properties. Perhaps a word of warning might be in order. An obvious quest is to find planetary systems like our own which could, perhaps, contain planets that might harbour life. So far, to many astronomers' surprise, the vast majority of solar systems found have been very unlike our own, which might lead one to the conclusion that solar systems like ours are very rare. I have even heard this point of view put forward by an eminent astrobiologist. But I do not believe one should draw this conclusion. For reasons that will become apparent, the techniques largely used to date would have found it very difficult to detect the planets of our own Solar System, so it should not be surprising that we have so far failed to find any other similar solar systems. As new techniques are used, this situation will improve, but it will be some time before we have any real idea how often solar systems like our own have arisen in the Galaxy. The story of the discovery of the first planet to orbit a sun-like star is very interesting in its own right, but, in order to appreciate its nuances, we need first to understand how this, along with many of the planets so far detected, have been discovered.

The visual detection of planets orbiting normal stars

It has been long thought that the detection of planets by direct imaging was not feasible due to the fact that the light reflected from the planet would be lost in the glare of the light from the star. However, in the infrared, stars are less

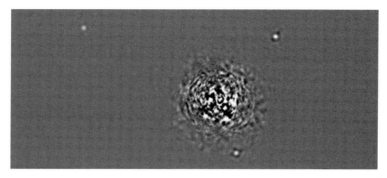

Figure 12.1 An infrared image taken by the Keck Telescope showing three planets in orbit around the star HR 8799. Image: C. Marois *et al.*, NRC Canada.

bright than in the visible and the brightness difference is reduced, so making detection easier. In fact, a planetary-sized body had been detected in orbit around a brown dwarf. This was achieved using one of the 8-metre telescopes of the VLT in Chile with the use of adaptive optics to correct for atmospheric turbulence. This technique is very effective in the infrared and allows telescopes to achieve higher resolution and so can allow planets at small angular distances from their star to be seen.

Planets observed in the infrared

In November 2008, a team of astronomers using the 10-metre Keck and 8-metre Gemini-North telescopes announced the discovery of three planets orbiting the star HR 8799, 129 light years distant (Figure 12.1). Again, the observations were made in the infrared and, in addition, an occulting device was used to remove much of the light from the star.

The three planets are several times the mass of Jupiter and even the closest has an orbital radius equal to the Sun–Neptune distance of ~30 AU.

A planet discovered in visible light by the Hubble Space Telescope

The first detection of a planet in visible light was also announced in November 2008. The discovery was based on two observations, in 2004 and 2006, of the star Fomalhaut, which lies at a distance of 25 light years from our Sun.

Again, an occulting disc was used to largely eliminate the light from the star. As seen in Figure 12.2, beyond the scattered 'starlight noise' is seen a prominent

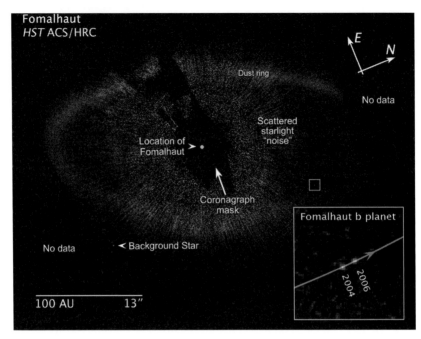

Figure 12.2 A composite of two images of the Fomalhaut star system taken by the Hubble Space Telescope. Image: STScI, ESA, NASA.

dust ring. To the lower right of the star's position a faint object was imaged, which was seen to move in the two years between the observations. This is the planet Fomalhaut b, which is ~3 Jupiter masses and orbits Fomalhaut at a distance of ~17 billion km or 73 AU. It is thus the first planet ever to have been imaged in visible light.

However, researchers analysing 2011 data from the Atacama Large Millimeter Array (ALMA) disputed its status as a planet and even its existence, but in October 2012 a team lead by Thayne Currie at the University of Toronto announced the first independent recovery of Fomalhaut b and revived the claim that Fomalhaut b is a planet. They re-analysed the original Hubble data using new, more powerful algorithms for separating planet light from starlight and confirmed that Fomalhaut b does exist.

It is now thought by some that the detected light is not that reflected from the surface of a planet, but rather light scattered by a large circumplanetary disc around a massive, but unseen, planet. The controversy about its existence has even led to its nickname of a 'zombie planet', although this term has not appeared in any scientific paper.

The radial velocity (Doppler wobble) method of planetary detection

Our own Solar System gives us a good insight into this method and its strengths and weaknesses. Astronomers often use the phrase 'the planets orbit the Sun'. This is not quite true. Imagine a scale model of the Solar System with the Sun and planets having appropriate masses and positions in their orbits from the Sun. All the objects are mounted on a flat, weightless sheet of supporting material. By trial and error, one could find a point where the model could be balanced on just one pin. This point is the centre of gravity of the Solar System model. The centre of gravity of the Solar System is called its 'barycentre', and *both* the Sun and the planets rotate about this position in space.

As Jupiter is more massive than all the other planets combined, its mass and position have a major effect on the position of the barycentre, which will thus lie a distance from the centre of the Sun in the approximate direction of Jupiter. In fact, if Jupiter were the only planet orbiting the Sun, the barycentre of the Solar System would always lie outside the Sun. When all of the major planets lie on one side of the Sun, as happened in the 1980s – allowing the Voyager spacecraft missions to the outer planets – the barycentre is further from the Sun's centre, and when Jupiter is on the opposite side to the other planets it is nearer the Sun's centre. On average, the barycentre is at a distance of ~1.25 solar radii from the Sun's centre, varying between extremes of ~0.3 and 2 solar radii.

Suppose that we observed the Solar System from a point at a great distance in the plane of the Solar System. We would not see the planets – their reflected light would be swamped by the light from the Sun – but, at least in principle, we could detect their presence. Due to the Sun's motion around the barycentre of the Solar System, it would at times be moving towards us and at other times moving away from us. If we could precisely measure the position of the spectral lines in the solar spectrum we could measure the changing Doppler shift and convert that into a velocity of approach or recession (Figure 12.3). The Solar System as a whole might, of course, be moving either away or towards us so we would see a cyclical change in velocity about a mean value.

Again, for the sake of simplicity, let us assume that our Solar System has only one planet: Jupiter. The Sun would be seen to rotate around the barycentre once every 11.86 years, the period of Jupiter's orbit. Given our calculation of the position of the barycentre, we can thus calculate the speed of the Sun in its orbit about the barycentre, which turns out to be 13 m/s so that the difference between the maximum and minimum velocities would be 26 m/s.

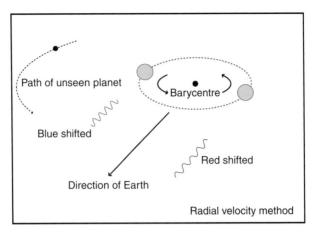

Figure 12.3 The change in wavelength of a spectral line as the star orbits the barycentre of its solar system.

The current precision in Doppler measurements is of order of 2–3 m/s, but the hope is that, in time, this might improve to ~0.5 m/s. Very high resolution spectrometers are used to observe a star whose light is first passed through a cell of gas to provide reference spectral lines to allow the Doppler shift to be measured.

The measurement accuracy of this method would thus be sufficient to detect the presence of Jupiter in orbit around the Sun. However, in order to be reasonably sure about any periodicity in the Sun's motion one would need to observe for at least half a period and preferably one full period. So observations have to be made on a time scale of many years in order to detect planets far from their suns. This is the major reason why few planets in large orbits have yet been detected – the observations have simply not been in progress for a sufficiently long time. But there is one other limitation that you might have realised: should we observe a distant solar system from directly above or directly below, then we would see no Doppler wobble and hence could not detect any of its planets. Unless we have additional information that can tell us the orientation of the orbital plane of a distant solar system, we can only measure the minimum mass of a planet, not its actual mass. If, for example, we later observed such a planet transit across the face of its sun then we would know that the plane of its solar system included the Earth so that the derived mass is the actual mass of the planet, rather than a lower limit.

A single planet in a circular orbit will give rise to a Doppler curve which is a simple sine wave. If the orbit of the planet is elliptical, a more complex but regularly repeating Doppler curve results. In the case of a family of planets, the Doppler curve is complex and will not repeat except on very long time scales. It can, however, still be analysed to identify the individual planets in the system.

In a manner similar to the way in which we calculated the orbital motion of the Sun due to Jupiter, one could calculate the Sun's orbital velocity due to the Earth. This is 0.1 m/s, well below the current and predicted future sensitivity of the radial velocity method, so other methods are required for the detection of Earth-like planets. As other techniques – discussed below – come to fruition and longer periods of observation are analysed by the radial velocity method, solar systems like our own are beginning to be found – but, as yet, we cannot say how common they are.

The discovery of the first planet around a sun-like star

In 1988, Canadian astronomers Bruce Campbell, G. A. H. Walker and S. Yang suggested from Doppler measurements that the star Gamma Cephei might have a planet in orbit about it. The observations were right at the limit of their instrument's capabilities and were largely dismissed by the astronomical community. Finally, in 2003, its existence *was* confirmed but, unfortunately, this was many years after the first confirmed discovery of a planet around a main sequence star.

Two American astronomers, Paul Butler and Geoffrey Marcy, were the first to make a serious hunt for extra-solar planets. They began observations in 1987 but, assuming that other planetary systems were similar to our own, did not expect that any planets could be extracted from the data for several years so their data was archived for later analysis. They would have thus been somewhat shocked when the discovery of a planet orbiting a star called 51 Pegasi was announced by Michel Mayor and Didier Queloz on 6 October 1995.

The star 51 Pegasi, or 51 Peg for short, lies just to the right of the square of Pegasus and is a type G5 star, a little cooler than our Sun, with a mass of 1.06 solar masses. Mayor and Queloz were studying the pulsations of stars, which also cause a Doppler shift in the spectral lines as the star 'breathes' in and out. With a sensitivity of only 15 m/s they had not really expected to discover planets but, greatly to their surprise, they found a periodicity in the motion of the star 51 Peg having a period of 4.23 days and a velocity amplitude of 57 m/s. The plot is very close to a sinusoid showing that the orbit is very nearly circular (Figure 12.4).

From the velocity of the star and the period of the orbit we can first calculate the circumference and hence the radius of the star's motion: 4.23 days is 365,472 seconds, so the circumference is 57 × 365,472 m or 20,831,904 m, giving a radius of 3,315,500 km. This is thus the distance from the centre of the star to

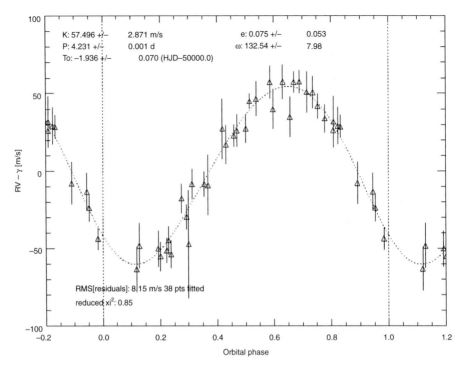

Figure 12.4 Radial velocity measurements of 51 Pegasi made by Korzennik and Contos using the Advanced Fiber Optic Echelle Spectrometer on the 1.5-metre telescope at the Whipple Observatory near Tucson, Arizona.

the barycentre of the system. You may remember that one can calculate the mass of the Sun given the orbital period and the distance of the Earth from the Sun along with a knowledge of the universal constant of gravitation, G (which has been found by experiment). In just the same way, the mass of the star 51 Peg is found to be $1.06 \times 2 \times 10^{30}$ kg. One can then find the distance of the planet from the star, which came to 0.052 AU. This is well within the distance of 0.39 AU at which Mercury orbits our Sun and only about 10 times the radius of the star.

As the star and planet will be balanced about the barycentre, it is then easy to find the mass of the planet, which is ~0.47 that of Jupiter. However, this would only be the mass of the planet if the plane of its orbit included the Earth and will thus be the planet's minimum mass. One can show that, for random orientations, the mass of a planet will on average be about twice the minimum mass, so the planet in orbit around 51 Peg is likely to be very similar in mass to Jupiter.

When Butler and Marcy learnt about this discovery, they realised that not only could they confirm its presence from several years of observations of 51 Peg in their database – which they did just six days later – but that if other massive planets with short periods existed around the stars that they had been observing, they should be able to rapidly find these as well. This hope was borne out and, over the following years, they became the world's most prolific planet hunters.

No-one had expected that a gas giant would be found so near its star, but many of the planets first discovered were similar in size and separation from their sun. It is not thought that giant planets can form so close to a star, so at some time in their early history it is assumed that they must have migrated inwards through their solar system. In doing so, they would very likely eject smaller (terrestrial type) planets that had formed nearer the star from the solar system and consequently these solar systems are thought unlikely to harbour life.

The radial velocity method has proven to be highly successful, but even with a hoped-for velocity precision of 0.5 to 1 m/s at best, the method will never be able to detect Earth-mass planets, no matter how close they are to their sun.

Gravitational microlensing

This is a method that has the potential to achieve the detection of Earth-mass planets. Einstein's General Theory of Relativity was proven by the observed movement of star positions due to the curvature of space close to the Sun. This effect gives rise to what is called gravitational lensing, specifically gravitational microlensing, as the effects are on a very small scale. In the same way that a convex lens can concentrate the light from a distant object into the eye and so make it appear brighter, if a distant star passes behind one of intermediate distance, the brightness of the distant star will undergo a temporary increase, which can last for many days. The peak brightness can be up to 10 times that normally observed.

If the lensing star has a planet in orbit around it, then that planet can produce its own microlensing event, and thus provide a way of detecting its presence. For this to be observed, a highly improbable alignment is required, so that a very large number of distant stars must be continuously monitored in order to detect planetary microlensing events. Observations are usually performed using networks of robotic telescopes (such as those forming the OGLE collaboration) that continuously monitor millions of stars towards the centre of the Galaxy in order to provide a large number of background stars.

If one of the telescopes finds that the brightness of a star is increasing, then the whole network, spaced around the world for continuous observation, will provide unbroken monitoring. The presence of a planet is shown by a very short

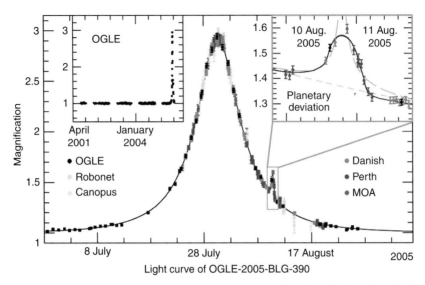

Figure 12.5 Observations by the OGLE consortium showing the microlensing caused by a planet of 5.5 Earth masses. Image: European Southern Observatory.

additional brightening appearing as a spike on the flanks of the main brightness curve.

On 25 January 2006, the discovery of OGLE-2005-BLG-390Lb was announced based on the observations shown in Figure 12.5. This planet is estimated to have a mass of ~5.5 Earth masses and orbits a red dwarf star at around 21,500 light years from Earth, towards the centre of the Milky Way Galaxy. The planet lies at a distance of 2.6 AU from its sun. At the time of its discovery, this planet had the lowest mass of any known extra-solar planet orbiting a main sequence star. The planet has surface temperature of ~220 °C below zero. It is likely to have a thin atmosphere with a rocky core buried beneath a frozen icecap.

A disadvantage of the method is that the chance alignment that allowed the lensing event that led to the planet's detection is highly unlikely ever to be repeated. Also, the detected planets will tend to be many thousands of light years away, so making any follow-up observations by other methods virtually impossible. However, if enough background stars can be observed over long periods of time, the method should finally enable us to estimate how common Earth-like planets are in the Galaxy.

In early 2008, the gravitational microlensing method detected two gas giant planets, similar to Jupiter and Saturn, orbiting a star 5,000 light years away in a planetary system with striking similarities to our own Solar System. The discovery suggests that giant planets do not live alone but are more likely to be found in family groups. The mass of the nearer planet is 0.71 times that of Jupiter and it lies 2.3 times as far from its host star as the Earth is from the Sun. The second planet is less massive, 0.27 times the mass of Jupiter, and twice as far away from the host star. Despite their host star being only half as massive as the Sun, the planetary system otherwise bears a remarkable similarity to our Solar System. Both the ratio of the masses of the two giant planets (close to 3:1) and the ratio of their distances from the host star (1:2) are remarkably similar to those of Jupiter and Saturn. The ratio between the orbital periods of 5 years and 14 years, respectively, also closely resembles that between Jupiter and Saturn (2:5). The system resembles our own Solar System more closely than any previously observed. Whilst there are more than 1,000 planets now known to be orbiting other stars (plus many more unconfirmed candidates), there are far fewer solar systems known to have multiple planets – but this number will surely rise as smaller planets fall within our detection methods.

By the end of 2013, a total of 24 planetary systems had been found by gravitational microlensing, including two multiple planet systems giving a total of 26 extra-solar planets. These ranged from 0.01 to 9.4 times the mass of Jupiter, that is, from 2.5 (Gliese 581 e) up to nearly 3,000 Earth masses. Their distances from their suns ranged from 0.2 to 8.3 AU. A listing of the current tally of microlensing planets can be found at http://exoplanet.eu/catalog/?f=%22microlensing%22+IN+detection.

Astrometry

The science of accurate positional measurement, called astrometry, is the oldest method that has been employed in the search for planets, and in the 1950s and 1960s the discovery of several planets was claimed using this method. Sadly, none has since been confirmed and by the end of 2007 no planets had been discovered by this method, though the Hubble Space Telescope (HST) has confirmed the existence of a planet in orbit around Gliese 876.

The method consists of observing, with great precision, how a star's position varies over time. All star systems are moving around the Galaxy (our Sun in ~230 million years) and over short periods of time will move in essentially straight lines. But, as I hope you might realise from our discussions above, if the star is in a binary or planetary system, the point that moves in a straight line is not the centre of the star but the barycentre of the system. So, as the star moves in a tiny circular or elliptical orbit about the barycentre, it will follow a wiggly path across the sky. As yet, except in the case of Gliese 876, it has not been possible to produce astrometric observations of sufficient accuracy to allow the presence of a planet to be detected.

Gliese 876, which lies at a distance of 15.6 light years, was first detected with the radial velocity method in 1998. The radial velocity measurements combined with two years of the HST astrometric measurements allowed the orientation of the plane of the planetary orbits to be determined so that the actual, not just the minimum, mass of the planets in that system has been found.

As described in Chapter 10, the Gaia spacecraft, which was launched in December 2013, will have sufficient positional accuracy (7 microarcseconds) to detect by this method planets orbiting relatively nearby brighter stars. Sadly, NASA's Space Interferometry Mission (SIM), which would have attained a measurement precision of 1 microarcsecond and have the capability to detect planets at considerably greater distances than 30 light years was cancelled in 2010.

The astrometric method nicely complements the radial velocity and transit methods in that it is more sensitive to planets at larger orbital distances. However, such planets will have long orbital periods so observations over many years would be required for their detection.

Planetary transits

As the number of known close-orbiting gas giants increased, there became a reasonable chance that the plane of some of their orbits would include the Earth and so, once each orbit, the planet might occult the star, giving a measurable drop in its brightness, as illustrated in Figure 12.6.

Let us estimate the brightness drop if a Jupiter-sized planet occulted our Sun as seen from a great distance. The Sun has a diameter of ~10 times that of Jupiter, so that its cross-sectional area will be ~100 times that of Jupiter. When

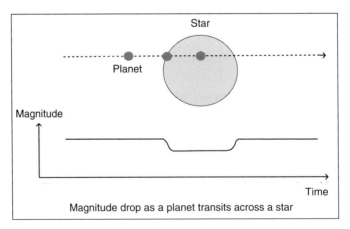

Figure 12.6 The effect of a planetary transit on the brightness of a star.

Jupiter occults the Sun, the effective area will drop from 100 to 99 – a ratio of 0.99 – and give a drop in brightness of 1%.

With care, such accuracy in measurement is achievable, and on 5 November 1999 two teams detected the transit of a planet, previously discovered by the radial velocity method, in orbit around the star HD 209469. During the transit, the brightness of the star dropped by 1.7%.

In 2002, a planet OGLE-TR-56b was discovered by the transit method and later confirmed using the radial velocity method. Then, in 2006, the HST made a survey of 180,000 stars up to 26,000 light years away towards the central bulge of our Galaxy. The survey discovered 16 candidate extra-solar planets of which three have since been confirmed. If all 16 were confirmed, it would imply that there would be of order 6 billion Jupiter-sized planets in the Galaxy. Five of the newly discovered planets were found to orbit their suns with periods of less than one day. The candidate with the shortest period – just 10 hours – is only 1.2 million km from its relatively small, red dwarf sun and has an estimated surface temperature of 1,400 K. It must be at least 1.6 times the mass of Jupiter in order to prevent the tidal forces from the star splitting the planet apart.

The transit method has the problem that transits can only be observed when the planet's orbit is nearly edge on. About 10% of planets in close orbits would show transits, but the fraction is far smaller for planets with large orbits as the alignment has to be more precise – only ~0.5% of Earth-like planets in orbit around stars similar to our Sun would cause transits. Two space missions called Kepler and COROT are equipped with giant CCD arrays to enable them to observe very large fields of view, so being able to continuously monitor many stars.

The transit method does have two significant advantages. The first is that, as a planet will take some time to fully cover its star, the size of the planet can be determined from the light curve. When combined with the planet's mass, determined by the radial velocity method, the density of the planet can be determined and so we can learn about its physical structure.

The second advantage is that it is possible to study the atmosphere of a planet. When the planet transits the star, light from the star passes through the atmosphere of the planet. By carefully studying the star's spectrum during the transit, absorption lines will appear that relate to elements in the planetary atmosphere.

The extra-solar planet HD 209458b, provisionally nicknamed Osiris, was the first planet observed transiting its sun. Observations by the HST first discovered a tail of evaporating hydrogen which may, in time, completely strip the planet of gas leaving a 'dead' rocky core. More recent HST observations have shown that the planet is surrounded by an extended envelope of oxygen and carbon believed to be in the shape of a rugby ball. These heavier atoms are caught up in the flow of the escaping atmospheric atomic hydrogen and rise from the lower atmosphere rather like dust in a whirlwind.

The Kepler space observatory

Kepler is a space observatory in solar orbit launched by NASA with the aim of discovering Earth-like planets orbiting other stars using the method of planetary transits. The spacecraft, launched on 7 March 2009, is named after the Renaissance astronomer Johannes Kepler. It was designed to survey a region of the Milky Way in the constellation Cygnus and continuously monitor over 145,000 main sequence stars using a 95 megapixel CCD array analysing the light collected by its 1.4-metre primary mirror. Three hundred, 6-second exposures are summed to give an effective exposure of 30 minutes. These images are then pre-analysed in the spacecraft to reduce the data that has to be downloaded to Earth.

Its aim was to discover dozens of Earth-sized extra-solar planets in or near their habitable zones and estimate how many of the billions of stars in our Galaxy have such planets. Its initial planned lifetime was 3.5 years, and it can easily be seen why three to four years would be needed to detect a planet, like our Earth, orbiting its sun in one year. It would need at least two and, to be sure, three transits of the planet. One might be lucky and the first transit could occur on the first day of observation, so only a year and one day would be needed to observe the second transit one year later, with two years and one day needed to observe a third, confirmatory observation. However, the first transit might only occur at the end of the first year and so three years would be needed to observe three transits. Even if only requiring two transits, only half of those with orbital periods of 2 years could be found with an observing period of 3.5 years.

When the first observations were analysed, it was found that the noise levels in the system were higher than expected – partly due to a greater intrinsic variability in the stars themselves than had been expected. This meant that to fulfil all the mission goals more transits would be needed to reliably detect a planet, and it was planned to extend the mission to 7.5 years, taking observations up to 2016.

The detection of planets depends on seeing very small changes in brightness (Kepler could detect changes of 80 parts per million) so that variable stars would not be worthwhile candidates. Within a few months, about 7,500 stars from the initial target list were found to be variable and so dropped from the target list to be replaced by new candidates.

As in the case of the HST, the telescope pointing was controlled by a set of reaction wheels (a form of gyroscope). You may remember (and this is discussed in Chapter 19) that those in the HST were replaced during the service missions and so it was able to be kept operational for many years. Those in Kepler could not be replaced and, sadly, on 11 May 2013 the second of four reaction wheels failed, which brought the initial collection of science data to an end.

As of July 2013, Kepler had found 134 confirmed exoplanets in 76 stellar systems, along with a further 3,277 unconfirmed planet candidates. In November 2013, further results were published bringing the total number to more than 3,500. Only 167 of the candidates had then been confirmed, but it is likely that the vast majority of these are real.

Of these candidates:

- 674 are less than 2.5 Earth diameters,
- 1,076 are between 2.5 and 4 Earth diameters,
- 1,457 are between 4 and 12 Earth diameters,
- 229 are between 12 and 30 Earth diameters,
- 102 are greater than 30 Earth diameters.

Based on Kepler space mission data, it is believed that there could be as many as 40 billion Earth-sized planets orbiting in the habitable zones of sun-like stars and red dwarf stars within the Milky Way Galaxy. Eleven billion of these estimated planets may be orbiting sun-like stars. It is estimated that one in five sun-like stars hosts a rocky planet in its habitable zone and that the nearest such planet may be only 12 light years away.

NASA has been considering further uses for Kepler. Its photometric sensitivity had dropped from 80 parts per million to 300 parts per million, but this is still far better than can be achieved from Earth. Possible uses that were being considered included searching for asteroids and comets, looking for evidence of supernovae, and finding exoplanets through gravitational microlensing.

There is still a further year of observations to be analysed and it is expected that significantly more Earth-sized planets will be found in the data – the number of such candidates increased by 78% from January to November 2013 as the analysis techniques have improved and more such planets will have made a third transit. The Kepler results are quite encouraging in that it may very well be that solar systems like ours *are* quite common despite what might appear at first sight.

At the second Kepler Science Conference in November 2013 the Kepler Science team proposed an extended mission called 'K2'. They showed that if Kepler is orientated appropriately to the Sun, its two remaining reaction wheels could stabilise the telescope's pointing provided the star fields were in the plane of the ecliptic. Each targeted field would only be able to be observed for 40 (and occasionally up to 70) days and only about 10,000–20,000 relatively bright stars (rather than ~150,000) stars could be monitored during this period. This latter restriction actually has the useful consequence that, as the planets would be orbiting brighter stars, follow-up observations from the ground using the radial velocity technique could be used to confirm their presence.

Suggestions for further reading:

Mirror Earth: The Search for Our Planet's Twin [Kindle Edition] by Michael D. Lemonick (Walker & Co.).
Transiting Exoplanets by Carole A. Haswell (Cambridge University Press).

13

Are we alone? The search for life beyond the Earth

In Chapter 4, the search for evidence for past life on Mars was discussed and, in Chapter 6, the possibility that life might exist in the subterranean ocean on Europa was considered. In addition, perhaps surprisingly, it is not impossible that we could find evidence of life on planets that are not too far distant in the Galaxy. Over the coming years the search for extra-terrestrial planets might find somewhere that a rocky planet exists within its sun's habitable zone and we might then sense whether life might exist there.

Let us put ourselves in the place of an advanced civilisation not too far distant in the Galaxy. Could they tell if life existed on Earth? The answer is, in fact, yes: they could, and the detection would be based on the taking of an infrared spectrum of the atmosphere of our planet. If they took spectra of Mars or Venus they would find a flat spectrum with a single deep absorption band due to the presence of carbon dioxide in the atmospheres. But that of our Earth would look very different. The presence of water vapour in our atmosphere would lower the outlying parts of the spectrum and there would be three absorption bands, not one. Along with that due to carbon dioxide, they would find a band due to methane, which would be a marker either for life (think cows) or for volcanic activity. But, more significantly, they would find a band due to ozone. Ozone can only exist in an atmosphere if there is free oxygen and, as oxygen is highly reactive, unless it is being replenished by some means, any present will soon disappear. The means by which oxygen is being replenished in our atmosphere is by the action of photosynthesis – a feature of plant life on Earth.

In the same way, should we find evidence of water vapour and ozone in the atmosphere of an exoplanet we could be pretty sure that some form of life might exist there.

I am often asked why we tend to restrict our search for other life forms to locations that are similar to the conditions on Earth. Why should we impose the facts of our own existence on other life forms? My justification for this is twofold. You will have seen that in the stars, somewhat more massive than our Sun, nitrogen is created as hydrogen is fused into helium and, in the later stages in the lives of stars like our Sun, first carbon and then oxygen are created. Thus, the life forms on our planet are very largely composed of what are the most common elements in the Universe. Further, it is generally recognised that carbon has the most complex chemistry of any element and has a complete subject, organic chemistry, devoted to it. So, our life forms are based on the most common elements linked by the chemistry of carbon. Is it not likely that the vast majority of other life forms will use a similar chemistry?

SETI

SETI, the Search for Extra-Terrestrial Intelligence, has now been actively pursued for close on 50 years without success. However, this does not imply that we are alone in the Milky Way Galaxy for, although most astronomers now agree that intelligent civilisations are far less common than once thought, we cannot say that there are none. But it does mean that they are likely to be at greater distances from us and, as yet, we have only seriously searched a tiny region of our Galaxy. It will not be until the mid 2020s that an instrument, now on the drawing board, will give us the capability to detect radio signals of realistic power from across the whole Galaxy. It is also possible that light, rather than radio, might be the communication carrier chosen by an alien race, but optical-SETI searches seeking out pulsed laser signals have only just begun.

The story so far

The subject may well have been inspired by the building of the 76-metre Mark I radio telescope at Jodrell Bank in 1957. In 1959 two American astronomers, Giuseppe Cocconi and Philip Morrison, submitted a paper to the journal *Nature* in which they pointed out that, given two radio telescopes of comparable size to the Mark I, it would be possible to communicate across interstellar distances by radio. They suggested a number of possible nearby, sun-like stars that could be observed to see if any signals might be detected. This list included Tau Ceti and Epsilon Eridani, both about 10–12 light years distant. They also pointed out that the radio spectral lines of H and OH, whose frequencies would be known to all civilisations capable of communicating with us, lie in a very quiet part the radio spectrum and could act as markers at either end of a band of

Figure 13.1 Frank Drake with the 25-metre telescope at Green Bank where he carried out Project Ozma. Image: SETI Institute.

frequencies that might be used for interstellar communication. This band of frequencies has become known as the 'water hole' (as H + OH = H_2O).

The following year Frank Drake, the father of SETI, using a 25-metre telescope at Green Bank, West Virginia (Figure 13.1), spent six hours every day for two months observing Tau Ceti and Epsilon Eridani in what was called Project Ozma – after L. Frank Baum's imaginary land of Oz. They did detect two brief signals in what should be a protected band for radio astronomy, but it is believed that these were transmitted by the, then, top secret U2 spy-plane!

The 'Wow!' signal

Since then there have been nearly a hundred serious SETI searches. In 1977 a telescope called 'Big Ear' operated by Ohio State University (Figure 13.2), which had been carrying out an all-sky SETI survey since 1974, picked up a signal that appeared to have all the right characteristics. It is called the 'Wow!' signal as the astronomer analysing the data wrote the word in the margin of the computer printout. Sadly, in follow-up observations, no signal has ever been picked up from the same region of sky.

To make a radio message as easy as possible to detect over interstellar distances it would almost certainly be in the form of a very slow 'Morse-code' type signal with a bandwidth of 1 Hz or less – in contrast to an audio transmission requiring a

182 A Journey through the Universe

Figure 13.2 The Big Ear Telescope and the 'Wow!' signal received in 1977. Images: The Ohio State University Radio Observatory and the North American AstroPhysical Observatory (NAAPO). Wikimedia Commons.

bandwidth of several kHz. To detect such signals requires highly specialised receivers with millions of channels covering the band of frequencies being searched. Paul Horowitz at Harvard, a leader in this field, developed receivers to simultaneously analyse 80 million channels each with a bandwidth of 0.5 Hz. These were used to search the whole of the 'water hole' using the 25-metre Harvard–Smithsonian telescope at Oak Ridge in projects META and BETA.

Projects SERENDIP and Phoenix

Two significant searches have used the 305-metre Arecibo Telescope in Puerto Rico. The first of these, Project SERENDIP, still continues whilst the second, Project Phoenix, terminated in 2003. SERENDIP, under the auspices of the University of California, Berkeley, is using the Arecibo dish in 'piggy-back' mode with a dedicated feed system observing the sky close to wherever other astronomers are pointing the telescope. Though the SETI observers have no control over what part of the sky is being observed, over a few years most of the sky accessible to the telescope will be observed, much of it several times over. SERENDIP is thus looking for signals that are seen on more than one occasion from the same location in the sky. A small part of these data, relating to a narrow band of radio frequencies close the 1,400 MHz hydrogen line, has been analysed by home computers across the world in what is known as SETI@home. After a few years a number of signals with appropriate characteristics had been detected several times and a special observing session was set up to observe these in detail. However, no signals appeared in the data to confirm a real detection.

This does highlight a real problem; a signal from ET might be transitory and one really needs to make an immediate confirmation that any signal has an extra-terrestrial origin. This was the premise of Project Phoenix, which arose out of the NASA SETI project when the American Congress cut funding. This had been managed for NASA by the SETI Institute, who then raised private funds to continue the targeted search part of the NASA programme and observe around 800 nearby sun-like stars. In Project Phoenix, two telescopes were used to make simultaneous observations so that any signals originating within our Solar System could be eliminated and there would be an immediate confirmation of any extra-terrestrial signal. Initially pairs of telescopes in Australia and the USA were used, but NASA had helped pay for a major upgrade to the world's largest radio telescope at Arecibo in Puerto Rico and had ~30 weeks of observing time allocated to use it to carry out SETI observations in Project Phoenix. By chance, at a conference on large radio telescopes in 1996, I happened to be sitting next to their project scientist, who told me about the proposed use of Arecibo and that they would need a very large radio telescope to operate in tandem with it. I immediately suggested that they use the 76-metre Lovell Telescope at my own observatory, Jodrell Bank – still then the fourth largest radio telescope in the world. This came to pass and the receiver system was installed on the telescope in the summer of 1998 with observations beginning that autumn.

Due to their separation across the Atlantic, any local interference at either telescope could be immediately discounted. In addition, as a result of the rotation of the Earth and the change in received frequency introduced by the Doppler effect, a signal from beyond our Solar System would be received at Jodrell Bank at a precisely calculable frequency that is approximately 2 kHz lower in frequency than that received at Arecibo. Thus, when Arecibo detected a possible alien signal, the receiver at Jodrell, offset in frequency by the required amount, attempted to confirm the signal. This enabled the elimination of any signals received from Earth itself or satellites orbiting nearby in the Solar System. The system was proven each day by observing the very weak signal from the Pioneer 10 spacecraft, then more than 10 million km from Earth and far beyond Pluto. In the five years of observations (with about six weeks of observations per year) 820 sun-like star systems were observed – some of which

Figure 13.3 A plaque at Green Bank Observatory commemorating the Drake equation. Image: Ian Morison.

we now know have planetary systems. It hardly needs saying that no positive signals were detected.

The Drake equation

The lack of success prompts one to ask what the likelihood is that in the Galaxy there exist other advanced civilisations who would be attempting to contact us. If we do not expect there to be any other civilisations then there would not be a lot of point in searching. This problem was first addressed by an eminent group of scientists at a meeting organised by Frank Drake at Green Bank in 1961. As an agenda for the meeting he came up with an equation that attempts to estimate the number of civilisations within our Galaxy who might be attempting to communicate with us. Known as the 'Drake equation' (Figure 13.3), it has two parts. The first part attempts to calculate how often intelligent civilisations arise in the Galaxy and the second is simply the period of time over which such a civilisation might attempt to communicate with us once it has arisen.

Some of the factors in the equation are reasonably well known, such as the number of stars born each year in the Galaxy, the percentage of these stars (like our Sun) that are hot enough, but also live long enough, to allow intelligent life to arise, and the percentage of these that have solar systems. But others are far harder to estimate. For example, given a planet with a suitable environment it seems likely that simple life will arise – it happened here on Earth virtually as soon as the Earth could sustain life. But it then took several billion years for multi-cellular life to arise and finally evolve into an intelligent species. So it appears that a planet must retain an equable climate for a very long period of time. The conditions that allow this to happen on a planet may not occur very often. Our Earth has a large Moon which stabilises its rotation axis, its surface is recycled due to plate tectonics and this releases carbon dioxide, bound up into carbonates, back into the atmosphere. This recycling has helped keep the Earth warm enough for liquid water to remain on the surface and hence allow life to flourish. Jupiter's presence in our Solar System has reduced the number of comets hitting the Earth; such impacts have given the Earth much of its water but too high an impact rate might well impede the evolution of an intelligent species. It could well be, as some have written, a 'rare Earth'. How many might there be amongst the stars?

It was widely assumed that once multi-cellular life had formed, evolution would drive life towards intelligence, but this tenet has been challenged in recent years – a very well adapted, but not intelligent, species could perhaps remain dominant for considerable periods of time, preventing the emergence of an intelligent species.

The final factor in this part of the equation is the percentage of those civilisations capable of communicating with us who would actually choose to do so. Our civilisation could, but currently does not, attempt to communicate. Indeed, there are some who think that it would be unwise to make others aware that here on Earth we have a nice piece of interstellar real estate! Any attempts at communication are very long term with the round travel time for a two-way conversation stretching into hundreds or thousands of years. It would be hard at present to obtain funding for such a programme. Estimates of 10% to 20% are often cited for this factor. This may well be optimistic.

The topic of 'leakage' radiation from, for example, radars and TV transmitters is often mentioned as a way of detecting advanced civilisations that do not choose to communicate with us. But this is, in my view, unlikely. Any signals that could be unintentionally detected over interstellar distances are, by definition, wasteful of energy. Already, on Earth, high-power analogue TV transmitters are being replaced with low-power digital transmissions, satellite transmissions are very low power and fibre networks do not radiate at all. The 'leakage' phase is probably a very short time in the life of a civilisation and one that we would be unlikely to catch. It could be that airport radars and even very high power radars for monitoring (their) 'near-Earth' asteroids might exist long term and give us some chance of detecting their presence, but we should not count on it.

When all these factors are evaluated and combined the average time between the emergence of advanced civilisations in our Galaxy is derived. If we find it hard to estimate how often intelligent civilisations arise it is equally hard to estimate the length of time, on average, such civilisations might attempt to communicate with us. In principle, given a stable population and power from nuclear fusion, an advanced civilisation could survive for a time measured in millions of years. Often a period of 1,000 years is chosen for want of anything better. This length of time is critical in trying to estimate how many other civilisations might be currently present in our Galaxy. If, for example, a civilisation arises once every 100,000 years – a not unreasonable estimate – but, typically, civilisations only attempt to communicate for 1,000 years, it is unlikely that more than one will be present at any given time. If, however, on average, they remain in a communicating phase for 1 million years then we might expect that nine other civilisations would be present in our Galaxy now.

In what has been the most sensitive search yet undertaken, Project Phoenix, each star was only observed for 1.5 hours, so for us to have had any chance of detecting a signal it would have to be effectively continuous. This would require considerable effort on the part of any other civilisation. If they were nearby in the Galaxy, they might know from analysis of the Earth's infrared spectrum that some form of life existed here, but unless intelligent life is very common they are likely to be too remote to be able to highlight our own Solar System as a possible target.

When the Drake equation was first evaluated, the estimates of other civilisations were quite high; numbers in the hundred thousands or even a million were quoted. Nowadays, astronomers who try to evaluate the Drake equation are far less optimistic. Many estimates are in the tens to hundreds and there are a minority of astronomers who suspect that, at this moment in time, we might be the only advanced civilisation in our Galaxy. I have to say that I am amongst that number. One reason is that new research indicates that the transition from single- to multi-cellular life is very unlikely, so, though I am happy to believe that simple life will be quite common across the Galaxy, intelligent life may well be very rare and I suspect that we are the only intelligent civilisation in our Galaxy at the present time.

The truth is we just do not know. It was once said with great insight that 'the Drake equation is a wonderful way of encapsulating a lot of ignorance in a small space'. Absolutely true, but an obvious consequence is that we *cannot* say that we are alone in the Galaxy. SETI is our only hope of finding out if other intelligent civilisations exist.

Optical SETI

Stars are not strong emitters of radio waves; it is thus not difficult to generate a signal that can be detected at great distance in the presence of the radio noise produced by a star – specifically that at the centre of a solar system from which the signal was being transmitted. (This would also be in the beam of the radio telescope that was attempting to detect the signal.) When SETI was first mooted it was believed that it would be impossible to outshine a star in the visible part of the spectrum but Charles Townes, having invented the laser, immediately realised that a laser might be able to generate very high intensity pulses that could, for brief instants of time, easily outshine a star. Thus the basis of optical SETI was laid. Laser systems close to what are required are now being developed for 'Star Wars' type weapons and nuclear fusion power plants, so we can easily envisage that advanced civilisations would have them.

Dan Werthimer and Geoff Marcy at the University of California, Berkeley and Paul Horowitz at Harvard University have been pioneers in O-SETI. The Berkeley pulsed laser search is directed by Dan Werthimer and plans to observe 2,500 nearby, largely sun-like, stars, looking for very short bright pulses that might last a billionth of a second or so, transmitted by a powerful pulsed laser operated by a distant civilisation. A second Berkeley search is for laser signals that are on for a large fraction of the time. This search, directed by Geoff Marcy, is a 1,000-star programme to search for ultra narrow band signals in the visible part of the spectrum (analogous to radio-SETI). They plan to search through thousands of extremely high resolution spectra for very sharp lines. Much of the data has already been taken in an ongoing (and highly successful) planet search.

Horowitz's first detector system piggy-backed onto the Harvard University's 61-inch telescope whilst it was carrying out a survey of 2,500 nearby sun-like stars. First light was in October 1998. The very sensitive detectors used to look for nanosecond time scale pulses are prone to false triggering, so the light beam is split into two and both detectors have to detect an event for it to be significant. During the first 27 months of observations of the Harvard system two detections were made on average every three nights. The detection events appeared to be uncorrelated with stellar magnitude and did not exhibit any periodicity. In fact, there was no clear evidence that they originated from light entering the telescope from the direction of the targeted star.

The Harvard group then combined their targeted search efforts with a group at Princeton to make simultaneous observations at two separate sites. By November 2003, 16,000 observations totalling 2,400 hours had been made but no pulses were detected simultaneously at the two sites (having, of course, first taken into account an appropriate time delay due to their separation).

In 2003, the telescope at Harvard was decommissioned, and so the targeted search ceased. Since then, the group have planned and commissioned an all-sky survey using a custom-built 72-inch (1.8-metre) telescope that came into operation in April 2006 and is shown in Figure 13.4. The telescope is equipped with an array of 1,024 light sensors that observes an area of sky 0.2 degrees wide by 1.6 degrees high. The telescope is a transit instrument observing a given declination on the celestial sphere as it passes due south. During one day of observations a 360-degree round strip of sky, 1.6 degrees high, will thus be observed. The next night the altitude of the telescope will be adjusted to observe an

Figure 13.4 The 1.8-metre Harvard 'Optical SETI' Survey telescope. Image: Harvard Observatory Optical SETI Project.

adjacent strip so that after about 200 clear nights the whole of the sky visible at Princeton will have been observed and the survey repeated.

Could we find evidence of other civilisations in our immediate locality?

As light and radio waves travel at the fastest speed possible through space, other possible means of contact have tended to be ignored. But if speed is not important, then the sending of spacecraft across the Galaxy as a one-way communication medium might be a sensible way of making contact – perhaps to give us the benefit of the knowledge of a highly advanced civilisation. Indeed, four spacecraft, Pioneers 10 and 11 and Voyagers 1 and 2, which are now coasting through space beyond our Solar System, carry messages from our civilisation that would tell anyone who recovered them a little about ourselves, the star system that we inhabit and even (very cleverly) the time that the spacecraft left our Earth.

Much like Arthur C. Clarke's lunar monolith in *2001: A Space Odyssey*, it is just possible that an alien craft might have landed on the Moon (where it would not suffer the consequences of the erosion we have on Earth) waiting for us to discover it. Or perhaps an alien spacecraft might be discovered in solar orbit.

Fanciful, but not impossible. It is also possible that evidence of a past extra-terrestrial civilisation might be found without there being any intent on their part. We are now producing quite significant amounts of space debris. Particles less than about one micrometre in size, perhaps of exotic alloys, will be ejected from our Solar System by radiation pressure and could, far into the future, land on the surface of an airless moon. When our Sun reaches the end of its life, intense solar winds could eject even larger particles into the interstellar medium. Might we find such material from another civilisation within the dust making up the lunar regolith?

For these possibilities to be at all likely, many civilisations must have reached high technical competence in the distant past. If one such civilisation came into existence every 100,000 years then, in the period since there have been sufficient heavier elements within the interstellar medium to allow planets to form and the intelligent life to have evolved – perhaps 4 billion years – 40,000 advanced civilisation might have come and gone. Could one of them have left any evidence of their existence?

The future of radio SETI

It has been a long-term dream of SETI astronomers to have a large dedicated telescope of their own. This dream has been realised, in part, with the partial construction of the Allen Telescope Array (ATA) at Hat Creek in California. The ATA was conceived as a combined project of the SETI Institute and the Radio Astronomy Laboratory at the University of California, Berkeley, to construct a radio telescope that will search for extra-terrestrial intelligence and *simultaneously* carry out astronomical research. (The Berkeley collaboration ended in April, 2012, and the project is now managed by SRI International, an independent, non-profit research institute.) This is, as one might suspect, not a telescope 'as we know it' and is exploiting the great advances in computing technology to build a highly flexible instrument where, as Jill Tarter of the SETI Institute points out, 'steel is being replaced by silicon'.

The cost of building a large single-dish antenna tends to rise as the cube of the diameter. The equivalent area could, in principle, be made up of an array of smaller antennas, in which case the cost only rises as the square of the diameter. However, the task of combining the signals from the individual elements must also be taken into account. With the reducing cost of electronics – from the receivers on each antenna, their fibre-optic links to the central processing system and the correlators that combine the data – this small diameter/large number approach has become both feasible and cost effective. But, in addition, there is a far more fundamental reason why this approach is particularly appropriate for SETI purposes. A large single antenna is only sensitive to signals received from a very small area of the sky defined by its 'beamwidth'. For a 120-metre antenna observing in the region of the 'water hole' this would be of order 7–8 arcminutes. (The use of a multi-beam receiver system can increase this by a small factor.) Let us suppose that, as in the ATA, the same effective area is made up by combining the signals from 350 7 × 6-metre antennas. These small antennas will have a beamwidth of ~120/7 times greater (beamwidth scales directly with diameter) giving ~146 × 125 arcminutes – over 2 degrees! At the heart of the array, the signals from all antennas are combined together to form a beam of comparable size to the single 120-metre antenna and having the same sensitivity. So nothing is lost. But there is much to gain. If, in additional electronics, the signals from each antenna are combined in a slightly different way then a second narrow beam can be formed anywhere within the overall beam of the small antennas, as shown in Figure 13.5. But if one can form a second beam, then with further electronics one can form a third, a fourth and so on. So the ATA will have multiple beams and could thus observe many stars simultaneously whilst the Berkeley group are observing pulsars or other astronomical objects in the same area of sky.

The first phase comprising 42 antennas was commissioned in the autumn of 2007 and began its SETI observations with a survey of the Galactic Centre. However, in April 2011, due to funding shortfalls, the ATA was placed in operational hibernation; then, having found some short-term funding, operation of the ATA was resumed in December that year.

In many respects the ATA is a technology demonstrator for what would be the ultimate radio SETI instrument – at least for many decades. A major, 38 million

Figure 13.5 (Left) The multiple beams formed with the Allen Telescope Array. (Right) Allen Telescope Array antennas. Image (right): SETI Institute.

euros, design study is now underway to finalise the technology for what is called the Square Kilometre Array (SKA). As its name implies, it aims to have a total collecting area of one square kilometre – a million square metres and the equivalent of about 90 120-metre single dishes in size. It would thus be nearly two orders of magnitude more sensitive than the ATA. Like the ATA it will also use the small diameter/large number approach, though the individual antennas might be somewhat larger, perhaps 16 m in height. There will be a central core of antennas with outlying 'stations' located on a number of 'spiral arms' radiating from the centre of the array, their separation increasing away from the array centre. The overall diameter of the whole array will be more than 3,000 km so it needed to be located in a large, sparsely populated, area. The Northern Cape in South Africa has been chosen as the location of the central array.

The SKA will be a 17-country collaboration and require funding of ~650 million euros to build Phase 1 comprising ~10% of the full array, so it is a truly major project. Construction of Phase 1 of the SKA is scheduled to begin in 2017 to allow initial observations by 2020. The headquarters of the project are at the Jodrell Bank Observatory in the UK. Like the ATA it will have the ability to form multiple beams so, for example, whilst a search might be being made for pulsars, beams could also be in use searching for any alien signals.

Should we be disheartened that no signals have as yet been detected? Not really, for as Peter Backus of the SETI Institute has made clear, of the hundred or so searches that have taken place since 1960, only SERENDIP and Phoenix (because of their use of the giant Arecibo Telescope) have had the sensitivity to detect signals from beyond our immediate locality in space. The use of the SKA will extend the search further and so have a realistic chance of detection if, as many astronomers now believe, intelligent life is thinly spread around the Galaxy.

There can be no better way to end this chapter than by quoting from Cocconi and Morrison's 1959 paper: 'The probability of success [in our search for extraterrestrial life] is difficult to estimate; but if we never search, the chance of success is zero.'

Two more books:

> *The Eerie Silence: Are We Alone in the Universe?* by Paul Davies (Allen Lane).
> *Beyond Contact: A Guide to SETI and Communicating with Alien Civilizations* by Brian McConnell (O'Reilly Media).

14

Our island Universe

The Milky Way

On a dark night with transparent skies, we can see a band of light across the sky that we call the Milky Way. (This comes from the Latin – Via Lactea.) The light comes from myriads of stars packed so closely together that our eyes fail to resolve them into individual points of light. This is our view of our own galaxy, called the Milky Way Galaxy or often 'the Galaxy' for short. It shows considerable structure due to obscuration by intervening dust clouds. The band of light is not uniform; the brightness and extent is greatest towards the constellation Sagittarius, suggesting that in that direction we are looking towards the Galactic Centre. However, due to the dust, we are only able to see about one-tenth of the way towards it. In the opposite direction in the sky the Milky Way is less apparent, implying that we live out towards one side of the Galaxy. Finally, the fact that we see a band of light tells us that the stars, gas and dust that make up the Galaxy are in the form of a flat disc (Figure 14.1).

A detailed image of the centre of the Galaxy in the constellations Sagittarius and Scorpius is shown in Figure 14.2.

The major visible constituent of the Galaxy, about 96%, is made up of stars, with the remaining 4% split between gas ~3% and dust ~1%. Here 'visible' means that we can detect the constituents by electromagnetic radiation: visible, infrared or radio. As will be discussed in detail in Chapter 21, 'The invisible Universe', we suspect that there is a further component of the Galaxy that we cannot directly detect called 'dark matter'.

Open star clusters

Amongst the general star background we can see close groupings of stars that are called clusters. These are of two types: open clusters and globular clusters.

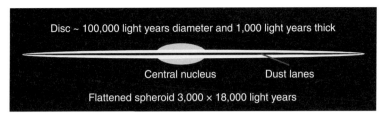

Figure 14.1 Cross section of our Galaxy.

Figure 14.2 Looking towards the heart of the Milky Way Galaxy showing the dust lanes that limit our view towards the centre. Also shown in colour in Plate 14.2. Image: Ian Morison.

Open clusters are a consequence of the formation of a group of stars in a giant cloud of dust and gas and are thus naturally found along the plane of the Milky Way – the disc of our Galaxy. Over time the stars will tend to drift apart but, whilst they are young, we will see the stars relatively closely packed together. Prime examples observable in the northern hemisphere are the Hyades and Pleiades clusters in Taurus and the Double Cluster in Perseus. Figure 14.3 shows the Perseus Double Cluster and the Pleiades Cluster. The Pleiades are passing through a dust cloud whose particles are scattering blue starlight.

Globular clusters

The globular clusters are, in contrast, very old stars in tight spherical concentrations (~200 light years across) of 20,000 to 1 million stars, as seen in

Figure 14.3 The Perseus Double Cluster (left) and Pleiades Cluster (right). Images: Ian Morison.

Figure 14.4 M5, a globular cluster in Serpens. Image: Ian Morison.

Figure 14.4. In the northern hemisphere the most spectacular is M13 in Hercules, whilst 47 Tucanae is a jewel of the southern hemisphere. (An even more spectacular object in the southern hemisphere is Omega Centauri, widely stated to be a globular cluster but now suspected to the nucleus of a small galaxy whose outer stars have been stripped away by the gravitational forces of our own galaxy.) They date from the origin of the Galaxy and were formed in the initial star formation period of our Galaxy but their precise origin and role in the evolution of the Galaxy is still unclear. Globular clusters orbit the centre of our Galaxy and form a roughly spherical distribution, helping to form what is known as the galactic halo. We know of 150 globular clusters associated with our Galaxy and perhaps a further 20 may be present but obscured by dust. Their spherical shape is due to the fact that they are very tightly bound by gravity, a further consequence of which is that the stars near their centres are very tightly packed.

The interstellar medium, and emission and dark nebulae

Together the gas and dust make up what is called the interstellar medium or ISM. All but 2% of the gas is made up of hydrogen and helium largely produced in the 'Big Bang' at the origin of the Universe. Originally the composition was ~75% hydrogen and 25% helium but, as helium and other elements were created in stars and then ejected into space, the percentage of hydrogen fell to ~70% and that of helium increased to 28%. The remaining gases are made up from other elements, such as carbon, oxygen and nitrogen, created in the stars and forming gases such as CO (carbon monoxide) and CN (cyanogen). The dust particles are composed of tiny lumps of material containing carbon, silicon and iron, which have sizes of 10^{-7} to 10^{-6} m across – similar in size to smoke particles and comparable in size to the wavelength of visible light, which they absorb and scatter so creating the dark dust lanes limiting our view along the plane of the Milky Way.

Most of the ISM is not apparent to our eyes, but in some regions we can see either 'emission nebulae' of glowing gas or 'dark nebulae' where a dust cloud appears in silhouette against a bright region of the Galaxy. Perhaps the most spectacular example of an emission nebula is the Great Nebula in Orion, or more simply the Orion Nebula – a region of star formation where the hydrogen gas is being excited by the ultraviolet light emitted by the very hot young stars – forming the 'Trapezium' at its heart, as seen in Figure 14.5. This type of emission nebula is called an H II ('H-two') region as it contains ionised hydrogen where the electrons have been split off from the protons by the ultraviolet photons emitted by very hot stars. The protons and electrons can then recombine to form neutral hydrogen atoms (H I – 'H-one') and the electrons drop through the allowed energy levels to the lowest energy state, emitting photons of various wavelengths as they do so. One of these transitions gives rise to a bright red emission line at 656.3 nm, so in photographs these regions look a lovely pinky-red colour.

An example of a dark nebula is the 'Coal Sack', seen against the background of the Milky Way close to the Southern Cross. Often the two types of nebulae are seen together, as in the Eagle Nebula in Serpens and the Horsehead Nebula in Orion shown in Figure 14.6, where the dark pillars of dust and the 'horse's head' respectively are seen against the bright glow of excited gas clouds.

Figure 14.5 The Orion Nebula. Image: Ian Morison.

Figure 14.6 (Left) The Eagle Nebula. Image: Hubble Space Telescope, STScI, ESA, NASA. (Right) The Horsehead Nebula. Image: European Southern Observatory.

Size, shape and structure of the Milky Way

In Chapter 10, the method of parallax that enables us to measure the distances to the nearby stars was described. This technique has enabled the distances out to a number of RR Lyrae stars to be determined. These variable stars pulsate with a period that is a function of their luminosity and this allows them to serve as 'standard candles' that may be used to measure astronomical distances. If the distance to a nearby RR Lyrae star is known from parallax methods, and one with the same period is observed at a great distance, the inverse square law can then be used to determine its distance. As RR Lyrae stars are typically 50 times more luminous than our Sun they can be used to determine distances out to some of the nearer globular clusters. In a similar way, Cepheid variable stars (up to ~10,000 times more luminous than our Sun and which will be discussed in detail in Chapter 19) can be used to extend distance measurements across the Galaxy.

In a groundbreaking series of observations up to 1919, Harlow Shapley was able to use observations of RR Lyrae stars and Cepheid variable stars to measure the distances of 100 of the globular clusters associated with our Galaxy. He found that they formed a spherical distribution, whose centre should logically be the centre of the Galaxy, and so deduced that our Galaxy was ~100,000 light years across and that our Sun was ~30,000 light years distant from the centre. Figure 14.7 shows a cross section of the Galaxy with the positions of the globular clusters observed by Harlow Shapley.

We now believe that the Sun is 27.7 light years from the Galactic Centre and, using spectroscopic measurements to observe its motion relative to the globular clusters, we can calculate that the Sun is moving around the centre of the Galaxy

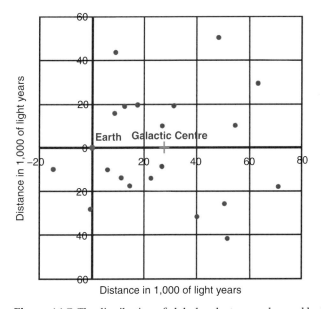

Figure 14.7 The distribution of globular clusters as observed by Harlow Shapley.

at about 230 km/s, taking ~220 million years to travel once around it. It appears that the central parts of the Galaxy rotate like a solid body so that the rotation speed increases as one moves out from the centre. Measurements of the speed at which stars and gas rotate around the centre of the Galaxy as a function of distance produce what is called the 'galactic rotation curve'. Its shape is not what one would expect and this will be further discussed in Chapter 21 about dark matter.

But what of its structure? Neutral hydrogen (H I) emits a radio spectral line with a wavelength of 21 cm. Radio observations of this line along the plane of the Milky Way show that the gas in the disc is not uniformly dense but is concentrated into clouds whose velocity away from or towards us can be determined using the Doppler shift in its observed wavelength. These data can be used to plot out the positions of the gas clouds, and when this is done a pattern of spiral arms emerges – indicating that we live in a typical spiral galaxy thought to be quite similar to the nearby Andromeda Galaxy.

Observations of the hydrogen line

A neutral hydrogen atom consists of a proton and an electron. As well as their motion in orbit around each other, the proton and electron also have a property called spin. This is actually a quantum-mechanical concept, but is analogous to the spin of the Earth and the Sun about their rotation axes. Again using this analogy, the spin may be clockwise or anticlockwise. Thus their two spins may be oriented either in the same direction or in opposite directions. In a magnetic field (as exists in the Galaxy) the state in which the spins of the electron and proton are aligned in the same direction has slightly more energy than one where the spins of the electron and proton are in opposite directions. Very rarely (with a probability of 2.9×10^{-15} s^{-1}) a single isolated atom of neutral hydrogen will undergo a transition between the two states and emit a radio spectral line at a frequency of 1420.40575 MHz. This frequency has a wavelength of ~21 cm so that it is often called the 21-cm line. A single hydrogen atom will only make such a transition in a time scale of order 10 million years but, as there are a very large number of neutral hydrogen atoms in the interstellar medium, it can be easily observed by radio telescopes. This radio spectral line was first detected by Professor Edward Purcell and his graduate student Harold Ewen, at Harvard University in 1951, using a simple horn antenna.

Let us consider what we would see if our Galaxy had a number of spiral arms. As well as delineating concentrations of stars and dust, they will also contain a higher concentration of hydrogen gas. As all the material in the Galaxy is

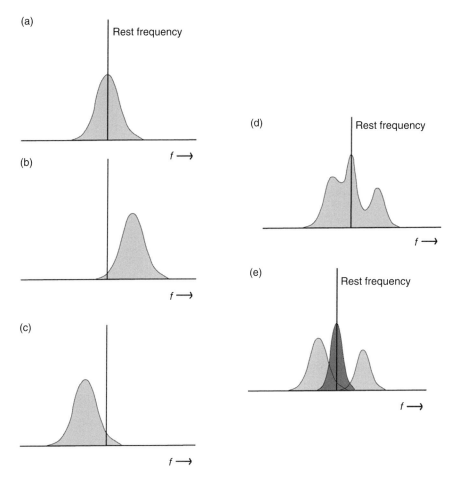

Figure 14.8 Hydrogen line profiles.

rotating around its centre, when we look in different directions along the galactic plane the material in the spiral arms – in some directions we might see several – will have differing velocities away from or towards us. The hydrogen line from these different arms will thus be Doppler shifted to higher or lower frequencies. Due to motion of the gas within an element of an arm that we observe, any given arm will have a line profile that is approximately a Gaussian, as shown in part (a) of Figure 14.8. This shows the gas close to us in our local spiral arm so its average velocity relative to us is zero. If the beam of our radio telescope receives radiation from the hydrogen gas in a spiral arm coming towards us, we will see that Gaussian profile shifted to higher frequencies, as in (b), whilst if moving away we will observe it at lower frequencies, as in (c). In general, should we look in some arbitrary direction we may observe several

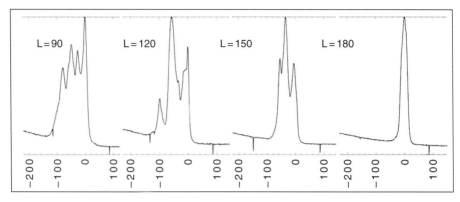

Figure 14.9 Galactic hydrogen line profiles. Images: Christine Jorden, Jodrell Bank Observatory, University of Manchester.

arms and so get a more complex profile, as shown in (d), which can be 'dissected' into the individual profiles from each arm, as shown in (e).

Figure 14.9 shows some hydrogen line profiles observed using a 6.4-metre radio telescope at the Jodrell Bank Observatory. Each has the galactic latitude (L) at which it was observed: latitude 0 would mean that our telescope would be pointing directly at the centre of the Galaxy, latitude 180 directly away from it – looking out from our Galaxy. You can see that in this latter direction, there is only one peak in the spectrum, which is centred on zero velocity with respect to us. The majority of this emission comes from hydrogen in our local spiral arm, which is, of course, at rest with us. Some will come from outer arms but, as these will be moving at right angles to the line of sight, they will show no Doppler shift. As we lie within our own spiral arm, there will always be a peak at zero relative velocity but, as we look along the Milky Way in other directions, we see other peaks in the spectrum, corresponding to other spiral arms. For example, at L = 90 we see three other spiral arms.

By observing such hydrogen line profiles along the galactic plane, and using a model of the rotation curve of the Galaxy, it is possible to locate the position of each observed arm along the line of sight through the Galaxy and build up a picture of the spiral structure of our own galaxy. Figure 14.10 shows a section of the plane of the Milky Way, with vertical relief showing the brightness of hydrogen. The local arm appears like a range of mountains on the right of the plot, with the Perseus arm curving out to its left. The more distant outer arm can just be seen appearing like 'foothills' even further to the left.

The fact that our Galaxy has a spiral structure is somewhat of a puzzle. In its life, our Sun has circled the Galactic Centre about 20 times, so why haven't the spiral arms wound up? The solution is hinted at by a visual clue. Spiral arms seen

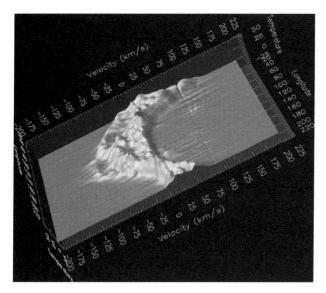

Figure 14.10 A relief map of a section of the Milky Way obtained with the 6.4-metre radio telescope at Jodrell Bank Observatory. Image: Tim O'Brien, Jodrell Bank Observatory, University of Manchester.

in other galaxies stand out because they contain many bright blue stars – remember, a single very hot star can outshine 50,000 suns like ours! But very hot bright stars must be young as they have very short lives, so the spiral structure we see now is not that which would have been observed in the past. As B. Lindblad first suggested, it appears that the spiral arms are transitory and caused by a spiral density wave rotating round the Galactic Centre – a ripple that sweeps around the Galaxy moving through the dust and gas. This compresses the gas as it passes and can trigger the collapse of gas clouds, so forming the massive blue stars that delineate the spiral arms. The young blue stars show us where the density wave has just passed through, but in its wake will be left myriads of longer lived (and less bright) stars that form a more uniform disc.

An infrared map of the Milky Way Galaxy

Infrared images from NASA's Spitzer Space Telescope, combined in the artist's impression shown in Figure 14.11, indicate that the Milky Way's elegant spiral structure is dominated by just two arms wrapping off the ends of a central bar of stars. The Galaxy's two major arms are attached to the ends of a thick central bar – so we now believe that our Galaxy is a barred spiral – with the Perseus Arm extending from the top right of the bar and sweeping round below it in the image and the Scutum–Centaurus Arm extending from the lower left

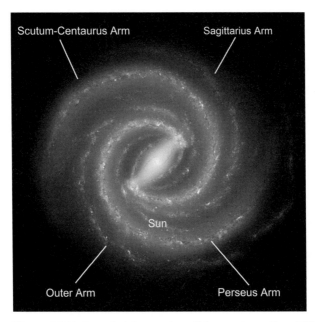

Figure 14.11 An artist's impression of the structure of our Milky Way Galaxy derived from infrared images taken by the Spitzer Space Telescope. Image: NASA/JPL-Caltech.

and sweeping round above it. These two arms contain the highest densities of both young and old stars whilst two less distinct minor arms (Outer and Sagittarius), located between the major arms, are primarily filled with gas and pockets of star-forming activity. Our Sun lies near a small, partial arm called the Orion Arm, or Orion Spur, located between the Sagittarius and Perseus arms.

A super-massive black hole at the heart of our Galaxy

Whereas at optical wavelengths the centre of our Galaxy is obscured by dust, at radio wavelengths we are able to peer deep into its heart. Astronomers have discovered a very compact radio source in the constellation Sagittarius called Sgr A*, which we believe marks the position of a super-massive black hole at the centre of our Galaxy. How can we be sure that this exists? In the same way that one can calculate the mass of the Sun by knowledge of the orbital velocity of the Earth and its distance from the Sun, we can estimate the mass of Sgr A* by measuring the speeds of stars in orbit around it at very close distances. For example, one of the 8-metre VLT telescopes at Paranal Observatory in Chile has observed a star in the infrared as it passed just 17 light hours from the centre of the Milky Way (three times the distance of Pluto from the Sun). This convincingly showed that it was under the gravitational influence of an object that had an enormous gravitational field yet must be extremely compact – a super-massive black hole. Its mass is now thought to lie between 3.2 and 4 million solar masses confined within a volume one-tenth the size of the Earth's orbit. This will be discussed further in Chapter 17 about black holes.

Our place in the Universe

In every direction that we look out into space we see galaxies having a wide variety of different shapes. Containing between hundreds of millions and hundreds of billions of stars, they are the fundamental building blocks of the Universe. Edwin Hubble devised the most commonly used system for classifying galaxies, grouping them according to their appearance in photographic images. He arranged the different types of galaxies in what became known as the Hubble sequence. They fell into three groups: elliptical, spiral (which were sub-divided into normal and barred spirals) and irregular galaxies. (It should be pointed out that galaxies do not evolve along the Hubble sequence so this is not, perhaps, a very useful name.)

Other galaxies

Galaxies, called originally 'white nebulae', have been observed for hundreds of years, but it was not until the early part of the last century that the debate as to whether they were within or beyond our Galaxy was settled – essentially when observations of Cepheid variables enabled their distances to be measured. They are, of course, objects outside our Galaxy and can now be observed throughout the Universe. Galaxies can be divided into a number of types and then sub-divided further to produce the classification scheme first devised by Edwin Hubble. As more and more galaxies were discovered, it became apparent that galaxies form groups (up to about a hundred galaxies) or clusters (containing hundreds to thousands of galaxies).

Elliptical galaxies

These, as their name implies, have an ellipsoidal form rather like a rugby ball. They range from those that are virtually circular in observed shape, called E0 by Hubble, to those, called E7, that are highly elongated. Giant elliptical galaxies may contain up to 10,000 billion solar masses in a volume some nine times that of our own galaxy ~300,000 light years across. These are the most massive of all galaxies but are comparatively rare. Far more common are elliptical galaxies containing perhaps a few million solar masses within a volume a few thousand light years across. Elliptical galaxies account for about one-third of all galaxies in the Universe.

Spiral galaxies

Like our own galaxy, these have a flattened spiral structure. The first observation of the spiral arms in a galaxy was made by the Third Earl of Rosse. During the 1840s he designed and had built the mirrors, tube and mountings for a 72-inch reflecting telescope, which for three-quarters of a century was the largest optical telescope in the world. With this instrument, situated at Birr Castle in Ireland, Lord Rosse made some beautiful drawings of astronomical objects. Perhaps the most notable was of the Whirlpool Galaxy. It was the first drawing to show the spiral arms of a galaxy and bears excellent comparison to modern day photographic images. (The galaxy is interacting with a second galaxy, NGC 5195, seen at the right of Figure 9.2.)

Spiral galaxies make up the majority of the brighter galaxies. Hubble classified them first into four types, S0, Sa, Sb and Sc. S0 galaxies, often called 'lenticular' galaxies, have a very large nucleus with hardly visible, very tightly wound spiral arms. As one moves towards type Sc, the nucleus becomes relatively smaller and the arms more open. In many galaxies the spiral arms appear to extend from either end of a central bar. These are called 'barred spirals' and are delineated SBa, SBb and SBc. Our own Milky Way Galaxy was thought to be an Sb or Sc but there is now some evidence that it has a bar, so making it an SBb or SBc.

Irregular galaxies

A small percentage of galaxies show no obvious form and are classified as 'irregulars'. One nearby example is the Small Magellanic Cloud (SMC). Its companion, the Large Magellanic Cloud (LMC), is usually classified as irregular too, though it shows some features of a small barred spiral. Such small galaxies are not very bright so we cannot see many but they may, in fact, be the most common type.

Groups and clusters of galaxies

Most galaxies are found in groups typically containing a few tens of galaxies or clusters that may contain up to several thousand. Our Milky Way Galaxy forms part of the 'Local Group', which contains around 40 galaxies within a volume of space 3 million light years across. Our Galaxy is one of the three spiral galaxies (along with M31 and M33) that dominate the group and contain the majority of its mass. The Andromeda Galaxy, M31, shown in

Figure 14.12 M31, the Andromeda Galaxy, along with two dwarf elliptical galaxies, M32 (left of centre) and M110 (below right). Also shown in colour in Plate 14.12. Image: Peter Shah.

Figure 14.12 and Plate 14.12, and our own are comparable in size and mass and their mutual gravitational attraction is bringing them towards each other so that, in a few billion years, they will merge to form an elliptical galaxy. M33 in Triangulum, Figure 14.13, is the third largest galaxy in the group. The group also contains many dwarf elliptical galaxies, such as the two orbiting M31 and visible in Figure 14.12. M32 is a small, type E2, elliptical galaxy seen to the left of the nucleus of M31, just outside its spiral arms, whilst NGC 205 (M110) is a more elongated, type E5 or E6, elliptical that is seen to the lower right of the nucleus of M31. The Local Group contains several large irregular galaxies, such as the Magellanic Clouds, and at least 10 dwarf irregulars to add to the total. There may well be more galaxies within the group, hidden beyond the Milky Way, which obscures over 20% of the heavens.

Observations of the region lying in Virgo just to the west of Leo show many hundreds of galaxies and it is thus called 'the realm of the galaxies'. Sixteen of these are bright enough to have been catalogued by Charles Messier using a

Figure 14.13 M33 in Triangulum. Also shown in colour in Plate 14.13. Image: Ian Morison.

Figure 14.14 The heart of the Virgo Cluster of galaxies with M86 above centre and M84 to its right. The Eyes Galaxies (NGC 4435 and NGC 4438) lie to the upper left. Image: Digitised Sky Survey.

telescope of just a few centimetres aperture. In this direction we are looking towards the heart of a galaxy cluster containing some 2,000 members called, due to the constellation in which we observe it, the Virgo Cluster (Figure 14.14). Two other nearby clusters are the Coma Cluster and the Hercules Cluster. Galaxy clusters typically contain 50 to 1,000 galaxies within diameters of 6 to 35 million light years and have total masses of 10^{14} to 10^{15} solar masses.

Superclusters

Small clusters and groups of galaxies appear to make up structures on an even larger scale. Known as superclusters, they have overall sizes of order 300 million light years (100 times the scale size of our Local Group). Usually a supercluster is dominated by one very rich cluster surrounded by a number of smaller groups. The Local Supercluster is dominated by the Virgo Cluster. The Virgo Supercluster, as it is often called, is in the form of a flattened ellipse about 150 million light years in extent with the Virgo Cluster at its centre and our Local Group near one end. In the same general direction, but further away, lies the Coma Cluster containing over 1,000 galaxies. It is the dominant cluster in the Coma Supercluster at a distance of 330 million light years. Two other nearby superclusters lie in the directions of the constellations Perseus/Pisces and Hydra/Centaurus at distances of 150 and 230 million light years respectively.

These very great distances are beyond that where the Cepheid variable distance scale can be used. In Chapter 19, entitled 'Hubble's heritage' it will be seen how relatively recent observations of Type Ia supernovae are being used to measure distances of this magnitude. However, it is worth mentioning a very simple method that had been used prior to this new technique. One could reasonably assume that the largest galaxies within a cluster are probably of

Figure 14.15 The Hubble Ultra Deep Field. Image: STScI, ESA, NASA.

the same size. One might take the third largest to avoid the possibility that the largest might be more extreme than in most clusters. If their angular sizes are measured on a photographic plate then one can compare them to a galaxy, hoped to be similar in size, whose distance we do know. In a similar vein, one could take the magnitude of the galaxy and use that to estimate its distance.

The most distant galaxies ever observed are shown in the Hubble Ultra Deep Field in Figure 14.15. Some are observed at a time just 0.7 billion years after the origin of the Universe, just 5% of its present age.

The overall structure of the Universe is like a sponge with regions that are virtually empty – called voids – surrounded by the clusters and superclusters of galaxies. The three-dimensional pattern is called the 'cosmic web'. It results from the effects of gravity acting on very small fluctuations in initial density of the Universe which are caused by the clumping of dark matter – as will be discussed in Chapter 21, 'The invisible Universe'. Those regions that were slightly denser attract material to them from the less dense regions – which thus become emptier and so result in the voids that we see. The concentrations of matter around the voids then give rise to the clusters and superclusters of galaxies.

More on the Milky Way:

> *The Milky Way: An Insider's Guide* by William H. Waller (Princeton University Press).
> *Revealing the Heart of the Galaxy: The Milky Way and its Black Hole* by Robert H. Sanders (Cambridge University Press).

15

Wonders of the southern sky

From northerly latitudes, there is a part of the heavens that we never see and a part that has such a low elevation when it rises in the south that we do not see it well. This region covers the heart of our Milky Way Galaxy, containing many beautiful clusters and nebulae, our two nearest galactic neighbours, the Magellanic Clouds, a star that may well be the next nearby supernova, a pair of colliding galaxies, and a globular cluster containing 'musical' stars. This chapter aims to introduce you to this most beautiful part of the heavens.

Our Milky Way

Stretching through Sagittarius and Scorpius, the central region of the Milky Way as seen in Figure 15.1 is a truly beautiful sight. We can see rich star fields crossed by intricate dust lanes and punctuated by bright nebulae and star clusters. The nebulae include the Lagoon and Trifid nebulae and the bright open clusters include M6 and M7 in Sagittarius and the Northern Jewel Box in Scorpius. There are also 'dark' nebulae such as the Pipe Nebula that lies on the borders of Sagittarius and Scorpius and the Coal Sack in Crux.

The region of the Milky Way that cannot be observed from the northern climes extends from Scorpius along the Milky Way to Canis Major. The first prominent constellation moving in this direction is Centaurus, three of whose objects, Alpha Centauri, Omega Centauri and Centaurus A, are discussed below. Somewhat hemmed in by Centaurus is one of the smallest constellations, Crux, often called the Southern Cross. It contains a beautiful open cluster, the Jewel Box, and one of the most prominent dark nebulae known as the Coal Sack. Next along the Milky Way is Carina, home to an open cluster called Mel 101, or the Southern Pleiades, along with the Eta Carinae Nebula in which lies the star Eta

Figure 15.1 The Milky Way as seen from Chile. Image: European Southern Observatory, Wikimedia Commons.

Carinae. This is likely to be the next nearby supernova and is discussed below. Next is Vela, which harbours a supernova remnant at whose heart is a rotating neutron star, the Vela Pulsar. Finally, crossing Puppis, we come to Canis Major.

The heart of the Milky Way

The Galactic Centre is the rotational centre of the Milky Way Galaxy. It lies in the direction of the constellations Sagittarius, Ophiuchus and Scorpius. When we look at images of the Milky Way we see extensive regions where the light is obscured by dark clouds of interstellar dust (containing much silicon and carbon) thus the Galactic Centre cannot be studied at visible, ultraviolet or soft X-ray wavelengths. We can, however, observe it at the extremes of the electromagnetic spectrum at gamma-ray, hard X-ray, infrared, sub-millimetre and radio wavelengths. Its direction and distance were first found by Harlow Shapley who, around 1918, measured the distances to 100 of the globular clusters (spherical clusters of ~1 million stars dating from the formation of our Galaxy) associated with our Galaxy. He found that they formed a spherical distribution, whose centre should logically be the centre of the Galaxy, and so deduced that the heart of the Galaxy was ~28,000 light years from our Solar System. By observing the Sun's motion relative to the globular clusters, we can calculate that the Sun is moving around the centre of the Galaxy at about 230 km/s, taking ~220 million years to travel once around it.

The Galactic Centre contains many millions of stars. Many of these are old, red main sequence stars (still burning hydrogen to helium in their cores), but it is also rich in massive stars that appear to have been born a few million years ago in a burst of star formation. Currently, star formation does not seem to be occurring at the Galactic Centre but studies predict that in ~200 million years there will be a further 'starburst' event there, with many stars evolving rapidly and producing supernovae at a rate 100 times that observed now. It is thought that the Milky Way undergoes a starburst event of this sort every 500 million years or so.

On either side of the Galactic Centre is a bar, composed primarily of red stars, which is thought to be about 27,000 light years long. It is inclined at an angle of ~44 degrees to the line between the Sun and the centre of the Galaxy. The bar is surrounded by a ring in which most of the Milky Way's star formation is concentrated. Many galaxies show such a ring and ours, as seen from the nearby Andromeda Galaxy, would be the brightest single feature of the Galaxy.

A super-massive black hole at the Galaxy's heart

Whereas at optical wavelengths the centre of our Galaxy is obscured by dust, at radio wavelengths we are able to peer deep into its heart, and astronomers have discovered a very compact radio source called Sgr A*, which we believe marks the position of a super-massive black hole at the centre of our Galaxy. How can we be sure that this exists? In the same way that we can calculate the mass of the Sun by knowledge of the orbital velocity of the Earth and its distance from the Sun, we can estimate the mass of Sgr A* by measuring the speeds of stars in orbit around it at very close distances. For example, one of the 8-metre VLT telescopes at Paranal Observatory in Chile has observed a star in the infrared as it passed just 17 light hours from the centre of the Milky Way (three times the distance of Pluto from the Sun). This convincingly showed that it was under the gravitational influence of an object that has enormous gravity and yet must be extremely compact – a super-massive black hole. Its mass is now thought to lie between 3.2 and 4 million solar masses confined within a volume one-tenth the size of the Earth's orbit.

The constellation Centaurus

Alpha Centauri

Also known as Rigel Kent, Alpha Centauri appears as a single star to the unaided eye and is the brightest star in the constellation Centaurus (as one might well suspect from its Beyer designation) and the fourth brightest star in the heavens. A telescope reveals that it is, in fact, a close binary system known as the 'Alpha Centauri AB' system, often abbreviated to 'α Cen AB' (Figure 15.2). Alpha Centauri A is the more massive star of the system. It is about 10% more massive than the Sun and about 23% larger and rotates once

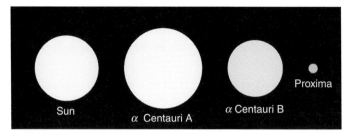

Figure 15.2 The Alpha Centauri star system.

every 22 days – a very similar speed to our Sun. Its companion, Alpha Centauri B, is slightly smaller and less luminous than our Sun with a mass of 0.9 solar masses. There is a third component, 'α Cen C', that is located 2.18 degrees away in the sky. Together, all three components make a triple star system, referred to as 'α Cen AB-C'. Alpha Cen C is normally called Proxima Centauri as it is the nearest star to our Sun at a distance of 4.2 light years, 0.21 light years closer than α Cen AB. Proxima Centauri is a red dwarf star which can be observed in a small telescope and has a mass about one-eighth that of our Sun. As all three stars are moving in the same approximate direction through space, it is generally assumed that Proxima Centauri is gravitationally bound to α Cen AB, orbiting them with a period of several hundred thousand years, though it is possible that it is just passing α Cen AB on a hyperbolic trajectory.

Omega Centauri

Omega Centauri, seen in Figure 15.3, has long been regarded as a globular cluster – a compact spherical group of typically a million stars that dates from the formation of the Galaxy. It had been listed as a star in Ptolemy's catalogue and, still thought to be a star, was given the designation ω Centauri by the German astronomer Johann Bayer his 1603 star atlas, *Uranometria*. Bayer assigned a lower-case Greek letter, such as alpha (α), beta (β), gamma (γ), etc., to each star he catalogued, combined with a form of the Latin name of the star's parent constellation. Omega Centauri was first recognised as a globular cluster by the English astronomer John William Herschel in the 1830s and, if so, is both the brightest and the largest known globular cluster associated with our Galaxy. We know of only one globular cluster, Mayall II in the Andromeda Galaxy, that is brighter and more massive.

Omega Centauri is located about 15,800 light years from Earth and contains several million stars. At its centre, the stars are so crowded that they are, on average, only 0.1 light years apart. It is thought to date from the time when our Milky Way Galaxy was formed, some 12 billion years ago, and is one of the few

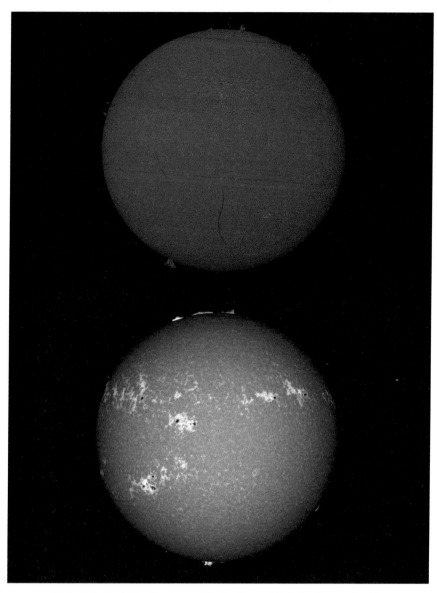

Plate 2.5 (Top) The Sun imaged in the light of H-alpha. Image: Ian Morison. (Bottom) The Sun imaged in the light of calcium K. Image: Greg Piepol.

Plate 2.6 The Aurora Borealis. Image: US Air Force, by Senior Airman Joshua Strang, Wikimedia Commons.

Plate 4.7 Computer-generated image of Mars. Image: MSSS, JPL, NASA.

Plate 6.4 Jupiter (top) imaged by Damian Peach and Saturn (bottom) imaged by the Cassini spacecraft. Image: CICLOPS, JPL, ESA, NASA.

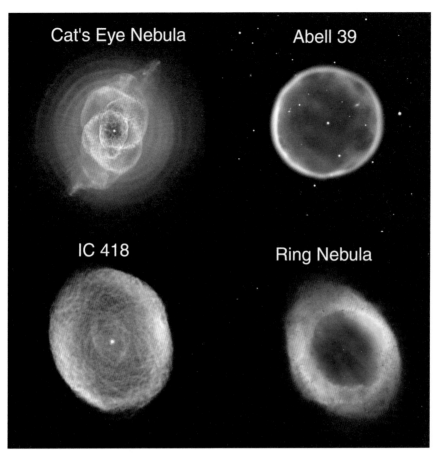

Plate 11.1 Hubble Space Telescope images of planetary nebulae. Images: Hubble Space Telescope, STScI, ESA, NASA.

Plate 14.2 The centre of our Milky Way Galaxy. Image: Ian Morison.

Plate 14.12 The Andromeda Galaxy, M31. Image: Peter Shah.

Plate 14.13 M33 in Triangulum. Image: Ian Morison.

Figure 15.3 Omega Centauri. Image: European Southern Observatory, Wikimedia Commons.

globular clusters visible to the naked eye, appearing about as large as the full Moon. However, unlike other globular clusters, it contains several generations of stars and it is now thought that Omega Centauri (as is now also thought to be the case with Mayall II) may well be the core of a dwarf galaxy whose outer stars have been disrupted and absorbed by our Milky Way Galaxy.

A central black hole?

Perhaps giving further weight to this idea, in 2008 astronomers claimed to have found evidence of an intermediate-mass black hole at the centre of the cluster. The observations, made with NASA's Hubble Space Telescope and the Gemini Observatory on Cerro Pachón in Chile, showed that the stars closer to the core are moving faster than the stars further away (as one might expect). But the speed at which they orbit the centre implies the gravitational pull of an unseen massive, dense object – most probably a black hole. They predict a mass of ~40,000 solar masses for the black hole, which lies between the masses of those resulting from the collapse of massive stars such as Eta Carinae (several solar masses) and those found at the heart of large galaxies, which can range from ~4 million solar masses as at the centre of our Milky Way Galaxy up to several billion solar masses within giant elliptical galaxies. As it is now thought

Figure 15.4 Crux, the Southern Cross, with the Coal Sack to its lower left. Image: Wikimedia Commons.

that black holes are found at the core of all galaxies, this is further evidence of a galactic, rather than a cluster, origin.

The constellation Crux: the Southern Cross

Alpha and Beta Centauri are often called the 'pointers' as they direct one towards the small cross (Figure 15.4) that makes up the constellation Crux – the smallest, but one of the most distinctive, of the 88 constellations in the heavens. It can be seen in April from northerly locations with latitudes less than +25 and, due to precession of the Earth's axis, will just be seen from the southern UK (and well seen in the USA) in about 10,000 years' time. Three of the four stars making up the cross, Acrux, Mimosa and Delta Crucis, are very young, hot, giant stars – about 10 to 20 million years old and part of the same moving star cluster. The principal star in the constellation is the binary system, Acrux or Alpha Crucis. This pair of stars is 320 light years away from our Sun, and each is approximately twice its size. The intrinsically brightest star in Crux is actually Beta Crucis (Mimosa), a blue-white giant, five times the Sun's diameter, which is ~580 light years away, and 8,000 times more luminous than our Sun.

Very close to Beta Crucis is a beautiful cluster called the 'Jewel Box', which contains about a hundred visible stars and is about 10 million years old. It contains many highly luminous blue-white stars along with a central red supergiant that makes a beautiful colour contrast. It was named by Sir John Herschel, who likened it to 'a gorgeous piece of fancy jewellery'. Just to the south of the

Jewel Box is a pear-shaped region of obscuring, or dark, nebula 7 degrees long by 5 degrees wide. Called the 'Coal Sack', it is a dense region of dust and gas about 2,000 light years from us that hides the light from more distant stars. Well seen in binoculars, it is the most prominent dark nebula along the plane of the Milky Way.

The constellation Carina

Carina contains three objects well worthy of note:

Mel 101, the Southern Pleiades, is the name of a brilliant and striking open cluster whose appearance and great brightness makes it comparable to the well-known Pleiades cluster in Taurus. It is one of the brightest open clusters in the southern sky, lying at a distance of ~480 light years and containing about 60 stars. It is thought to be about 50 million years old. Theta Carinae is the brightest star within the cluster and easily visible to the unaided eye but the remaining stars would usually require binoculars to be seen.

The Eta Carinae Nebula, also known as the Carina Nebula, is a large bright nebula that surrounds several open clusters of stars which are significant in that they contain two of the most massive and luminous stars in our Milky Way Galaxy, Eta Carinae and HD 93129A. The nebula lies at an estimated distance of between 6,500 and 10,000 light years from Earth and contains many hot, blue O-type stars. It is one of the largest diffuse nebulae in our skies – both four times as large as and brighter than the Orion Nebula. Within the nebula is a much smaller feature, known as the Homunculus Nebula (from the Latin meaning Little Man), that surrounds the star Eta Carinae and is believed to have been ejected from that star during an enormous outburst in the 1840s, which briefly made Eta Carinae the second brightest star in the sky.

Eta Carinae is, perhaps, the most interesting single star in the southern hemisphere. It was first catalogued in 1677 by Edmond Halley. Lying at a distance of 7,500 light years it is a very massive variable star, which is highly unstable and expected to explode in a supernova in the astronomically near future. Once thought to be a giant single star, Eta Carinae is now believed to be part of a system that contains at least two stars with a combined mass of 100 solar masses and about 4 million times brighter than our Sun. They orbit each other with a period of 5.52 years. Its brightness is highly variable and peaked in 1843, when it became the second brightest star in the sky (just brighter than Canopus but less bright than Sirius). It is now just visible to the unaided eye given a dark and transparent sky. Such stars are quite rare, with perhaps only a few dozen in a galaxy like ours.

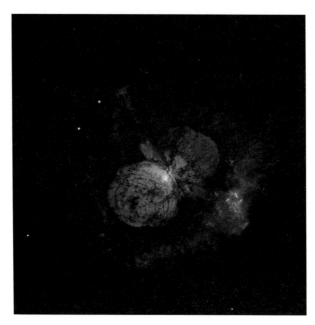

Figure 15.5 Eta Carinae and its surrounding nebula as imaged by the Hubble Space Telescope. Image: STScI, ESA, NASA.

Eta Carinae is surrounded by a huge, billowing pair of gas and dust clouds as shown in the Hubble Telescope image, Figure 15.5. These are 10 billion miles across, so about the size of our Solar System, and were formed when Eta Carinae suffered a giant outburst about 160 years ago. Over a period of time it then released nearly as much visible light as a supernova explosion, but survived! The two lobes produced in the explosion move outwards at about 2.4 million km/h. The reason for Eta Carinae's large outbursts is not yet known; the most likely possibility is that they are caused by a build-up of radiation pressure from the star's enormous luminosity. After the 1843 outburst, Eta Carinae's brightness faded away, and between about 1900 and 1940 it was invisible to the unaided eye. This was probably due to it being surrounded by the dust clouds that form the expanding lobes. Eta Carinae suddenly doubled its brightness in 1998–9, so becoming visible again.

Eta Carinae is expected to explode as a supernova or hypernova some time within the next million years or so. As its current age and evolutionary path are uncertain, it could explode within the next several millennia or even in the next few years. A similar outburst to that shown by Eta Carinae in 1841–3 was observed during 2004 in the star SN2006jc, which lies some 77 million light years away in the constellation of Lynx. First thought to be a supernova it, like Eta Carinae, survived even though it lost about 0.01 solar masses of material into space – about 20 times Jupiter's mass. However, it then exploded as a Type Ib supernova on 9 October 2006 – just two years later. As a result, some suggest that Eta Carinae could explode in our lifetime or even in the next few years. However, others say it is likely that Eta Carinae is at an earlier stage of evolution, and that the material in its core will still be able to support nuclear fusion (and hence prevent core collapse) for some time to come.

If Eta Carinae were to become a hypernova it would probably eject gamma-ray bursts aligned to the rotational axis of the resulting black hole. Happily, as seen in the Hubble image, the rotation axis does not currently point towards the Earth; however, as part of a binary system, this axis could change and, if it were pointing towards Earth at the time of the explosion, calculations show that the intercepted energy (in the form of gamma rays) would be equivalent to 1 kilotonne of TNT per square kilometre over the entire hemisphere facing the star. Terrestrial life forms will be protected by the atmosphere, but gamma rays could destroy spacecraft or satellites and the ozone layer. In any event, it will probably be so bright that it could be seen in daylight and one would be able to read a book at night by its light!

The constellation Vela

The most interesting object within Vela is the Vela supernova remnant shown in Figure 15.6. It is the result of a stellar explosion some 11,000–12,300 years ago at a distance of about 800 light years. In 1968, a neutron star in the form of a pulsar was associated with it and so provided the first direct observational proof that supernovae form neutron stars, as has been described in Chapter 11. The Vela Pulsar spins at a rate of one rotation every 89 milliseconds – 11 times a second – and as well as being a radio pulsar can also be observed in visible light and is the brightest persistent source of gamma rays in the sky. Pulsars are usually ejected at high speed from the exploding star and the Vela Pulsar is moving through space at about 1,200 km/s.

The Magellanic Clouds

The two Magellanic Clouds, seen in Figure 15.7, are members of our Local Group of galaxies with the large Magellanic Cloud (LMC) being the fourth largest after (in order) the Andromeda Galaxy or M31, our own Milky Way Galaxy and M33 in Triangulum. They are both classified as irregular dwarf galaxies, though the LMC has some characteristics of a barred spiral. They lie

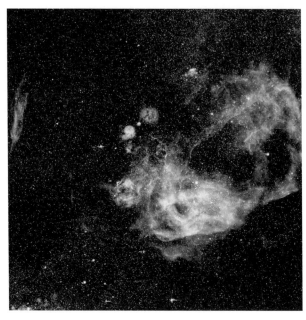

Figure 15.6 The Vela supernova remnant. Image: Southern H-Alpha Sky Survey Atlas (SHASSA) of the Cerro Tololo Inter-American Observatory (CTIO), La Serena, Chile. Wikimedia Commons.

Figure 15.7 The Large (LMC) and Small (SMC) Magellanic Clouds. The globular cluster 47 Tucanae is just to the left of the SMC. Image: European Southern Observatory, Wikimedia Commons.

at distances of ~160,000 light years (LMC) and ~197,000 light years (SMC) and are separated by ~75,000 light years. Until recently, they were thought to be orbiting our Milky Way Galaxy but new research seems to indicate that this is not the case and that they may be just 'passing by' at a speed of 480 km/s. Only the Sagittarius dwarf elliptical galaxy and the Canis Major dwarf galaxy lie closer to the Milky Way. The LMC was first recorded in AD 964 from a latitude of 12°15′ north in Southern Arabia. The pair were later reported by Antonio Pigafetta during the circumnavigation by Ferdinand Magellan from 1519 to 1522 and were, much later, named after him. They are visible as faint 'clouds' in the southern night sky, hence their name. Their makeup differs from our Galaxy in that a higher fraction of their mass comprises hydrogen and helium and so contains a lesser fraction of heavier elements. Their stars range from the very old to the very young, indicating a long history of stellar formation.

As seen from the Magellanic Clouds, the Milky Way would be a spectacular sight and span about 36 degrees across the sky – the width of over 70 full Moons. Because they lie well above the Milky Way's galactic plane, any observers there would get an oblique view of the entire Galaxy – far better than ours due to the interstellar dust that obscures our own view through its plane.

The Small Magellanic Cloud (SMC)

The SMC is a dwarf irregular galaxy with a diameter of about 7,000 light years. It contains several hundred million stars with a total mass of approximately 7 billion solar masses. (Remember that our Sun is well above average in terms of stellar masses – there are few more massive stars than our Sun and many less massive.) Its declination is -73 degrees so it can only be viewed from the southern hemisphere and the lower latitudes of the northern hemisphere. It appears as a hazy, light patch in the night sky about 3 degrees across in the constellation of Tucana. Due to its very low surface brightness, it can only really be seen from a dark sky location.

The SMC played a key role in setting up the 'distance ladder' that enabled the distances of what were first called 'white nebulae' (galaxies) to be determined – a story that is recounted in detail in Chapter 19.

The Large Magellanic Cloud (LMC)

The LMC has a mass of about 10 billion solar masses – about one-tenth the mass of the Milky Way, and a diameter of about 14,000 light years. It contains a very prominent bar in its centre, suggesting that it may have previously been a barred spiral galaxy which has been distorted by tidal interactions with both the Milky Way and the SMC. The LMC, like many irregular galaxies, is rich in gas and dust, and contains the Tarantula Nebula, the most active star-forming region in our Local Group of galaxies. The Tarantula Nebula (also known as 30 Doradus) is an H II region (a region of hydrogen gas excited by the ultraviolet light from very hot young stars) in the Large Magellanic Cloud. It was originally thought to be a star, hence the star name 30 Doradus, but in 1751

Nicolas Louis de Lacaille recognised its true nature. It is easily seen in binoculars which, given its distance of about 170,000 light years, means that it is extremely luminous – so bright that if it were as close to Earth as the Orion Nebula (our nearest H II region) the Tarantula Nebula would cast shadows. It is, in fact, the largest (about 650 light years across) and most active region of star formation that we know of within our Local Group of galaxies. Most of the energy that excites the nebula comes from a compact (35 light years across and of 45,000 solar masses) star cluster at its heart.

Supernova 1987A

Supernova 1987A was first seen in the outskirts of the Tarantula Nebula on 23 February 1987. With a peak visual magnitude of −3, it was bright enough to be visible to the unaided eye and was the closest observed supernova since SN1604, Kepler's supernova, seen in the constellation Ophiuchus. Its brightness peaked in May and slowly declined in the following months, giving modern astronomers their first chance to observe a nearby supernova. It was first discovered by Ian Shelton and Oscar Duhalde at the Las Campanas Observatory in Chile on 24 February 1987. Ian had processed the image he had taken immediately, but an earlier image (the first to show the supernova) taken by Albert Jones in New Zealand was not processed until following morning, so Ian became the discoverer.

The progenitor star was soon identified as Sanduleak −69° 202a, a blue supergiant – this was surprising as it was not then thought that a blue supergiant would produce such a supernova event. It is now thought that the progenitor star was in a binary system, whose stars merged about 20,000 years prior to the explosion, producing the blue supergiant. Most supernovae grow dimmer with the passage of time as they release their energy but, surprisingly, the X-ray and radio emissions from 1987A grew brighter with time as the shock wave from the explosion excited a dense cloud of gas and dust surrounding the star that had been ejected some time earlier.

The distance to SN1987A

Images of SN1987A taken by the Hubble Space Telescope now show three bright rings around it – material from the stellar wind of the progenitor that has been excited by the ultraviolet flash from the supernova explosion. These rings did not 'turn on' until the ultraviolet light was able to reach them several months later, and so give us a measure of their radius in light days. The rings are large enough for their angular size to be measured accurately by the Hubble Space Telescope, with the inner ring shown in Figure 15.8

Figure 15.8 The inner ring around supernova SN1987A. Image: Hubble Space Telescope, STScI, ESA, NASA.

being ~0.8 arcseconds in radius. Knowing the angular size and diameter of the inner ring enables its distance to be calculated by simple geometry, giving about 168,000 light years. These important observations have resulted in an estimate of the distance from Earth to the centre of the LMC of 52.0 ± 1.3 kpc. This has given a new value for setting the zero point of the Cepheid distance scale, which has greatly improved our knowledge of galactic distances and, as a consequence, the value of Hubble's constant – the scale factor of the Universe.

The globular cluster 47 Tucanae

Cluster 47 Tucanae, often just called 47 Tuc, is a globular cluster, 120 light years across, located at a distance of ~16,700 light years in the constellation Tucana. It can be seen with the unaided eye very close (in direction) to the Small Magellanic Cloud and was discovered by Nicolas Louis de Lacaille in 1751. Under very dark skies, it appears roughly the size of the full Moon. Assuming, as indicated above, that Omega Centauri is *not* a globular cluster, it is the brightest globular cluster in the sky and is noted for having a very bright and dense core where the stars are very tightly packed, as seen in Figure 15.9.

Search for planets within 47 Tucanae

The Hubble Space Telescope has recently made a search for large, close-orbiting planets within 47 Tuc by searching for stellar occultations as their transits temporarily block some of the light of their parent stars. The HST observed ~34,000 stars but found no light curves that could be convincingly

Figure 15.9 The globular cluster 47 Tucanae. Image: European Southern Observatory, Wikimedia Commons.

interpreted as being due to a planet occulting a star. Intriguingly, these observations showed that such 'hot Jupiters' must be much less common (at least 10 times) in 47 Tuc than around stars in our own neighbourhood. It could well be that the dense stellar environment is unhealthy for even such close planets, or that planet formation today is very different from the time ~12 billion years ago when 47 Tuc was formed.

Millisecond pulsars in 47 Tucanae

Cluster 47 Tuc is of very great interest to radio astronomers as it contains at least 23 so-called 'millisecond pulsars'. These are pulsars where the passage of a passing star has enabled the neutron star to 'pull' material from the outer envelope of the passing star (or a companion in a binary orbit) onto itself. This also transfers angular momentum, so spinning the pulsar up to give periods in the millisecond range – hence their name. The fastest known pulsar is spinning at just over 700 times per second – with a point on its equator moving at 20% of the speed of light and close the point where it is thought theoretically that the neutron star would break up!

The dense core of 47 Tuc is a perfect home for millisecond pulsars whose periods range from 8 down to 2 milliseconds, so the fastest are spinning on their axis around 500 times per second. If the pulse train derived from a single millisecond pulsar is amplified and applied to a loudspeaker cone, the periods are such that the sound appears as a tone rather than a sequence of regular pulses. One must, however, be cautious to note that it is *not* sound waves that are being received by our radio telescopes! The sounds may be found at www.jb.man.ac.uk/research/pulsar/Education/Sounds/.

Centaurus A

Centaurus A is a lenticular or elliptical galaxy with a superimposed dust lane about 11 million light years away in the constellation Centaurus and is shown in Figure 15.10. The fifth brightest galaxy in the sky, it is exceedingly interesting as it appears to be the result of a merger between two smaller galaxies. First identified as a 'peculiar galaxy' by John Herschel in 1847, it was included in the 1966 *Atlas of Peculiar Galaxies* as one of the best examples of a 'disturbed' galaxy with dust absorption. The Galactic Bulge comprises mainly old red stars, but over a hundred star-formation regions (like the Orion Nebula or the Tarantula Nebula in the Large Magellanic Cloud) have been identified in the disc, giving rise to more recent star formation. It appears that Centaurus A is devouring a smaller spiral galaxy – a process that usually initiates an intense period of star formation.

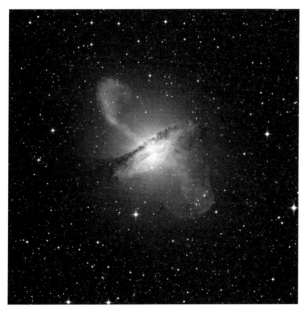

Figure 15.10 Centaurus A. Image: European Southern Observatory, Wikimedia Commons.

These are galaxies where some processes going on within them make them stand out from the normal run of galaxies, particularly in the amount of radio emission that they produce. It was mentioned earlier that, at the heart of our Galaxy, lies a radio source called Sgr A*, one of the strongest radio sources in our Galaxy. However, this would be too weak to be seen if our Milky Way Galaxy was at a great distance and our Galaxy would therefore be termed a 'normal' galaxy. However, there are some galaxies that emit vastly more radio emission and shine like beacons across the Universe. Because most of the excess emission lies in the radio part of the spectrum, these are called radio galaxies. (The name Centaurus A implies that it is the brightest radio galaxy in the constellation.) Other galaxies produce an excess of X-ray emission and, collectively, all are called active galaxies. Though relatively rare, there are obviously energetic processes going on within them that make them interesting objects for astronomers to study. Centaurus A is one of the closest radio galaxies to Earth so has been extensively studied by professional astronomers.

The active galactic nucleus

We believe that the cause of the bright emissions lies right at the heart of the galaxy in what is called an active galactic nucleus – or AGN. It was mentioned earlier that we believe that there is a super-massive black hole at the centre of our Galaxy. We now believe that black holes, containing up to several billion solar masses, exist at the centres of all large elliptical and spiral galaxies. In the great majority of galaxies these are quiescent, but in some, matter is currently (as we observe now) falling into the black hole fuelling the processes that give rise to the X-ray and radio emission. A pair of relativistic jets that are responsible for the emissions at X-ray and radio wavelengths extract energy from the vicinity of the super-massive black hole at the centre of the galaxy. By taking radio observations of the jet separated by a decade, astronomers have determined that the inner parts of the jet are moving at about one-half of the speed of light. X-rays are produced further out as highly energetic particles are created as the jet collides with the gas in the outer parts of the galaxy. Centaurus A is the nearest galaxy to Earth that contains a super-massive black hole actively powering a jet. It is thought that the black hole at its heart has a mass of 100 million solar masses.

Let us consider what happens as a star begins to fall in towards the black hole. As one side will be closer to the black hole than the other, the gravitational pull on that side will be greater than on the further side. This exerts a force, called a tidal force, which increases as the star gets closer to the black hole. The final effect of this tidal force will be to break the star up into its constituent gas and dust. A second thing also happens as the material falls in. It is unlikely that a star would be falling directly towards the black hole and would thus have some rotational motion – that is, it would be circling around the black hole as well as gradually falling towards it. As the material gets closer it has to conserve angular momentum and so speeds up – just like an ice skater bringing her arms in towards herself. The result of the material rotating in close proximity at differing speeds is to produce friction, so generating heat that causes the material to reach temperatures of more than a million degrees K. Such material gives off copious amounts of X-ray radiation, which we can observe, but only if we can see in towards the black hole region. This is surrounded by a torus (or doughnut) of material called the accretion disc, which contains so much dust that it is opaque. But if, by chance, this torus lies roughly at right angles to our line of sight, then we can see in towards the black hole region and observe the X-ray emission.

Nuclear fusion of hydrogen can convert just under 1% of its rest mass into energy. What is less obvious is that the act of falling into a gravitational potential well can also convert mass into energy. In the case of a super-massive black hole, energy equivalent to at least 10% of the mass can be released before it falls within the event horizon, giving the most efficient source of energy that we know of. This energy release often results in the formation of two opposing jets of particles moving away from the black hole along its rotation axis. Moving at speeds close to that of light, these 'bore' a hole through the gas surrounding the galaxy and in doing so the particles will be slowed down – or decelerated. They then produce radiation across the whole electromagnetic spectrum, which allows us to observe the jets. If one of the jets happens to be pointing towards us, the observed emission can be very great and so these objects can be seen right across the Universe.

Three more books about the southern sky:

> *Treasures of the Southern Sky* by Robert Gendler (Springer).
> *A Walk through the Southern Sky* by Milton Heifetz and Wil Tirion (Cambridge University Press).
> *The Southern Sky Guide* by David Ellyard and Wil Tirion (Cambridge University Press).

16

Proving Einstein right

This chapter aims to give an understanding of Einstein's General Theory of Relativity and show how, over nearly 100 years, it has stood up to all the observational tests that have been made.

Einstein's Special Theory of Relativity postulates that nothing can travel *through* space faster than the speed of light, 3×10^5 km/s. The word 'through' has been highlighted as the expansion of space *can* carry matter apart faster than the speed of light. Think of a currant bun: when brought out of the oven one would hope that it would be larger with the currants further apart from each other than before it went in. They have not moved through the dough but have been carried apart by the expansion of the dough. In the same way, the expansion of space can carry objects within it away from each other at speeds faster than that of light.

Perhaps a thought experiment will help to make it clear that, if this is the case, Newton's Law of Gravitation cannot be totally correct. Suppose that the Sun could suddenly cease to exist. Under Newton's theory of gravity, the Earth would instantly fly off at a tangent. Einstein realised that this could not be the case. Not only would we not be aware of the demise of the Sun for 8.32 minutes – the time light takes to travel from the Sun to the Earth – the Earth must continue to feel the gravitational effects of the Sun for just the same time, and would only fly off at a tangent at the moment we ceased to see the Sun. This assumes, of course, that whatever carries the information about the gravitational field of the Sun will also propagate at the speed of light. Thus, something has to propagate through space to carry the information about a change in gravity field. Einstein thus postulated the existence of gravitational waves that would carry such information. As we will see later, the existence of such gravitational waves has already been shown indirectly and it is likely that, before long, direct evidence of their existence will be gained.

In 1915, Einstein published his General Theory of Relativity often called 'general relativity', which is essentially a theory of gravity. Objects in our Universe exist in a four-dimensional space-time continuum which combines the three co-ordinates of space with a fourth co-ordinate, time. For simplicity, in what follows, the term 'space' will be used. In the absence of mass, Einstein's theory predicts that space is 'flat'. This is a rather unfortunate term as it seems to imply a two-dimensional plane surface. In fact, it simply means that light will travel in straight lines, so two initially parallel beams of light remain parallel. In 'flat' space a triangle in any orientation will have inscribed angles that add up to 180 degrees. Euclidian geometry holds true! (Personally, I would like to start a movement to stop calling space 'flat' and use the terms 'Euclidian' or 'zero curvature' space instead!)

If a mass is now introduced into flat space it makes the space positively curved, so that two initially parallel beams of light will converge and the inscribed angles of triangles will add up to more than 180 degrees. A simple, two-dimensional analogy is a flat stretched sheet of rubber. Ball bearings rolled across it will travel in straight lines. If a heavy ball is now placed on the rubber sheet, it will cause a depression and should a ball bearing come close, it will follow a curved path. In just the same way, the space around our Sun is positively curved and the Earth is simply following its natural path through this curved space – there is no force acting on it.

Imagine a spherical world where the inhabitants believe totally (but incorrectly) that the surface is flat. In the region near their north pole the icy surface is virtually frictionless. The inhabitants use a hovercraft-like transport so that, once moving, the craft experience no frictional forces. Two craft, 10 km apart, and at identical distances from the pole, set off at the same time and at the same speed heading due north on parallel paths across the ice. As they believe that the surface is flat, they will expect to remain this distance apart as they travel across the ice. They will thus be somewhat surprised (and possibly hurt) when they collide at the north pole! In order to maintain their belief that the surface of their world is flat they would have to postulate a force, which they might call 'gravity', to explain why their craft were drawn together. In the same way, *we* postulate the force of gravity to explain what *we* observe in the (incorrect) belief that three-dimensional space is flat, not curved, in the vicinity of mass.

Gravity is a force that we 'invent' to explain what we observe happening (such as the planets going around the Sun) in the belief that space is flat when it is, in fact, positively curved.

The first 'test' of Einstein's theory was its application to the orbit of Mercury. As Kepler's first law of planetary motion tells us, the orbit of Mercury should be

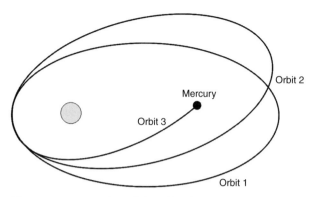

Figure 16.1 Precession of the orbit of Mercury.

an ellipse with the Sun at one focus. The point of closest approach, called its perihelion, would remain fixed in space if the Sun was a sphere and there were no other planets. However, the oblateness of the Sun and perturbations caused by the other planets cause the orbit to 'precess' – think of the patterns produced by a 'spirograph' (Figure 16.1). Accurate observations showed that the observed value of this precession, 5,599.7 arcseconds per century, disagreed with that calculated from Newton's theory by 43.0 arcseconds per century. The application of Einstein's theory provided a correction term of 42.98 ± 0.04 arcseconds – exactly that required to remove the anomaly!

It was realised that Einstein's theory could be tested by observing the positions of stars when viewed close to the Sun. His theory predicted that the positions of the nearest stars would be shifted by just 1.75 arcseconds – close to the limitations in measurement accuracy due to the atmosphere.

Two significant solar eclipses

We cannot usually measure the positions of stars close to the Sun except during a total eclipse of the Sun, and thus the eclipses of 1919 and 1922, which followed the publication of Einstein's theory, played a significant role in the history of science. In essence the plan was simple. Prior to a solar eclipse, images of the sky where the Sun would lie during totality would be taken to give the stars' nominal positions. The same images would then be taken during totality – the only time when stars can be seen close to the Sun's position – and the positions of the stars compared.

Following two expeditions to the eclipse of 29 May 1919, Sir Arthur Eddington proclaimed the eagerly anticipated result to the world. Einstein was right. With hindsight, the confirmation was not as conclusive as Eddington portrayed though, to be fair, he did make all the photographic plates available for anyone else to analyse should they so wish.

Sir Arthur Eddington led the British eclipse expedition to the Atlantic island of Principe, whilst a second set of observations were made from Sobral in Brazil. The telescopes used thus had to be portable and this limited their accuracy. The control images obviously had to be taken at night when it would have been colder than during the day time. Even disregarding these problems, the experiment was not easy. The anticipated deflection of 1.6 arcseconds has to be compared with the typical size of a stellar image as observed from the ground (due to atmospheric turbulence) of 1 to 2 arcseconds. The data from the observations were not quite as conclusive as was implied at the time. The telescope at Principe was used to take 16 plates, but partial cloud reduced their quality. Two usable plates from the telescope on Principe, though of a poor quality, suggested a mean of 1.62 arcseconds.

Two telescopes were used at Sobral (Figure 16.2) where conditions were superb; sadly, however, the focus of the main instrument shifted, probably due to the reduction in temperature during the eclipse, and the stellar images were not clear. They were thus difficult to measure and produced a result of ~0.93 arcseconds. A smaller 10-cm instrument did, however, produce eight clear photographic plates and these showed a mean deviation of 1.98 ± 0.12 arcseconds. If all the data had been included, the results would have been inconclusive, but Eddington, with some justification, discounted the results obtained from the larger Sobral telescope and gave extra weight to the results from Principe (which he had personally recorded). On 6 November that year the Astronomer Royal and the President of the Royal Society declared the evidence was decisively in favour of Einstein's theory. However, there were many scientists who, at the time, felt there were good reasons to doubt whether the observations had been able to accurately test the theory.

A more positive test of the theory came from observations made by William Campbell's team from the Lick Observatory, who viewed the 1922 eclipse from Australia. They determined a stellar displacement of 1.72 ± 0.11 arcseconds. Campbell had believed that Einstein's theories were wrong, but when his experiment proved exactly the opposite, he immediately admitted his error and thereafter supported relativity. (One tends to believe an experiment when the results do not agree with the expectations of the observer!)

Gravitational lenses

If the Sun's mass can produce a small shift in the position of a distant object, so also should the mass of a galaxy. Occasionally, a galaxy will be close to

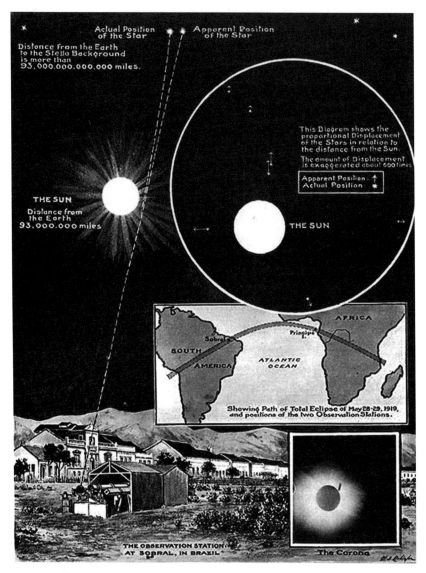

Figure 16.2 The 1919 solar eclipse experiment.

the line of sight of a more distant object. The mass of the galaxy distorts the space around it forming a 'gravitational lens'. Depending on the relative positions, this lens can form multiple images of the distant object or even spread its light or radio emission into an arc or ring – called an Einstein ring.

In 1977, observations using the Lovell Telescope at Jodrell Bank discovered two quasars whose positions were ~6 arcseconds apart and close to that of a foreground galaxy, as seen in Figure 16.3. Quasars are very distant bright radio

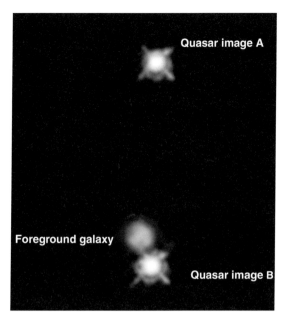

Figure 16.3 Hubble Space Telescope image of the 'Double Quasar'. Image: STScI, ESA, NASA.

sources which appear like stars on photographic plates – hence their full name 'quasi-stellar object', which means 'looking like a star'. Now called the 'Double Quasar', it was soon realised that we were observing two images of the same object as a result of a gravitational lens produced by the foreground galaxy. But there is a subtle difference. The path length through space between us and the quasar is longer for one of the images by a distance of 417 light days. We thus, simultaneously, see it at two times in its existence – separated by 417 days! Time and space *do* interact, showing why space-time is implicit in Einstein's theory. The prediction by Einstein's theory of gravitational lenses was thus proved.

(One might wonder how the time difference has been measured. Quasars are giant galaxies which at their heart have a super-massive black hole. Stellar material falling in towards the black hole provides the energy source of the quasar, and as the rate at which material is consumed varies so does the energy output of the quasar. The effect is that the brightness varies with time. Suppose the brightness of the image of the quasar whose light has travelled the greater distance is seen to increase rapidly by 10%. Then we would see that the brightness of the image whose light has travelled the lesser distance would increase by the same amount some time later, as determined by the difference in path length. By comparing the brightness curves of the two images, a 'match' was found when the time difference was 417 days.)

232   A Journey through the Universe

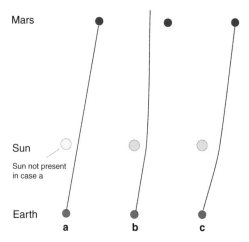

Figure 16.4 Representation of the Shapiro delay.

In the 1960s Irwin A. Shapiro realised that there was another, and potentially far more accurate, way of testing Einstein's theory. Shapiro was a pioneer of radar astronomy and realised that the time that a radar pulse would take to travel to and from a planet would be affected if the pulse passed close to the Sun. In Figure 16.4, (a) shows the direct path that a radar pulse would take to and from Mars if we could imagine that the Sun was not present and that, as a consequence, space was flat. Path (b) on the diagram shows that, due the curvature of space surrounding the Sun, a radar pulse sent along this precise path would curve away to the left and so not reach Mars. The pulse that *would* reach Mars, shown as path (c), has to take a path slightly to the right of its true position, so the curvature of space near the Sun would deflect it towards Mars. The echo would follow exactly the same path in reverse. As the pulse has had to follow a longer route to Mars and back it will obviously take longer than if the Sun was not present. The radar pulse will thus be delayed. The 'Shapiro delay', as it is called, can reach up to 200 microseconds and provided an excellent test of Einstein's theory.

Further tests, of even higher accuracy, using the Shapiro delay have been made by monitoring the signals from spacecraft as the path of the signals passed close to the Sun. In 1979, the Shapiro delay was measured to an accuracy of one part in a thousand using observations of signals transmitted by the Viking spacecraft on Mars. More recently, observations made by Italian scientists using data from NASA's Cassini spacecraft, whilst en route to Saturn in 2002, confirmed Einstein's General Theory of Relativity with a precision 50 times greater than previous measurements. At that time the spacecraft and Earth were on opposite sides of the Sun separated by a distance of more than 1 billion kilometres (~621 million miles). They precisely measured the change in the round-trip travel time of the radio signal as it travelled close to the Sun. A signal was transmitted from the Deep Space Network station in Goldstone, California, which travelled to the spacecraft on the far side of the Sun and there triggered a transmission that returned to Goldstone. Multiple frequency techniques enabled the effects of the solar atmosphere on the signal to be eliminated, so giving a very precise round-trip travel time. The Cassini experiment confirmed Einstein's theory to an accuracy of 20 parts per million.

The Global Positioning System satellite network

Though not specifically a 'test' of Einstein's theories, the Global Positioning System (GPS) network is a beautiful illustration to show that, if Einstein's two theories are not taken into account, the GPS network could not function. GPS essentially works by the transmission of highly accurate timing signals from a constellation of satellites orbiting the Earth. By 'knowing' where the satellites are when they transmit their time signals, a receiver on the ground can calculate the distance from each observed satellite and hence where on the surface of the Earth it must be. The timing signals are derived from hydrogen maser atomic clocks carried in each satellite. They orbit the Earth at a height of ~20,200 km whilst moving at a speed of ~14,000 km/s. Both these statements are significant. Einstein's Special Theory of Relativity shows that a moving clock, when observed from a body at rest, will appear to run slow. The result is that, if the hydrogen maser is set to give precise timing signals on the ground, it will appear to run slow when in orbit by 7.2 microseconds per day. One might thus set the clock to run fast on the ground so that, when in orbit, it runs at the correct rate.

But this would ignore Einstein's General Theory of Relativity. At a height of 20,200 km, the value of the acceleration due to gravity, g, is reduced by a quarter as compared to that measured on the Earth's surface. As a result of what is called the gravitational redshift, clocks run faster in weaker gravitational fields and this effect would make the clocks run fast by 45.9 microseconds per day compared to those on the ground. In order to run at the correct rate in orbit the clocks have to be made to run slow by ~38.7 microseconds per day when calibrated on the ground!

The binary pulsar

The next major advance in testing Einstein's theory came with the discovery, by Russell Taylor and Joseph Hulse in 1974, of the first 'binary pulsar'. Pulsars have been discussed in detail in Chapter 11 and it is the fact that pulsars are such accurate clocks that has made them such valuable tools with which to test Einstein's theory. In the binary pulsar system discovered by Taylor and Hulse, a 1.4 solar mass pulsar is orbiting a companion star of equal mass. It thus comprises two co-rotating stellar mass objects. General relativity predicts that such a system will radiate gravitational waves – ripples in space-time that propagate out through the Universe at the speed of light. Though gravitational wave detectors are now in operation across the globe, this gravitational radiation is far too weak to be directly detected. But there is a consequent effect that *can* be detected. As the binary system is losing energy as the result of its gravitational radiation, the two stars should gradually spiral in towards each other. The fact that one of these objects is a pulsar allows us to very precisely determine the orbital parameters of the system. Precise observations over the 40 years since it was first discovered, shown in Figure 16.5, indicate how the two bodies are

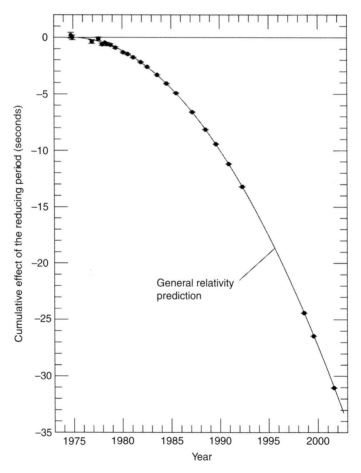

Figure 16.5 Observations of the first 'binary pulsar' by Taylor and Hulse. Image: Wikimedia Commons.

slowly spiralling in towards each other, exactly agreeing with Einstein's predictions! Taylor and Hulse received the Nobel Prize in Physics in 1993 for this outstanding work.

It is another pulsar system, this time where both objects in the system are pulsars, and called the 'Double Pulsar', that has produced one of the most stringent tests of general relativity to date. It was discovered in a survey carried out at the Parkes Telescope in Australia using receivers and data acquisition equipment built at the University of Manchester's Jodrell Bank Observatory. In analysis of the resulting data using a supercomputer at Jodrell Bank, the Double Pulsar was discovered in 2003. It comprises two pulsars of masses 1.25 and 1.34 solar masses spinning with rotation rates of 2.8 seconds and 23 milliseconds respectively. They orbit each other every 2.4 hours with an orbital major axis just less than the diameter of the Sun. The neutron stars are moving at speeds of 0.01% that of light and it is thus a system in which the effects of general relativity are more apparent than any other known system. At this moment in time, general relativity predicts that the two neutron stars should be spiralling in towards each other at a rate of 7 mm per day. Observations made across the world since then, including those using the Lovell Telescope at Jodrell Bank, have shown this to be exactly as predicted.

In fact, five predictions of general relativity can be tested in this unique system. One that has provided high precision is a measurement of the Shapiro delay. By good fortune, the orbital plane of the two pulsars is almost edge on to us. Thus, when one of the two pulsars is furthest away from us its pulses have to pass close to the nearer one on their way to our radio telescopes. They will thus have to travel a longer path through the curved space surrounding the nearer one and suffer a delay that is close to 92 microseconds. The timing measurements agree with theory to an accuracy of 0.05%. Einstein must be at least 99.95% right!

As the two neutron stars are gradually getting closer, at some point in the future they will coalesce to form what may well become a black hole. As they finally merge into one, what can only be called a gravitational wave 'tsunami' will be produced. The predicted strength of the gravitational wave produced by such events is sufficient for it to be detected by the gravitational wave detectors being developed on Earth, and it may not be too long before a gravitational wave is directly detected rather than being inferred.

The way that one could detect such a gravitational wave can perhaps be understood by a 'possible' way to detect a tsunami wave crossing an ocean. Suppose, in a thought experiment, two boats are spaced a kilometre apart and an accurate laser system measures the distance between them. Should a

Figure 16.6 The LIGO gravitational wave detector showing the two arms at right angles. Image: LIGO Observatory.

tsunami wave first reach one of them, the boat will carry out a circular motion as the wave passes beneath, thus making a small momentary change in the two boats' separation, which will be detected by the laser system. Some time later the wave will reach the second boat and the separation will again show a deviation. Note, however, that a tsunami wave coming side-on and reaching both boats simultaneously would not be detected as the boats' motion would be at right angles to the distance being measured. To overcome this one might well have three boats to make a right-angled triangle, and so waves reaching the boats from any angle could be detected.

This is exactly similar to the gravitational wave detectors such as 'LIGO', the Laser Interferometer Gravitational Wave Observatory, in North America, seen in Figure 16.6. LIGO uses a device called a laser interferometer, which measures the time it takes light to travel between suspended mirrors to very high precision. Two mirrors, 4 km apart, form one 'arm' of the interferometer, and two further mirrors make a second arm perpendicular to the first, forming an L shape. Laser light enters the system at the corner of the L and a beam splitter divides the light between the arms. The laser light reflects back and forth between the mirrors repeatedly before it returns to the beam splitter. Any deviations in the path lengths can be measured with extreme precision – movements as small as one thousandth the diameter of a proton can be measured! To achieve this, the mirrors and the light paths between them are housed in one of the world's largest vacuum systems, with a volume of nearly 8,500 cubic metres, evacuated to a pressure of only one-trillionth of an atmosphere. High-precision vibration-isolation systems are needed to shield the suspended mirrors from natural vibrations such as those produced by earth tremors.

To date, no gravitational waves have been directly detected. Gravitational wave detectors will not detect the merging of the two neutron stars in the Double Pulsar for they are predicted to merge in 84 million years' time. However, we believe that such binary systems are common and that such an event should happen on time scales of a few years within this galaxy. By 2015 the sensitivity of the LIGO systems will be greatly enhanced and then events happening over much of our local universe could be detected – the direct detection of gravitational waves cannot be far off.

However, though we are now showing that Einstein's theory holds true to high precision, this cannot be the whole story. One of the most perplexing problems in theoretical physics at the present time is the attempt to harmonise the general theory of relativity, which describes gravitation and applies to the large-scale structure of the Universe (including stars, planets, galaxies), with quantum mechanics, which describes the fundamental forces acting at the atomic scale of matter. It is commonly thought that quantum mechanics and general relativity are irreconcilable, but general relativity can be linked to massless particles called gravitons. There is no proof of their existence, but quantised theories of matter necessitate their existence and they would act as 'messenger particles' carrying information about changes in mass distribution in the same way that the other fundamental forces have messenger particles – for example, photons are the messengers of the electromagnetic force and gluons are the messengers of the strong force (which keeps groups of three quarks bound together to form protons and neutrons).

The graviton is an essential element of much modern theoretical physics and one major effort of the Large Hadron Collider, the world's largest particle accelerator and collider, is to provide evidence for their existence though it will not be able to detect them as such.

One problem is that the force of gravity is $\sim 10^{39}$ times weaker than the other forces that control the Universe. One idea is that gravity may in fact have an intrinsic strength similar to that of the other forces, but appears weaker because it operates in a higher-dimensional space. This provides a link with string theories where there may, in fact, be 11 dimensions in all. Six of these are tightly curled and form the fundamental particles – called strings. The way in which these vibrate defines the type of particle. Four further dimensions are those of space and time, which thus leaves one further dimension. Some think that gravitons can 'leak out' into this hidden dimension so that gravity appears to be far weaker than it actually is.

We have a lot to learn!

Suggestions for further learning:

> *Einstein: His Life and Universe* by Walter Isaacson (Pocket Books).
> *Why Does E = mc²? (And Why Should We Care?)* by Brian Cox and Jeff Forshaw (Da Capo Press).
> *Simply Einstein: Relativity Demystified* by Richard Wolfson (W. W. Norton & Company).
> *What Is Relativity?: An Intuitive Introduction to Einstein's Ideas, and Why They Matter* by Jeffrey Bennett (Columbia University Press).

17
Black holes: no need to be afraid

Black holes seem to have a reputation for travelling through the Galaxy 'hoovering up' stars and planets that stray into their path. It's not really like that. If our Sun were a black hole, we would continue to orbit just as we do now – we just would not have any heat or light. Even if a star were moving towards a massive black hole, it would be far more likely to swing past – just like the fact very few comets hit the Sun but fly past to perhaps return again. So, if you are reassured, then perhaps we can consider . . .

What is a black hole?

If one projected a ball vertically from the equator of the Earth with increasing speed, there would come a point, when the speed reaches 11.2 km/s, when the ball would not fall back to Earth but would escape the Earth's gravitational pull. This is the Earth's escape velocity. If either the density of the Earth was greater (so its mass increased) or its radius smaller (or both) then the escape velocity would increase. Pierre-Simon Laplace realised that, if the mass within a given volume were sufficiently high, the escape velocity would exceed that of light and so not even light could escape. Much later, the eminent American physicist John Wheeler came up with name 'black hole' for such an object.

Suppose that the density of the Earth was vastly increased but it was able to retain its present size so the escape velocity just reached the speed of light at the surface and so it became a black hole. The surface of the Earth would then become what is termed the 'event horizon' of the black hole. In fact, of course, the immense gravity would crush the Earth's material into what is often referred to as a singularity of zero size and infinite density at its centre. However, the event horizon would remain exactly the same size. The fundamental point is that the diameter of the event horizon bears no relationship to the size of the matter forming the black hole, only its mass. As we will see, we believe that virtually all (if not all) of the interior of a black hole is empty space.

Black holes have no specifically defined size or mass. Until recently we had only found evidence for black holes in two circumstances. The first, with masses of many millions of solar masses, are found at the heart of galaxies and are called 'super-massive black holes', whilst the second are believed to result from the collapse of giant stars of perhaps 20 solar masses or more whose stellar cores have a mass exceeding ~4 solar masses. This is the point at which we believe that neutron degeneracy pressure (which allows stellar cores in the range 1.4 to 3 solar masses to form neutron stars) can no longer prevent gravitational collapse. More recently, evidence has been building for what are called 'intermediate mass black holes' having a mass of perhaps 40,000 solar masses, which have been found at the centre of what have in the past been classified as globular clusters. Omega Centauri is one such cluster, but the evidence of a black hole coupled with the fact that it contains many more young stars than globular clusters implies that, instead, it might be the remnant core of a dwarf galaxy whose outer stars have been stripped off by the gravitational effects of our own galaxy.

A black hole should perhaps be better thought of in terms of Einstein's General Theory of Relativity. This states that a massive body distorts the space around it making it curved so that light, for example, no longer travels in straight lines but curves round the location of the mass. A black hole is simply when the mass is so great that space curves back on itself and so light can no longer escape.

A Schwarzschild black hole

In the simplest case in which the stellar remnant is not rotating, the event horizon has a radius, called the 'Schwarzschild radius', which is directly proportional to the mass.

The interior of an event horizon is forever hidden from us, but Einstein's theories predict that at the centre of a non-rotating black hole is a singularity, a point of zero volume and infinite density where all of the black hole's mass is located and where space-time is infinitely curved. I do not like singularities; in Einstein's view they are where the laws of physics are inadequate to describe what exists. We know that somehow Einstein's classical theories of gravity must be combined with quantum theory, and so relativity can almost certainly *not* predict what happens at the heart of a black hole. We have the name for the relevant theory – quantum gravity – but, as yet, it has to be formulated.

Particle physics tells us that nucleons are thought to be composed of up quarks and down quarks. It is possible that at densities greater than those that can be supported by neutron degeneracy pressure, quark matter could occur – quark-degenerate matter *may* occur in the cores of neutron stars and *may* also occur in hypothetical quark stars. Whether quark-degenerate matter can exist in these situations depends on the, poorly known, equations of state of both neutron-degenerate matter and quark-degenerate matter. Some theoreticians even believe that quarks might themselves be composed of more fundamental particles called 'preons' and, if so, preon-degenerate matter might occur at densities greater than that which can be supported by quark-degenerate matter. Could it be that the matter at the heart of a black hole is of one of these forms?

Let us just suppose that the matter at the heart of a 10 solar mass black hole is in the form of quark-degenerate matter. How big might it be? Neutrons have a diameter of ~10^{-15} m and it is suspected that quarks have a diameter no larger than ~10^{-18} m. So let us suppose that the quark's diameter is 10^{-18} m and so is 1,000 times less than the neutron's diameter. The diameter of a 1.4 solar mass neutron star is ~20 km. As the volume goes as the cube of the diameter, the volume of a quark mass of 1.4 solar masses would be $(10^3)^3$ smaller if there were one quark in the neutron, that is 10^9 times smaller or 20,000/1,000,000,000 m in diameter, 0.00002 m or 0.02 mm! As there are three quarks for each neutron, this has to be increased somewhat and, as the black hole would be about seven times more massive, the quark mass would be about two times larger or ~0.04 mm. That is pretty small, but it is not infinitely small. However if, as some think, quarks are point-like and have no size, then who is to say that there is not a singularity.

The more massive a black hole, the greater the size of the Schwarzschild radius: a black hole with a mass 10 times greater than another will have a radius 10 times as large. A typical 10 solar mass stellar black hole would have an event horizon whose radius was 30 km.

Kerr black holes

There is a theorem, called the 'no-hair' theorem, that postulates that all black hole solutions of the Einstein–Maxwell equations of gravitation and electromagnetism are completely characterised by only three observable properties; their mass, electric charge and angular momentum. Once matter has fallen into the event horizon, all other information (the word 'hair' is a metaphor for this) about the matter 'disappears' and is permanently lost to external observers. (This is somewhat contentious, as the theorem violates the principle that if complete information about a physical system is known at one point in time then it should be possible to determine its state at any other time.)

On the large scale matter is neutral, so it is not thought that black holes carry an electromagnetic charge but, on the other hand, the stars, dust and gas that might go to form a black hole have angular momentum – rotational energy – such as has a spinning star. Thus, in general black holes are thought to be spinning. This makes them both much more interesting, but at the same time far more complex! The solutions for a rotating black hole were first solved by Roy Kerr in 1963 and are thus called 'Kerr black holes'. The vast majority of black holes in the Universe are initially thought to be of this type, but there is a mechanism named after Roger Penrose that theoretically allows spinning black holes to lose angular momentum and so they might eventually turn into Schwarzschild black holes.

Like a Schwarzschild black hole, there is a singularity at the heart of a Kerr black hole surrounded by an event horizon, but beyond this is an egg-shaped region of distorted space called the ergosphere (Figure 17.1) caused by the spinning of the black hole, which 'drags' the space around it. (This is called frame dragging and gives a way of observing that a black hole is rotating.) The boundary of the ergosphere and the normal space beyond is called the static limit. An object within the ergosphere can gain energy from the hole's rotation and be ejected, so removing angular momentum from the black hole by what is called the Penrose process. But, of course, if an object crosses the event horizon it can never escape.

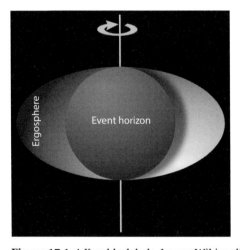

Figure 17.1 A Kerr black hole. Image: Wikimedia Commons.

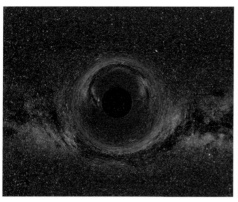 

Figure 17.2 Simulations of a nearby black hole between us and the Milky Way.

How can we discover them?

Could we see them?

Though no one has ever seen a black hole directly, it is, at least in principle, possible to 'see' one if one could get near enough. This is due to the fact that its mass distorts the space-time around it forming a gravitational lens which distorts what we see beyond it. The 'lens' acts rather like the base of a wine glass or the 'bubble' glass in the window of an old cottage and tends to convert point sources into arcs or even circles, which are called Einstein rings. The two images in Figure 17.2 indicate what one might observe if one was near enough to a black hole lying between us and the Milky Way. As the distorted image of the Milky Way surrounds a totally black circle, it could be said that we are 'seeing it', though of course we only see its silhouette.

By gravitational microlensing

The gravitational lensing effect of a large mass gives us a second way in which a black hole could be detected. As it acts rather like a convex lens or magnifying glass it will brighten the image of a star that lay behind it. This has been observed thousands of times when a nearby star passes in front of a distant one. In what is called a gravitational microlensing event, the brightness of the lensed star can increase by many times. (Should the foreground star have a planet in orbit around it that also passes in front of the distant star this can also cause it to brighten, so providing a method of detecting extra-solar planets.) If, as might be expected, there were black hole stellar remnants orbiting within the Galaxy, then these would also give rise to lensing events so enabling their presence to be detected.

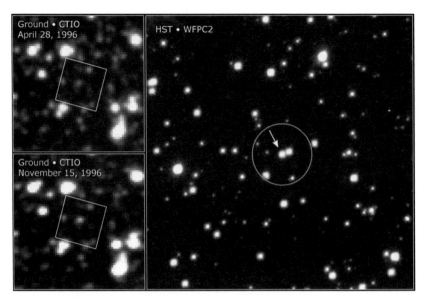

Figure 17.3 Observations of a microlensing event. Images: CTIO and STScI, ESA, NASA.

On the left-hand side of Figure 17.3 are two images of a star field observed with a ground-based telescope that show the brightening of a star due to a gravitational microlensing event. On the right is a Hubble Space Telescope image of the same field that clearly resolves the lensed star and so determines its true brightness. From the increase in brightness it is possible to calculate that the mass of the foreground object must be at least 6 solar masses. If it were a star, it would be visible and outshine the background star. As no foreground star is seen, one deduces that the lensing mass must be a black hole.

By determining the mass of an unseen stellar companion

If a stellar mass black hole, formed when a massive star ends its life in a supernova explosion, existed in isolation, it would be very difficult to detect except by gravitational microlensing as described above. However, many stars exist in binary systems. In a binary system in which one of the components is a black hole, its gravity can pull matter off the companion star forming an accretion disc of gas swirling into the black hole. As the gas spins up as it nears the black hole due to conservation of angular momentum, the differential rotation speeds give rise to friction and the matter in the accretion disc reaches temperatures of more than 1 million K. It thus emits radiation, mostly in the X-ray part of the spectrum. X-ray telescopes have now detected many such X-ray binary systems, some of which are believed to contain a black hole.

If the unseen companion object exceeds a calculated mass of ~4 solar masses, then it is likely to be a black hole. An excellent candidate in our own galaxy is Cygnus X-1 – so called because it was the first X-ray source to be discovered in the constellation Cygnus – and it is the brightest persistent source of high-energy X-rays in the sky. Usually called Cyg X-1, it is a binary star system that contains a supergiant star with a surface temperature of 31,000 K (with its spectral type lying on the O and B boundary) together with a compact object. The mass of the supergiant is between 20 and 40 solar masses and observations of its orbital parameters imply a companion of 8.7 solar masses. This is well above the 4 solar mass limit of a neutron star, so it is thought to be a black hole.

There follows a case study of the discovery of one of the best black hole candidates in our Galaxy and in which I had a small role.

In the summer of 1975, an X-ray satellite called Ariel IV, built and operated by Leicester University, detected one of the strongest sources of X-rays that had ever been observed. It lay in the constellation Monoceros, over to the left of Orion. Unfortunately, in those days, X-ray satellites were not able to give accurate positions to allow the star system that gave rise to the flare to be determined and thus enable follow-up observations at other wavelengths.

I was contacted by Professor Ken Pounds who asked if the Jodrell Bank telescopes could be used in an attempt to locate the source of the X-ray flare as it was likely that it would be producing significant radio emission as well. I immediately used the Mark II telescope to observe the region of sky indicated by the Ariel IV observations, but no bright radio source was observed. It appears that no other optical or radio telescopes were able to detect the source of the X-ray emission at this time either.

Unfortunately, a second telescope that could be combined with the Mark II to make a superb survey instrument was in use by some guest observers, but a week later we regained its use and were able to make a sensitive sky survey centred on the nominal position. In those days paper charts were used and the sky map produced stretched some 20 feet (6 m) across. By great luck, we had surveyed just sufficient sky to find the radio waves being emitted by the flaring object: it was within four inches of the edge of our chart – far from the nominal position. This was an exciting day! The object's precise position and our follow-up radio observations were published in the journal *Nature* adjacent to the original X-ray results. It has two names, Monoceros X-1 and A0620−00, the latter name being a shortened version of our position: Right Ascension of 06 hours, 22 minutes, 44.5 seconds and Declination of −00 degrees, 20 arcminutes, 45 arcseconds.

Extensive observations during the 1980s established that this was arguably the best stellar mass black hole candidate yet discovered; the X-ray flare having been caused by matter infalling from a companion star into an accretion disc surrounding the black hole. A0620−00 had previously flared in 1917, since when the density had slowly built up until it became unstable and exploded releasing a massive burst of X-rays. The flare observed in 1975 is the most intense ever observed. So why do we believe that a black hole was the cause of these X-ray outbursts? At the position we derived is seen a K-type star. These have typical masses of 0.5 to 0.8 solar masses. From Doppler measurements of its spectrum it is possible to deduce that it is orbiting an unseen object with a period of 0.32 days with an orbital velocity of 460 km/s. A pair of such measurements may be familiar as it is the way a planet can be inferred to be orbiting a star and which then allows one to calculate its distance from the star and its minimum mass. (This technique is described in Chapter 12.) In this case, when there are two massive co-orbiting objects, the observed motion only allows us to calculate their combined mass, which is ~10.5 solar masses. So, taking the maximum mass of 0.8 solar mass for the K-type star, this implies that the mass of the unseen companion must be ~9.7 solar masses. This is well above the minimum mass of a stellar-derived black hole. Lying at a distance of 3,500 light years, it is thought to be the nearest black hole to our Solar System.

Super-massive black holes

We now believe that at the centre of all large elliptical and spiral galaxies there exist black holes of vastly greater mass than those resulting from the evolution of individual stars – with the most massive thought to contain several billion solar masses. Initially, the evidence for them was indirect as they were believed to provide the energy to power what are called 'active galaxies'. These are galaxies where some processes going on within them make them stand out from the normal run of galaxies, particularly in the amount of radio emission that they produce. At the heart of our Galaxy lies a radio source called Sgr A*, one of the strongest radio sources in our Galaxy. However, this would be too weak to be seen at if our Milky Way Galaxy was at a great distance, and our Galaxy would therefore be termed a 'normal' galaxy. However, there are some galaxies that emit vastly more radio emission and shine like beacons across the Universe. Because most of the excess emission lies in the radio part of the spectrum, these are called 'radio galaxies'. Other galaxies produce an excess of X-ray emission and, collectively, all are called active galaxies. Though relatively rare, there are obviously energetic processes going on within them that make them interesting objects for astronomers to study.

We believe that the cause of their bright emissions lies right at their heart in what is called an 'active galactic nucleus', or AGN, where matter is currently falling into the black hole fuelling the processes that give rise to the X-ray and radio emissions.

These highly luminous objects were first discovered by radio astronomers in a series of experiments to measure the angular sizes of radio sources. In the early 1960s, the signals received with the 75-metre Mark I radio telescope at Jodrell Bank were combined with those from a smaller telescope located at increasingly greater distances across the north of England. It was discovered that a number of the most powerful radio sources had angular sizes of less than 1 arcsecond. So small, in fact, that they would appear as 'stars' on a photographic plate. They were thus given the name 'quasi-stellar object' (looking like a star) or 'quasar' for short. This meant that they were very hard to identify until their precise positions were known. The first quasar to be identified was the 273rd object in the Third Cambridge Catalogue of Radio Sources so it had the name 3C 273.

Though its image, taken by the 200-inch Hale Telescope, looked very like a star, a jet was seen extending ~6 arcseconds to one side. In the highly unusual spectrum of the object, Maarten Schmidt was able to find hydrogen emission lines that were shifted by ~16% to the red – vastly more than in any previously observed object. This implied that its distance was about 2,500 million light years – then the most distant object known in the Universe. But 3C 273 is one of the closer quasars to us and the most distant currently known lies at a distance of ~13 billion light years. So quasars are some of the most distant and most luminous objects that can be observed in the Universe. One reason that they appear so bright is that much of the emission is concentrated in two opposing beams and we tend to see those where the beams are pointing towards us – just as if the light from a 6-volt bulb is concentrated in a beam by a torch we could see it from far greater distances providing that the beam were pointed towards us. Even so, the energy outputs were greater than could reasonably be provided by nuclear fusion.

Nuclear fusion of hydrogen can convert just under 1% of its rest mass into energy. What is less obvious is that the act of falling into a gravitational potential well can also convert mass into energy. In the case of a supermassive black hole, energy equivalent to at least 10% of the mass can be released before it falls within the event horizon, giving the most efficient source of energy that we know of. If, as we suspect is usual, the black hole is rotating, the conversion efficiency can rise towards 30%. This energy release often results in the formation of two opposing jets of particles moving away from the black hole along its rotation axis. Moving at speeds close to that of light, these 'bore' a hole through the gas surrounding the galaxy and in doing so the particles will be slowed down – or decelerated. They then produce radiation across the whole electromagnetic spectrum that allows us to observe the jets. If one of the jets happens to be pointing towards us, the observed emission can be very great, and so these objects can be seen right across the Universe.

One can calculate the mass that would be needed to power a bright quasar with luminosity of order 10^{41} watts. Assuming that just 10% of the mass is converted into energy, such a black hole would need to 'consume' about 175 solar masses per year.

The active galaxy NGC 4261

In the section 'The active galactic nucleus' in Chapter 15, what happens as a star begins to fall in towards a black hole is described. Copious amounts of X-ray radiation are given off, but can only be seen if the accretion disc is roughly at right angles to the line of sight. This is the case in the active galaxy NGC 4261. The Hubble Space Telescope image on the right of Figure 17.4 shows a giant disc of cold gas and dust, about 300 light years across, that fuels a possible black hole at the core of the galaxy. This disc feeds matter into the black hole, where gravity compresses and heats the material as described above. Particles accelerated from the vicinity of the black hole produce two opposed jets of particles which, as they are decelerated, give off radio emission to form the two radio 'lobes' that we observe. The jets are aligned perpendicular to the disc, like an axle through a wheel – exactly as we would expect if a rotating black hole forms the 'central engine' in NGC 4261.

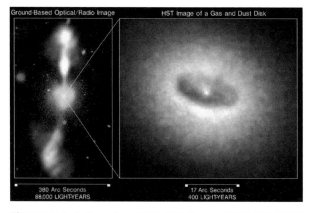

Figure 17.4 Active galaxy NGC 4261. Image: STScI, ESA, NASA.

The size of the active galactic nucleus

There is a simple observation of quasars that can give us an indication of the size of the emitting region around the black hole. It has been observed that the light and radio output of a quasar can change significantly over periods of just a few hours. Perhaps surprisingly, this can provide a reasonable estimate of its size as the following 'thought experiment' will show.

Suppose the Sun surface instantly became dark. We would see no change for 8.32 minutes due to the light travel time from the Sun to the Earth. Then we would first see the central region of the disc go dark as this is nearest to us and the light travel time from it is least. This dark region would then be seen to expand to cover the whole of the Sun's visible surface. This is because the light from regions of the Sun further from us would still be arriving after the light from the central region was extinguished. As the Sun has a radius of 695,000 km, the time for the change to occur would be given by this distance divided by the speed of light and this gives 2.31 seconds. It is thus apparent that a body cannot appear to instantaneously change its brightness and can only do so on time scales of order of the light travel time across the radiating body.

Suppose that an AGN is observed to significantly change its brightness over a period of 12 hours, that is 720 minutes. As light can travel 1 AU in 8.32 minutes (the light travel time from the Sun to the Earth) the scale size of the object must be of order 720/8.32 AU or ~86 AU.

Calculating the mass of super-massive black holes

In recent years astronomers have gained more direct evidence of the presence of super-massive black holes by measuring the speed at which stars or dust are rotating around the centre of the galaxies in which they reside. Two examples follow: firstly, the galaxy M84, which lies in the Virgo Cluster some 50 million light years distant and, secondly, our own Milky Way Galaxy.

The galaxy M84

At the right of Figure 17.5 is an observation taken with the Hubble Space Telescope Imaging Spectrograph of a strip across the centre of the galaxy M84 as shown in the image on the left.

The right-hand plot shows the Doppler shift in the spectra of the material rotating around the galactic centre. Moving down towards the centre, there is a

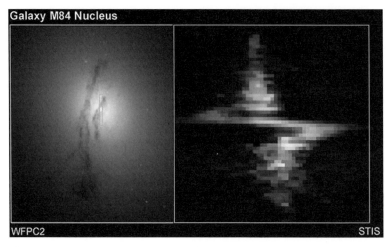

Figure 17.5 Hubble Space Telescope observations of the active galaxy M84. Image: STScI, ESA, NASA.

sudden blueshift, indicating rapid motion of the gas towards us. The Doppler shift indicates that the velocity towards us reaches a speed of about 400 km/s at a distance only 26 light years from the centre. Crossing the centre, the sign of the radial velocity rapidly reverses to give a redshift, indicating a similar speed away from us.

The most obvious interpretation of these data is that there is a large rotating disc around the nucleus of M84 that is seen in cross section – an interpretation strengthened by the fact that its nucleus is very active and emits jets of particles that give rise to strong radio emission.

The observations allow us to compute the mass of the central region in exactly the same way that we can calculate the mass of the Sun. From a knowledge of how fast the gas is rotating around the black hole and its distance away from it, along with an experimentally derived value of the universal constant of gravitation, G, we get a mass of ~300 million solar masses. It is expected that most of this mass will be in a black hole near the centre of the galaxy. So, given the mass, one can then compute the radius of the event horizon and this turns out to be 8.8×10^8 km – somewhat less than the size of the orbit of the planet Venus.

The Milky Way Galaxy

At the centre of our Galaxy is a strong radio source called Sagittarius A*. As we know that the regions surrounding super-massive black holes tend to emit strongly in the radio part of the spectrum, it has long been thought that such a black hole lay at the Galaxy's heart. In the infrared it is possible to eliminate the effects of the Earth's atmosphere and produce images that are limited only by the diameter of the telescope. In the case of the 10-metre Keck Telescopes on the peak of Mauna Kia in Big Island, Hawaii, this is ~1/25 arcsecond, roughly equivalent to that of the Hubble Space Telescope in the visible part of the spectrum. This resolution has allowed individual stars near the centre of the Galaxy to be imaged and, with observations taken over a period of 15 years, has enabled the orbits of a number of stars to be determined. As we have seen above, a knowledge of the periods and size of the stellar orbits enables the mass of the body that they are orbiting to be found and this is exactly what the team from UCLA (University of California, Los Angeles) have achieved.

Figure 17.6 shows the central, 1 arcsecond by 1 arcsecond region of our Galaxy as observed in the infrared by the Keck Telescope in 2010. Plotted on top of the image are the 15-year tracks of seven stars that are orbiting the Galactic Centre. One star, S0–2, orbits with a period of just 15.78 years and a second, S0–16, went to within 45 AU of the Galactic Centre, at which time it was moving at 12,000 km/s! The best fit to the data gives a mass for the central body of 4.1 million solar masses and is the best evidence yet for the presence of a super-massive black hole at the centre of our own galaxy.

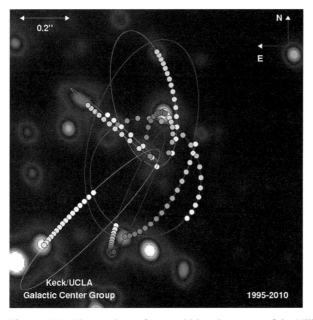

Figure 17.6 Observations of stars orbiting the centre of the Milky Way Galaxy. Image: Keck/UCLA Galactic Center Group.

Black holes are not entirely black

In the 1970s, Stephen Hawking showed that, due to quantum-mechanical effects, black holes can actually emit radiation – they are not entirely black! The energy that produces the radiation in the way described below comes from the mass of the black hole. Consequently, the black hole gradually loses mass and, perhaps surprisingly, the rate of radiation increases as the mass decreases, so the black hole continues to radiate with increasing intensity, losing mass as it does so until it finally evaporates.

The theory describing why this happens is highly complex and results from the quantum-mechanical concept of virtual particles – mass and energy can arise spontaneously provided they disappear again very quickly and so Heisenberg's uncertainty principle is not violated. In what are called 'vacuum fluctuations', a particle and an antiparticle can appear out of nowhere, exist for a very short time, and then annihilate each other. These could, for example, be two photons of opposite spin. Should this happen very close to the event horizon of a black hole, it is possible for one particle to fall across the horizon into the black hole whilst the other escapes. In order to preserve total energy, the particle that fell into the black hole must have had a negative energy – equivalent to negative mass – which thus reduces the mass of the black hole. The particle that escapes carries energy away from the black hole and can, in principle, be detected, so that it appears as if the black hole was emitting radiation. This radiation is called 'Hawking radiation'.

Black holes can be said to have an effective temperature, called the 'Hawking temperature', which is proportional to the surface gravity of the black hole. It turns out that the more massive the black hole, and hence the radius of its event horizon, the lower the surface gravity and the lower the effective temperature. Even for a stellar mass black hole this is exceedingly small, the order of 100 nanokelvin (10^{-7} K) and, as the effective temperature is inversely proportional to mass, vastly less for super-massive black holes.

Consider an object placed in a bath of radiation at a certain temperature – say in a room at 20 °C (293 K). Only if the object is hotter than this can it lose heat by radiation; if cooler it will absorb radiation and warm up. Black holes exist in a Universe whose space is now at an effective temperature of ~2.7 K due to the Cosmic Microwave Background (CMB) – the afterglow of creation left by the annihilation of matter and antimatter particles at the time of the Big Bang and discussed in Chapter 22. Unless a black hole has an effective temperature greater than this it cannot evaporate and will, in fact, gain energy and hence mass from the CMB photons that fall into it. It will thus grow with time rather than shrink. A temperature of 2.7 K is vastly higher than the effective temperatures of even solar mass black holes, so at this time in the Universe none of the black holes that we know of can be evaporating. To have a Hawking temperature higher than 2.7 K and so be able to evaporate, a black hole needs to be lighter than the Moon and would be an object with a diameter of less than a tenth of a millimetre.

Once a black hole begins to evaporate it loses mass and hence size. Its Hawking temperature increases so it begins to radiate more strongly and so lose mass more quickly. This is a runaway process, so a black hole will finally disappear in a blinding flash of radiation.

On the other hand, small black holes, should they exist, would evaporate in an instant. If there have ever been (perhaps at the time of the Big Bang) black holes whose mass was comparable to that of a car (which would have a diameter of ~10^{-24} m) they would evaporate in the order of a nanosecond, during which time they would outshine more than 200 of our Suns!

It is just possible that tiny black holes might be created in the collisions of particles in the Large Hadron Collider at CERN. Some have worried that these might grow and consume the world. But we believe that these would evaporate essentially as soon as they were created – on a time scale of 10^{-18} seconds! It is thought that very high energy gamma rays will have created vast numbers of such micro black holes during the Earth's history but, encouragingly, we are still here.

As will be described in Chapter 21, we now believe that the Universe is expanding at an ever increasing rate due to the pressure derived from the 'dark energy' that appears to make up 73% of the mass/energy of the Universe. As the temperature of the CMB scales inversely with size, it is falling at an increasing rate too. Eventually, in aeons, when the temperature of this relict radiation has fallen sufficiently and assuming Hawking's theory is correct, stellar mass black holes will finally begin to evaporate – on a time scale of 10^{100} years!

Might black holes be evaporating now?

Let us suppose that, at the time of the Big Bang, black holes in a range of masses (and so effective temperatures) were created. As the Universe cooled, the lighter ones could begin to evaporate followed by the heavier ones and now, as mentioned above, only those whose mass is comparable to that of the Moon. The final moments of such an evaporation would give rise to massive bursts of gamma rays, which our satellites could detect. In fact, we do observe what are termed gamma-ray bursts, but it is thought (as will be described in Chapter 20) that these are caused by the coalescing of two stellar remnants to form a black hole. The fact that we cannot show that they are caused by the evaporation of a black hole is regarded as a pity by Stephen Hawking for, if they were, he would almost certainly win the Nobel Prize!

More on black holes:

> *Black Holes, Wormholes and Time Machines* by Jim Al-Khalili (Institute of Physics).
> *Black Holes: A Very Short Introduction* by Katherine Blundell (Oxford University Press).
> *Black Holes and Beyond: An Introduction* by Werner Brueckner (CreateSpace).

18

It's about time

As the title of this chapter implies, it is about the enigmatic concept of time – perhaps the least understood topic in physics. It is almost impossible to give a definition of time, and perhaps the most understandable was provided by the physicist John Wheeler, whose definition was 'Time is nature's way of preventing everything happening at once'!

Let us first discuss how astronomers measured the passage of time until the 1960s.

Local Solar Time

For centuries, the time of day was directly linked to the Sun's passage across the sky, with 24 hours being the time between one transit of the Sun across the meridian (the line across the sky from north to south) and that on the following day. This time standard is called 'Local Solar Time' and is the time indicated on a sundial. The time such clocks showed would thus vary across the UK, as noon is later in the west. It is surprising the difference this makes. In total, the UK stretches 9.55 degrees in longitude from Lowestoft in the east to Manger Beg in County Fermanagh, Northern Ireland, in the west. As 15 degrees is equivalent to 1 hour, this is a time difference of just over 38 minutes.

Greenwich Mean Time

As the railways progressed across the UK, this difference became an embarrassment, and so London or 'Greenwich' time was applied across the whole of the UK. A further problem had become apparent as clocks became more accurate: due to the fact that, as the Earth's orbit is elliptical, the length of

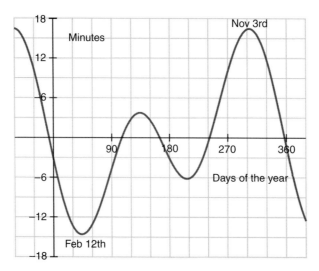

Figure 18.1 The Equation of Time: the difference between GMT and Local Solar Time at Greenwich Observatory.

the day varies slightly. Thus 24 hours, as measured by clocks, was defined to be the *average* length of the day over one year. This time standard became known as 'Greenwich Mean Time' (GMT).

The Equation of Time

The use of GMT has the consequence that, during the year, our clocks get in and out of step with the Sun. The difference between GMT and the Local Solar Time at Greenwich is called the 'Equation of Time' (Figure 18.1). The result is that the Sun is not always due south at noon (even in London) and the Sun can transit (cross the meridian) up to 16 minutes 33 seconds before noon and up to 14 minutes 6 seconds afterwards as measured by a clock giving GMT. This means that sunrise and sunset are not usually symmetrically centred on midday and this does give a noticeable effect around Christmas time. Though the shortest day is on 21 December, the Winter Solstice, the earliest sunset is around 10 December and the latest sunrise does not occur until around 2 January, so the mornings continue to get darker for a couple of weeks after 21 December whilst, by the beginning of January, the evenings are appreciably lighter.

Universal Time

Greenwich Mean Time was formally replaced by Universal Time (UT) in 1928 (though the title has not yet come into common usage) but UT was essentially the same as GMT until 1967 when the definition of the second was changed. Prior to this, one second was defined as one 86,400th of a mean day as determined by the rotation of the Earth. The rotation rate of the Earth was thus our fundamental time standard. The problem with this definition is that, due to the tidal forces of the Moon, the Earth's rotation rate is gradually slowing and, as a consequence, the length of time defined by the second was increasing. This was not something that physicists were happy with so, in 1967, a new definition of the second was made:

> **The second is the duration of 9,192,631,770 periods of the radiation corresponding to the transition between the two hyperfine levels of the ground state of the caesium-133 atom.**

Thus our clocks are now related to an atomic time standard that uses caesium beam frequency standards (as described later) to determine the length of the second.

But this has not stopped the Earth's rotation from slowing down, and so very gradually the synchronisation between the Sun's position in the sky and our clocks will be lost. To overcome this, when the difference between the time measured by the atomic clocks and by the Sun (as determined by the Earth's rotation rate) differs by around a second, a leap second is inserted to bring solar and atomic time back in step. This is usually done at midnight on New Year's Eve or 30 June. Since the time definition was changed, 22 leap seconds have had to be added, about one every 18 months, but there were none between 1998 and 2005, showing that the slowdown is not particularly regular. Leap seconds are somewhat of a nuisance for systems such as the Global Positioning System (GPS) network and there is pressure to do away with them, which is, not surprisingly, opposed by astronomers. If no correction were made and the average slowdown over the last 40 years of 0.56 of a second per year continues then, in 1,000 years, UT and solar time will have drifted apart by ~9 minutes.

Sidereal time

If one started an electronic stopwatch running on UT as the star Rigel, in Orion, was seen to cross the meridian and stopped it the following night when it again crossed the meridian, it would be found to read 23 hours, 56 minutes and 4.09 seconds, not 24 hours. This period is called the 'sidereal day' and is the length of the day as measured with respect to the apparent rotation of the stars.

Why does the sidereal day have this value? Imagine that the Earth was not rotating around its axis and we could observe from the dark side of the Earth facing away from the Sun. At some point in time we would see Rigel due south.

As the Earth moved around the Sun, Rigel would be seen to move towards the west and, three months later, would set from view. Six months after setting in the west, it would be seen to rise in the east, and precisely one year later we would see it due south again. So, in the absence of the Earth's rotation, Rigel would appear to make one rotation of the Earth in one year and so the sidereal day would be one Earth year. But, in reality, during this time, the Earth has made ~365 rotations so, in relation to the star Rigel (or any other star), the Earth has made a total of ~(365+1) rotations in one year and hence there are ~366 sidereal days in one year. The sidereal day is thus a little shorter and is approximately 365/366 of an Earth day.

The difference would be ~1/366 of a day or 1440/366 minutes, giving 3.93 minutes or 3 minutes 55.8 seconds. The length of the sidereal day given by this simplified calculation is thus approximately 23 hours, 56 minutes and 4.2 seconds, very close to the actual value.

Clocks

Sundials

These are perhaps the most fundamental clocks of all and keep, by definition, Local Solar Time. They are not, unfortunately, as useful in the UK as in more southerly countries. They are made in many forms: some with a horizontal flat face with gnomon pointing up to the pole star, some with vertical faces on the sides of buildings, and some where a band stretches around the gnomon. There is quite an art, and some mathematics, employed in their design, with a particularly interesting one, where a person acts as the vertical gnomon, called an analemmatic sundial.

In 1906, George James Gibbs invented what he called a 'Helio-chronometer' (Figure 18.2). Adjusted at the Pilkington–Gibbs factory for the latitude and longitude of the purchaser, it had a rotating disc on which the gnomon was located. By setting this to the day's date, the Equation of Time would be compensated for, so allowing GMT to be determined to better than a minute of time.

Water clocks

These are rather fun, and until the invention of the pendulum clock were the most accurate clocks. The simplest just filled a cylinder of constant cross section from a steady supply of water. A float might then rise with a toothed vertical arm mounted on it to rotate the hour hand of a clock face. As in all water clocks, the key to accuracy was to have a constant 'head' of water so that the water flow into the clock was constant. A good way to achieve this was to allow more water than required for the clock itself to pass into a reservoir,

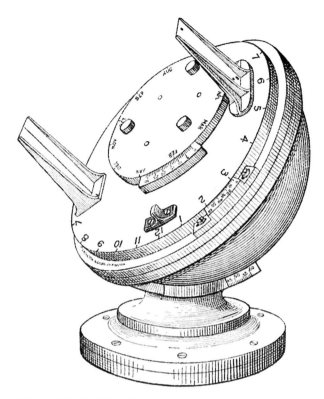

Figure 18.2 The Gibbs Helio-chronometer.

which thus continuously overflows and so is kept brim full. An exit pipe at some distance below the surface will thus have a constant head of water above it and so be at a fixed pressure. Later, water clocks were designed to provide a mechanical system to move an hour hand (and perhaps a minute hand) around a dial, and some were based on a water wheel, which would rotate at a constant speed to drive the hands through a series of gears.

Pendulum clocks

The idea of using a pendulum to keep time is attributed to Galileo who, as a student in 1602, watched a suspended lamp swing back and forth in the cathedral of Pisa. Galileo discovered that the period of swing of a pendulum (at least for relatively small swings) is independent of its amplitude – the so-called 'isochronism' of the pendulum. In 1603 a friend of his began to use a short pendulum to measure the pulse rate of his patients.

In 1641, at the age of 77 and totally blind, Galileo, aided by his son, turned his attention to using a pendulum to construct a clock but, although drawings were made and a clock was partly constructed in 1649, it was never completed. Galileo's work inspired the Dutch scientist Christiaan Huygens in 1657 to invent and patent a working pendulum clock. His first design used a verge escapement, which required quite a wide pendulum swing, causing its period to be somewhat variable. He made a second design that used gears to limit the swing and then, in his third design, used curved 'jaws' to effectively change the length of the pendulum depending on the swing, so correcting the problem to a large extent. Later, in 1670, clockmakers invented the anchor escapement, which reduced the pendulum's swing to 4–6 degrees. This allowed the clock's case to accommodate longer, slower pendulums – in particular, the 'seconds' pendulum (also called the Royal pendulum). The length of the pendulum is about one metre with each swing taking one second. The tall narrow clocks built using these pendulums became known as 'long case' or 'grandfather' clocks and, because of their increased accuracy, a minute hand began to be added after 1690.

One problem was that pendulum clocks were observed to slow down in summer due to the thermal expansion of the pendulum rod. This was solved by the invention of the mercury pendulum, which had a mercury vessel as its bob, and the gridiron pendulum, which used alternating rods of iron and zinc. More recently, pendulums made of invar, a steel–nickel alloy, and having an exceedingly low coefficient of expansion have been used. A key objective is to try to allow the pendulum to run as freely as possible and the most accurate pendulum clocks (called regulators) only give a sustaining pulse to the pendulum every 30 seconds. The pendulum swing (and hence period) will also be slightly affected by changes in the barometric pressure. A bellow device that changes its size as a function of pressure can be used to compensate this, but some of the very best regulator pendulums operated in a near vacuum. It should be noted that, as the period of a pendulum is a function of the gravitational pull of the Earth, they have to be calibrated for both their height above sea level and the latitude of their location.

For many years, regulators, located in observatories to allow astronomical calibration, served as the primary standards for national time distribution services. Initially, the US time standard used Riefler pendulum clocks, accurate to about 10 milliseconds per day. In 1929 it switched to the Shortt free pendulum clock (about 1 second per year) before phasing in quartz time standards in the 1930s.

Quartz clocks

A quartz clock uses an electronic oscillator that is regulated by a quartz crystal to keep time. They are at least an order of magnitude more accurate than good mechanical clocks. In most modern quartz clocks or watches, the quartz crystal resonator is in the shape of a small tuning fork, laser-trimmed to vibrate at 32,768 Hz. This frequency is equal to 2^{15} Hz. A power of 2 is chosen so a chain of 15 digital divide-by-2 stages can derive the 1 Hz signal that then drives the second hand of the clock or watch. A typical quartz wristwatch will gain or lose less than a half second per day at body temperature.

If a quartz watch is kept at a reasonably constant temperature it can be accurate to within 10 seconds per year. To improve accuracy, quartz chronometers that are to be used as time standards include a crystal oven to keep the crystal at a constant temperature. From the 1930s, quartz time standards replaced pendulum regulators in providing national time standards. In 1932, a quartz clock was able to measure the tiny weekly variations in the rotation rate of the Earth and that (due to the Moon's tidal forces) the rotation rate was slowing down.

Atomic clocks

Quartz time standards remained in use until the 1960s when they were replaced by atomic clocks. These are the most accurate time and frequency standards known, and use the precise microwave signal that electrons emit when they change energy levels in an atom. They provide accuracies of approximately 1 part in 10^{14}, which is $\sim 10^{-9}$ seconds per day.

The first accurate atomic clock was built by Louis Essen in 1955 at the National Physical Laboratory in the UK and used a beam of caesium-133 atoms passing through a cylinder that acts as a resonant cavity at the frequency emitted by the caesium atoms (Figure 18.3). Such caesium beam clocks provide the fundamental time standards of most nations, but are very expensive and are usually backed up with hydrogen maser atomic clocks, such as that at Jodrell Bank.

The hydrogen maser uses the fact that, when in a magnetic field, the lowest energy level of the hydrogen atom is split into two. In the upper energy level, the spins of the proton and electron are parallel whilst in the lower, antiparallel. A beam of hydrogen atoms is produced (having equal numbers in both states) and is passed through a special (hexapole) magnet which splits it into two beams with different spin states. The beam of hydrogen atoms in the higher state is passed into a resonant cavity which contains radiation at the frequency corresponding to the transition from the upper to the lower state – 1,420,405,752 Hz. This radiation stimulates the arriving atoms to radiate and build up the level of radiation in the cavity. A small probe extracts a small amount of energy from the cavity which is used to lock a crystal oscillator to a frequency with equal precision. This frequency can then be divided down to give a 'pulse' at 1 Hz to drive a clock.

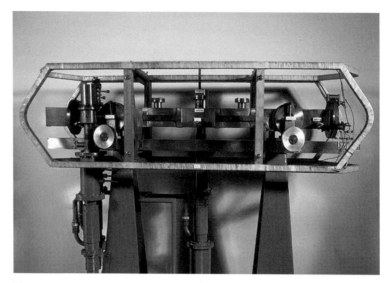

Figure 18.3 The first caesium beam atomic clock. Image: Science Museum/Science & Society Picture Library, UK.

The most common atomic clocks use rubidium atoms, and these are used in many commercial, portable and aerospace applications. These are inexpensive but are inherently less accurate. However, they can be periodically corrected by a GPS receiver to achieve long-term accuracy equal to the US national time standards.

One of the most accurate atomic clocks in continuous use today is NIST-F1, which is now the USA's primary time and frequency standard. It is a caesium fountain atomic clock which extracts the resonant frequency (9,192,631,770 Hz) of the caesium atoms when they are virtually stationary. Six infrared laser beams gently push the caesium atoms together into a ball, which slows down the movement of the atoms and cools them to temperatures near absolute zero. This beautifully removes the effect of the Doppler shift that affects atomic clocks that use atoms in motion. The precision given by NIST-F1 is now about 5×10^{-16}, which means it would neither gain nor lose a second in more than 60 million years and is about 10 times more accurate than the caesium beam atomic clock that served as the USA's primary time and frequency standard from 1963 to 1999.

Time transfer: not as simple as one might think

At one o'clock each day except Sunday, a gun is fired from the ramparts of Edinburgh Castle to give a time signal to the inhabitants of the city and ships

Figure 18.4 The time ball at the Royal Observatory, Greenwich. Image: Wikimedia Commons.

in the port of Leith. Should the clocks in Holyrood Palace, which is located at the other end of the Royal Mile, be set when the time signal is heard, they will actually be five seconds slow as this is how long the time takes to travel from the castle to the palace.

If the 'time transfer' is carried out by a light or radio signal the error will be vastly smaller. Above the roof of the (now) Royal Observatory, Greenwich, is mounted a 'time ball' as shown in Figure 18.4. This is raised just before 1 p.m. and drops on the hour as determined by the master clock at the observatory. Thus, marine chronometers on ships in the Pool of London could be set accurately and their 'rate' such as '+1 second per day' (meaning that the chronometer gained one second per day) measured. These chronometers made it possible for ships to measure their longitude by observing when the Sun was due south (and highest in the sky). Suppose when at sea, 10 days after setting the chronometer in London, the clock gave a time of 13:00 hours plus 5 seconds when the Sun was due south. By subtracting off the rate of, say, '+ 0.5 seconds per day' to give the true time of 13:00 hours showed that the ship was at a longitude of 15 degrees west, as 24 hours of time corresponds to 360 degrees of longitude, so 1 hour of time corresponds to 15 degrees.

Selling the time: the Belville family

On the wall beside the entrance to the Royal Observatory, Greenwich, is a clock installed in 1852 by Charles Shepherd, which was electrically operated and controlled by a master clock inside the main building. One could thus visit the observatory to find the precise time.

Members of the Belville family, first John then his wife Maria and finally their daughter Ruth, would take a chronometer up to the observatory and then travel round the City of London selling the 'time'. Ruth Belville became known as the 'Greenwich Time Lady'. She carried on this service (with her chronometer, which she called 'Arnold') until the 1930s.

Radio-controlled 'atomic clocks'

Such clocks and wristwatches are now widely available and are based on quartz watch movements but with additional circuitry to receive time signals from a number of longwave radio transmitters around the world, such as 'MSF' in the UK. These signals are used to correct the time displayed by the clock – often around midnight – and can even adjust for British Summer Time. They will normally be accurate to the second, which is good enough for most people. An interesting point is that such a clock in London will be about 1.6 milliseconds slow as it takes this long for the time signal to reach London from the transmitter in Cumbria.

Until the advent of 'GPS clocks', the timing system (built by the author) at Jodrell Bank Observatory used the MSF time signal transmitted from Rugby in central UK to synchronise its clock. In order to correct for the time delay, a portable atomic clock was sent to Rugby and synchronised. Brought back to the observatory, it was thus possible to use it to measure how long the MSF signal took to reach its location in Cheshire (around 200 microseconds) and the clock was set so that it provided its one-second 'ticks' at precisely this time *prior* to the arrival of the MSF time signal.

Pulsars: the best natural clocks in the Universe

In Chapter 11, the discovery and nature of pulsars was discussed and it was pointed out that they make very accurate clocks. Pulsars slowly radiate energy, which is derived from their angular momentum. This is so high that the rate of slowdown is exceptionally slow and so pulsars make highly accurate clocks and some may even be able to challenge the accuracy of the best atomic clocks. One of the best pulsar clocks known at the present time is 1713+07, which has been 'spun up' by matter falling onto it from a companion white dwarf star. It now has a pulse period of 4.57 milliseconds – spinning 218.8 times per second – and is currently slowing down at a rate of 200 nanoseconds in 12 years. That is a precision of better than one part in 10^{13}.

An absolute time standard: cosmic time

In 1905, Albert Einstein, then working in the Bern Patent Office, published his paper on the Special Theory of Relativity. Perhaps one of the most well-known aspects of this theory is that moving clocks appear to run slow when compared to a clock at rest with an observer – a phenomenon called 'time dilation'. This prediction has been proven by flying highly accurate atomic clocks around the world and has to be taken into account in the GPS used for navigation.

As time is relative, can we actually define a time standard with which to observe the evolution of the Universe? One could, perhaps, define what might be called 'cosmic time' as that measured by a clock that is stationary with respect to the Universe as a whole. But how would this time relate to clocks on Earth? We know that the Earth is moving around the Sun, and that the Sun is moving around the centre of our Milky Way Galaxy once every ~220 million years. But can we measure how fast the Solar System is moving with respect to the Universe? Perhaps surprisingly, we can.

Since 1965, observations have been made of what is called the Cosmic Microwave Background (CMB) – radiation that originated near the time of the origin of the Universe and which now pervades the whole Universe, as will be discussed in detail in Chapter 22. This radiation is very largely composed of a mix of long wavelength infrared and very short wavelength radio waves. For simplicity, just suppose that it is made up of only one wavelength and that the Solar System is moving in a certain direction with respect to this radiation. The Doppler effect will alter the apparent wavelength that we observe so that, when looking along the direction in which the Solar System is moving it will be blueshifted and appear to have a shorter wavelength. Conversely, in the opposite direction, the radiation will appear to be redshifted and have a longer wavelength. From very precise measurements of the CMB we now know that we are moving through space towards the constellation Leo at a speed of ~650 km/s. (2,340,000 km/h or about 0.22% of the speed of light.) This is thus our speed with respect to the Universe as a whole.

We can thus calculate how the time of a clock at rest with the Universe – measuring cosmic time – will differ from our clocks. Einstein produced a formula to give the ratio of the times of a moving clock relative to a static clock which, in this case, gives a ratio of 1.0000023. This is exceedingly small so, to a very good approximation, our clocks can be used to measure the time scale of the Universe.

Gravitational time dilation

Einstein in his General Theory of Relativity showed that clocks will tick more slowly in a strong gravitational field than in a weak one, giving a form of time dilation known as 'gravitational time dilation'. As the gravitational field gets stronger the time dilation gets greater as, for example, when approaching a black hole. At what is called the event horizon of the black hole – from within which not even light can escape – the time dilation seen by an observer in free space becomes infinite and time is effectively frozen!

When did time begin?

As will be described in Chapter 19, Edwin Hubble showed that the Universe was expanding. If one assumes a uniform rate of expansion related to what is called Hubble's constant, then one can extrapolate backwards until the Universe had no size – its origin and the beginning of time in our Universe. The result has to be modified to take into account that the rate of expansion has not been constant during the life of the Universe and, given the currently accepted value of his constant, the age of the Universe comes out as ~13.8 billion years.

If you have more time:

> *A Question of Time: The Ultimate Paradox* [Kindle Edition] by *Scientific American* Editors.
> *The Time Book: A Brief History from Lunar Calendars to Atomic Clocks* by Martin Jenkins and Richard Holland (Walker).
> *The Measurement of Time: Time, Frequency and the Atomic Clock* by Claude Audoin, Bernard Guinot and Stephen Lyle (Cambridge University Press).

19

Hubble's heritage: the astronomer and the telescope that honours his name

The astronomer

Edwin Hubble, the son of Virginia and John Hubble, was born at Marshfield, Missouri, USA, on 20 November 1889. His early interest in astronomy was indicated by the fact that, when just 12 years of age, his article about the planet Mars was published in a local paper! The family moved to Wheaton, Illinois, where his father, an insurance executive, had an office. Edwin went to Wheaton High School where not only did he do well at his academic studies but also excelled at athletics and football. (He held the Illinois High School high jump record for some time.) Following his graduation from Wheaton, he was awarded a scholarship to the University of Chicago, where he studied physics, astronomy and mathematics.

Hubble arrived at the university during the fall of 1906 and, with his height of 6 feet 1 inch (1.85 m) and fine physique, soon became a star of the gymnasium, track and sports field. In his third year he played in 6 out the 12 basketball games that brought Chicago the national universities title (Figure 19.1). He graduated in March 1910, having been vice president of his class.

He gained a prestigious Rhodes scholarship to study law at Queen's College, Oxford, partly as a result of the letter of commendation given him by Physics Nobel Laureate-to-be Robert Millikan, who recommended him as 'a man of magnificent physique, admirable scholarship, and worthy and lovable character'. Surprisingly, he chose to study law and Spanish – at, it seems, his father's insistence – but he would often visit the university observatory and learnt about the new field of celestial photography from its director, Herbert Hall Turner.

During his time at Oxford, it seems that Edwin picked up some affectations and, on his return, surprised his sisters with his attire. Their athletic brother was

Figure 19.1 Hubble as an athlete, champion basketball player and (at right) on his return from Oxford.

'dressed in a cape, knickers, and sported a walking stick. A signet ring graced his little finger, and he was wearing a wristwatch he had won for high jumping.' (It should be pointed out that in America 'knickers' are full breeches gathered and banded just below the knee!)

Back in the USA, he first taught physics and mathematics at the New Albany High School in New Albany, Indiana, and then, having passed the bar examination in 1913, became an attorney at Louisville, Kentucky. It is not at all obvious that Hubble actually practised to any great extent (if at all) and, by 1914, had tired of the law and in August of that year decided to return to the University of Chicago to study for a PhD in astronomy. Chicago was no place for an observatory, so the university had built its observatory on the north shore of Wisconsin's Lake Geneva, 75 miles (120 km) to the northwest of Chicago. Here, with money pledged by Charles Tyson Yerkes, the Chicago streetcar magnate, they had built the world's largest refracting telescope with an aperture of 40 inches.

However, Hubble was using the observatory's 24-inch reflector to photograph areas of the sky to study what were then known as 'white nebulae' (now known as galaxies). One object soon took his attention: he was amazed to find that one of his target objects, known as NGC 2261, was changing significantly on relatively short time scales. He described it as 'the finest example of a cometary nebula in the northern skies'. His maiden discovery was published in the *Astrophysical Journal*, where he cautiously stated, 'No attempt is here made to explain the phenomenon of illumination, the nebula must be very near.' (This is now known as Hubble's Variable Nebula and it changes its appearance noticeably in just a few weeks. It is a reflection nebula made of gas and fine dust fanning out from the star R Monocerotis (R Mon). About one light year across, it lies about 2,500 light years away in the constellation Monoceros. It

 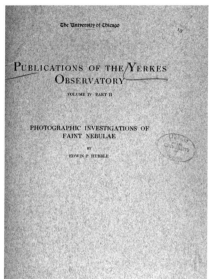

Figure 19.2 The 24-inch telescope at Yerkes Observatory, which Hubble used to image and study faint nebulae.

is thought that dense knots of opaque dust pass close to R Mon and cast moving shadows onto the reflecting dust forming the nebula.)

By the time his thesis (Figure 19.2) was completed, some 17,000 nebulae had been catalogued by other astronomers. In his thesis he wrote, 'Extremely little is known of the nature of the nebulae ... At least some of the great diffuse nebulosities, associated as they are with stars visible to the naked eye, seemed to lie within the stellar system (our Milky Way Galaxy) ... Yet others, most particularly the giant spirals which display no visible motion, apparently lie outside our system.' Though his thesis was somewhat shaky on technical grounds and rather confused in its theoretical interpretations, it laid the basis of the great discoveries that he was to make in the next 10 years.

Whilst finishing work for his doctorate early in 1917, Hubble was invited by George Ellery Hale to join the staff of the Mount Wilson Observatory, in Pasadena, California. However, after sitting up all night to finish his PhD thesis and taking the oral examination the next morning, Hubble enlisted in the infantry and telegraphed Hale, 'Regret cannot accept your invitation. Am off to the war.' Hubble was commissioned a captain, later rising to the rank of major, and was sent to France where he served as a field and line officer. He returned to the USA in the summer of 1919, and went immediately to join Hale at the Mount Wilson Observatory, where the 100-inch Hooker telescope had been completed some two years earlier. He remained on the staff of the observatory throughout his life apart from a further spell in the US Army during World War II, when he was chief of ballistics and director of the supersonic wind tunnels at the Aberdeen Proving Ground in Maryland. For his work there he received the Legion of Merit award.

Hubble's law

Before we can discuss what was, perhaps, the greatest discovery of the last century, we need to learn about two highly significant sets of observations. The first were made by Henrietta Leavitt whilst working at the Harvard College Observatory, where she became head of the photographic photometry department. Her group studied images of stars to determine their magnitude using a photographic measurement system developed by Miss Leavitt that covered a brightness range of over 7 million. Many of the plates measured by Leavitt were taken at Harvard Observatory's southern station in Arequipa, Peru, from which the Magellanic Clouds could be observed, and she spent much time searching the plates taken of them for variable stars. Amongst the many variable stars observed within them were 25 Cepheid variable stars. These stars are amongst some of the brightest – between 1,000 and 100,000 times the brightness of our Sun – and are named after the star Delta Cepheus, which was discovered to be variable by the British astronomer John Goodricke in 1784. These stars pulsate regularly, rising rapidly to a peak brightness and then falling more slowly. As they are very bright they can be seen at great distances. Leavitt determined the periods of 25 Cepheid variables in the Small Magellanic Cloud (SMC) and in 1912 announced what has since become known as the period–luminosity relation. She stated, 'A straight line can be readily drawn among each of the two series of points corresponding to maxima and minima (of the brightness of Cepheid variables), thus showing that there is a simple relation between the brightness of the variable and their periods.' As the SMC was at some considerable distance from Earth and was relatively small, Leavitt also realised, 'As the variables are probably nearly the same distance from the Earth, their periods are apparently associated with their actual emission of light, as determined by their mass, density, and surface brightness.'

The relationship between a Cepheid variable's luminosity and period is quite precise; a 3-day period Cepheid corresponds to a luminosity of about 800 times the Sun whilst a 30-day period Cepheid is about 10,000 times as bright as the Sun (Figure 19.3). So, for example, we might measure the period of a Cepheid variable in a distant galaxy and observe that it is 10,000 times fainter than a Cepheid variable having the same period in the Large Magellanic Cloud (LMC). We can then deduce that, from the inverse square law, it is 100 times further away than the LMC, that is 100×51.2 kpc (166,000 light years) giving a distance

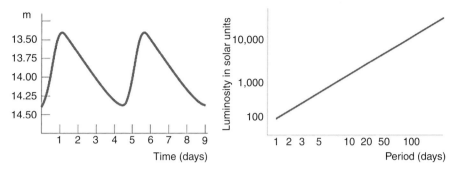

Figure 19.3 A Cepheid variable light curve and period–luminosity relation.

of 5,100 kpc (16,600,000 light years). Cepheid stars are thus the ideal standard candle to measure the distance of clusters and external galaxies. (As we do not know the precise location of the Cepheid variable within the cluster or galaxy there will be a small uncertainty, but this error is typically small enough to be irrelevant.)

There had long been arguments as to whether the 'white nebulae' were within or beyond our own Milky Way Galaxy. In 1912 Vesto Slipher at the Lowell Observatory published his observations of the spectral lines in M31, the Great Nebula in Andromeda, and found that they showed a shift towards the blue. Assuming that this was due to the Doppler shift, this indicated that Andromeda was moving towards us at a speed of 300 km/s – greater than any previously observed. Slipher wrote, 'The magnitude of this velocity, which is the greatest hitherto observed, raises the question whether the velocity-like displacement might not be due to some other cause (than the Doppler shift), but I believe we have at present no other interpretation for it.' Three years later he reported on the spectral shifts in the lines of a further 14 galaxies, all but three of which were receding from us at high speeds. This was perhaps an indication that these objects were not part of our own galaxy as the measured Doppler shifts of known objects within our Galaxy were far less.

But an opposing view was promoted by Harlow Shapley, who had used Cepheid variables to measure the size of the Galaxy and the place of our Sun within it. As a result, he was a highly respected astronomer, so many accepted his word that the nebulae were within our own galaxy. His key point was that novae were observed in these objects and if they were at great distances they would have to be unimaginably bright. (They were – they were supernovae!)

What was really needed was the measurement of the distance to one of these 'white nebulae'. Hubble knew that if he could locate a Cepheid variable in one and measure the period of its oscillation, he could compare its brightness with one of similar period that had been observed in the SMC. If, say, it appeared 100 times fainter, he would know that it would lie at a distance 10 times further away than the SMC, whose distance was known. The Andromeda Nebula was the obvious target and finally, on an image taken on the night of 5–6 October 1923, he found one and was thus able to calculate that Andromeda lay at a distance of 860,000 light years – well beyond the extent of our own galaxy, then thought to be about 300,000 light years in diameter. (You will note that these values are about three times smaller than those currently accepted. There are two main types of Cepheid variable and those observed in Andromeda were several times brighter than those observed by Henrietta Leavitt in the SMC. This reduces the calculated distance.)

His discovery, announced on 30 December 1924, profoundly changed our understanding of the Universe. Hubble (Figure 19.4) then went on to measure the distances to the galaxies whose redshifts had been measured by Vesto Slipher and combined his distance measurements with Slipher's velocity measurements to make what was, perhaps, the single most important discovery of the last century.

Hubble found that the more distant galaxies had greater velocities of recession and (roughly in the original data) the greater the distance, the greater the velocity, as shown in Figure 19.5. This has become known as Hubble's law: the velocity of recession being proportional to the distance.

Figure 19.4 Edwin Hubble at his desk and at the focus of the Hooker 100-inch telescope at Mount Wilson Observatory, which he used to measure the distances to the 'white nebulae'.

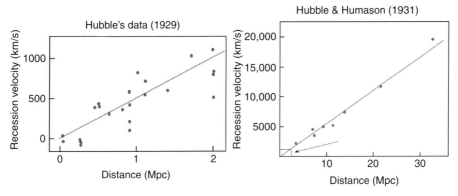

Figure 19.5 The Hubble plots of 1929 and 1931.

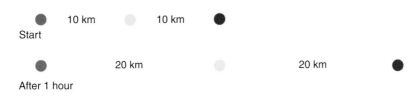

Figure 19.6 A schematic of an expanding one-dimensional universe.

Why was this so important? Imagine the very simple one-dimensional universe shown in Figure 19.6. Initially the three components are 10 km apart, as shown in the upper plot. Let this universe expand uniformly by a factor of 2 in one hour. As seen from the left-hand component, the middle component will appear to move 10 km in one hour whilst the right-hand component will appear to move 20 km – the apparent recession velocity is proportional to the distance. Thus the linear relationship between the velocity of recession and distance implied that the Universe was expanding and not 'static' as many believed at the time.

The speed of recession and distance were directly proportional and related by Hubble's constant, H_0. The value of the constant derived from his original data was ~500 (km/s)/Mpc. The use of the word 'constant' is perhaps misleading. It would only be a real constant if the Universe had expanded linearly throughout the whole of its existence. It has not, which is why the subscript is used: H_0 is the *current* value of Hubble's constant.

If one makes the simple assumption that the Universe has expanded at a uniform rate throughout its existence, then it is possible to backtrack in time until the Universe had no size – its origin – and hence estimate the age, known as the Hubble age, of the Universe. This is very simply given by $1/H_0$ and, using 500 (km/s)/Mpc, one derives an age of about 2 billion years:

$$\begin{aligned}
1/H_0 &= 1 \text{ Mpc}/500\,\text{km/s} \\
&= 3.26 \text{ million light years}/500\,\text{km/s} \\
&= 3.26 \times 10^6 \times 365 \times 24 \times 3600 \times 3 \times 10^5\,\text{s}/500 \\
&= 3.26 \times 10^6 \times 3 \times 10^5 \text{ years}/500 \\
&= 1.96 \times 10^9 \text{ years} \\
&= \sim 2 \text{ billion years}
\end{aligned}$$

This is obviously incorrect; the distance calibration data Hubble used were incorrect and, in addition, he was actually observing a brighter class of Cepheid variable, which led him to significantly underestimate the distance to the galaxies.

Hubble also devised the most commonly used system for classifying galaxies, grouping them according to their appearance in photographic images. He arranged the different groups of galaxies in what became known as the Hubble sequence. They fell into three groups: elliptical, spiral (which were sub-divided into normal and barred spirals) and irregular galaxies.

After World War II he supported the construction of the gigantic 200-inch telescope on Palomar Mountain south of Mount Wilson, which helped put him on the cover of *Time* magazine in 1948. He was honoured with making its first observation - that of NGC 2661, the Cometary Nebula, which he had first studied in 1914. Sadly, in 1953, shortly after the 200-inch was completed, Hubble died of a heart attack. Up to the time of his death, astronomers were not able to be considered for the Nobel Prize. This was remedied shortly afterwards but, as it is never awarded posthumously, he never became a Nobel Laureate – an honour that he so richly deserved.

The Hubble Space Telescope

In 1946, the astronomer Lyman Spitzer wrote the paper 'Astronomical advantages of an extra-terrestrial observatory' in which he discussed the two main advantages that a space-based observatory would have over ground-based telescopes. Firstly, the angular resolution would be limited only by the size of the mirror (assumed to be optically perfect), rather than by the turbulence in the atmosphere and, secondly, a space-based telescope could observe in the infrared and ultraviolet light wavebands that are strongly absorbed by the atmosphere. In 1965 he was appointed to chair a committee given the task of defining the scientific objectives for such a telescope. (In recognition of this work, the infrared space telescope 'Spitzer' is named after him.)

The Hubble Space Telescope (HST), as it is known, was funded in 1970 with a proposed launch date of 1983, and was built by NASA with contributions from the European Space Agency. It was to be launched by the space shuttle and, amongst many other problems, its launch was delayed by the Challenger disaster. The launch finally took place on 24 April 1990 when the HST was carried into orbit by Discovery. The low 559-km orbit meant that many objects could only be observed for up to 12 hours before they were hidden by the Earth, but this gave the HST the invaluable (and, as it turned out, critical) ability to be serviced by further shuttle flights throughout its working life. A total of five servicing missions from 1993 to 2009 have taken place; replacing parts with a limited working life – in particular the gyroscopes that are used to orientate the telescope towards the target objects – and upgrading the instruments to greatly improve their capabilities.

The task of figuring the primary 2.4-m mirror was given to the Perkin-Elmer Corporation. The HST optical design is a Ritchey–Chrétien variant of the Cassegrain reflecting telescope. This design gives a wide field of view with excellent image quality across the field and is now used by the majority of large optical telescopes. Its disadvantage is that the primary and secondary mirrors are hyperbolic in shape and are thus difficult to fabricate and test. As the HST was to be used at ultraviolet wavelengths, to achieve the desired diffraction-limited optics meant that its mirror needed to be polished to an accuracy of 10 nanometres, or about 1/65 the wavelength of red light. Construction of the Perkin-Elmer mirror began in 1979, using Corning ultra low expansion glass. To minimise weight, it consisted of inch-thick top and bottom plates sandwiching a honeycomb lattice. The mirror polishing was completed in 1981 and it was then washed using hot, deionised water before it received a reflective coating of aluminium under a protective coating of magnesium fluoride.

In low orbit, the spacecraft (Figure 19.7) has to withstand frequent passages from direct sunlight into the Earth's shadow, so a shroud of multi-layer insulation is used to keep the interior temperature stable. The optics are mounted in a graphite-epoxy frame to keep them precisely aligned. The HST's guidance system uses gyroscopes, controlled by three fine guidance sensors, to keep the telescope accurately pointed during an observation. The scientific operation of the HST is under the control of the Space Telescope Science Institute, in Baltimore, who schedule the telescope's observations and provide support for the astronomers who use the HST (with a European support institute in Germany). Scheduling the observations is a difficult task as the telescope must not be pointed near the Sun, Earth or Moon as scattered light would harm the observations and possibly damage the instruments. Due to its low orbit,

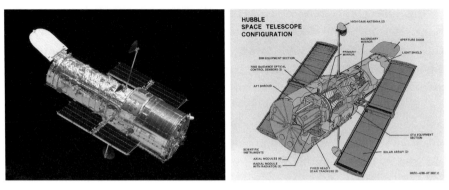

Figure 19.7 The Hubble Space Telescope. Note: The image at left shows the smaller solar panel arrays fitted on the fourth servicing mission. Image: STScI, ESA, NASA.

effectively in the upper atmosphere, its orbit changes over time so its position cannot be predicted with accuracy until a few days before a proposed observation.

A flawed mirror

During the initial tests of the optical system after the launch of the telescope it became obvious that there was a serious problem. Stellar images that should have been ~0.1 arcsecond across were more than 1 arcsecond across – no better than ground-based images! Image analysis showed that the primary mirror suffered from spherical aberration, as the mirror – though the most precisely figured mirror ever made – had been ground to the wrong shape. The result was that the image of a star was composed of two parts: a central core 0.1 arcseconds across containing ~15% of the total light (instead of a theoretical ~84%) with the remaining light forming a halo surrounding the core about 1 arcsecond across. It was still possible to image bright objects and spectroscopy was largely unaffected but faint object imaging (such as of distant galaxies) was virtually impossible. Techniques developed by radio astronomers to produce high-quality images from sparse arrays of telescopes came to the rescue to some extent and the HST was able to make productive observations during the three years before the first servicing mission could provide a solution to the problem.

Though the primary mirror was the wrong shape, it precisely followed the wrong shape and this was much better than a badly made mirror of the correct shape! Having found the error in the shape, it was then possible to design and make corrective optics, having an identical error but in the opposite sense, that were installed during the first HST servicing mission in 1993. (This is rather like the fact that the appropriate concave lenses of my glasses correct for the fact that my retinas are too far from my eye lenses and so enable me to have essentially perfect vision.)

It turned out that the device (called a null corrector) used by Perkin-Elmer to test the primary mirror had been incorrectly assembled, so giving rise to the error in the surface curvature. It appeared that they ignored tests made with two other less precise null correctors, both of which indicated that the mirror was suffering from spherical aberration. To incorporate the corrective optics (called COSTAR) into the HST, one instrument had to be initially sacrificed, but upgraded instruments installed in 2002 all incorporated their own corrective optics and so, in 2009, COSTAR was replaced with a new spectrograph.

It was known that the four gyroscopes used in the telescope pointing system would have a limited lifespan, and so the telescope had always been designed so that it could be regularly serviced, but the first servicing mission in December 1993 assumed great importance as the astronauts had to carry out extensive work over a 10-day period to install the corrective optics. The seven astronauts also replaced the solar arrays and upgraded the on-board computers before boosting the telescope into a higher orbit to compensate for the drag and resultant orbital decay experienced over the three years since its launch. Happily the mission was a complete success and in January 1994 NASA published the first perfectly sharp image from the corrected optics.

Further servicing missions were flown in February 1997, December 1999 and March 2002. The fourth mission replaced the solar arrays for the second time. The new arrays, whilst only two-thirds the size of the old arrays, provided 30% more power whilst significantly reducing the atmospheric drag, so lengthening the time before the HST would need to be boosted back into a higher orbit. This was a critical factor in extending its life. The next scheduled mission was planned for October 2005, when the shuttle crew would have replaced two broken gyroscopes, installed a new CCD camera, WFC3, to replace WFPC2 and installed a new spectrograph to take up the space no longer need by COSTAR. The Columbia disaster in February 2003 delayed all shuttle flights for two years and under a new safety regime (which mandated that all future shuttle flights would have to be able to reach the safety of the International Space Station (ISS) should an in-flight problem occur) any further servicing missions were ruled out as the shuttle was not capable of reaching both the HST and ISS in one mission. This would have meant that the HST would reach the end of its operational life in 2010, well before the launch of the James Webb Space Telescope (JWST), then scheduled for 2014 but now delayed until at least 2018.

This was a major concern to astronomers, given the great scientific impact of HST, and Congress asked NASA to look into ways that it could be saved. Robotic missions were considered and discarded as 'not feasible'. In 2005 the new NASA

Administrator, Michael D. Griffin, authorised the Goddard Space Flight Center to proceed with preparations for a manned Hubble maintenance flight, which was scheduled for October 2008. However, a failure of the main data handling unit in the HST in September 2008 meant that this was postponed until May 2009 so that a replacement could be carried to the HST. (The HST had a backup data system which was successfully activated, but should this have failed the HST would then be crippled.)

Two new instruments were installed, one of which, Wide Field Camera 3 (WFC3), increased Hubble's sensitivity in the ultraviolet and visible parts of the spectrum by up to 35 times. At the same time, the astronauts replaced the six battery packs, the three rate sensor units, one of the fine guidance sensors and positioning gyroscopes, as well as giving the HST a new thermal protective blanket. Hubble resumed operation in September 2009 and it is hoped that it will continue to be fully operational until the end of 2014 and perhaps longer.

Hubble was originally designed to be returned to Earth on board a shuttle. With the retirement of the shuttle fleet this was no longer possible, so NASA engineers developed the 'Soft Capture and Rendezvous System', which was attached to Hubble's aft bulkhead during its final servicing mission. This will enable the HST to be captured by a future spacecraft (probably robotic) and then safely 'deorbited' at the end of its long and productive life.

Hubble science

In its over 20-year lifetime, the HST has probably been the most productive scientific instrument ever built: it has targeted over 30,000 individual objects, and produced over 44 terabytes of data from which astronomers have published nearly 9,000 papers. In the following text it will thus be only possible to highlight some of the HST's most interesting results.

Observation of SN1987A

In February 1987, a supernova was observed in the Large Magellanic Cloud (LMC) and immediately became the focus of virtually every telescope capable of observing the southern sky. It was continuously monitored by the HST and has enabled a key step in the cosmic distance scale to be determined, that of the distance to the LMC. It appeared that at some time before the star exploded, it ejected a ring of gas that surrounded the star. This was not initially visible, but as the ultraviolet light from the explosion travelled outwards at the speed of light it reached the ring and caused the gas to glow. Had the ring been at right angles to the line of sight, it would all have lit up simultaneously, but it lies at an angle and thus, as a result, the nearest part of the ring was seen to brighten

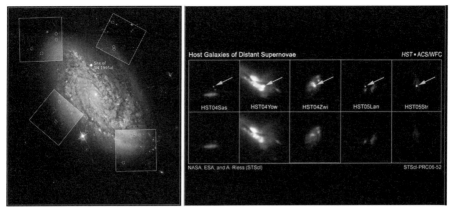

Figure 19.8 (Left) Cepheid variables in NGC 3012 also showing the site of the 1995 Type Ia supernova. (Right) Type Ia supernovae in distant galaxies. Images: STScI, ESA, NASA.

after 75 days and the furthest part after 390 days – this is just due to the light travel time across the ring delaying the observed brightening of the more distant parts.

From these observations it was deduced that the radius of the ring was 232.5 light days. But also, as the HST had sufficient resolution to image the ring, seen in Figure 8 of Chapter 15, it was measured to be 17.2 arcseconds across. Given both the angular diameter and the distance in light days across the ring (465 light days), the distance of the supernova can be calculated, so putting the LMC at a distance of 52 kpc or 170,000 light years. This has given a new value for the zero point of the Cepheid distance scale which, as described below, has enabled Hubble's constant to be measured more accurately.

Measurement of the distance scale of the Universe, Hubble's constant and the age of the Universe

Hubble's constant, which is a measure of the expansion rate of the Universe (and from which one can determine its age), depends on measuring two things about a distant galaxy: its speed of recession and its distance. The former is relatively easy to measure from the redshift seen in its spectral lines. The latter is far harder and it requires the observation of objects of known brightness. The HST's superb resolution has enabled astronomers to study Cepheid variables, described earlier in the chapter, out to far greater distances than before. Their period is closely related to their absolute brightness and so Cepheid variables can be used as 'standard candles'. As the period is easy to measure, the brightness can be deduced and hence, using the inverse square law, the distance found.

This, for example, enabled the distance to the galaxy NGC 3012 (Figure 19.8, left) to be determined to be 92 million light years. In 1995 a Type Ia supernova was observed in this galaxy. These supernovae have an exceedingly high and well-defined brightness and will be discussed further in Chapter 21. The HST has been able to observe them in very distant galaxies, and so by comparing their observed brightness (Figure 19.8, right) with that observed in NGC 3012, their distances could be determined. Allied to their redshifts this has enabled the error in the value of Hubble's constant to be reduced to less than 5% – a value of 74.2 ± 3.6 (km/s)/Mpc and this gave an age of the Universe of 13.75 ± 0.11 billion years. (Note: The HST measure of Hubble's constant is somewhat larger than that found by the Planck Spacecraft observations of the Cosmic Microwave Background, which will be discussed in Chapter 22, which gave a value of 68 ± 1.)

Our Solar System and exoplanets

Shoemaker–Levy 9

The HST imaged the train of comets that resulted from the breakup of a comet called Shoemaker–Levy 9 (discussed further in Chapter 8, 'Impact!'), which had first become captured by Jupiter and then went so close that it came within the Roche limit and broke up into ~25 separate pieces. These were predicted to impact onto Jupiter's surface over a period of one week in 1994. The impact site was beyond the limb, but the HST was able to image the plumes that arose from the impact sites and then the sites themselves as they came into view an hour or so later (Figure 19.9).

Pluto and the Kuiper Belt

In 2005, the HST discovered two new moons of Pluto, Nix and Hydra, and was able to show significant changes in the surface markings of the dwarf planet itself. Two further small moons, Kerberos and Styx, were discovered in 2011 and

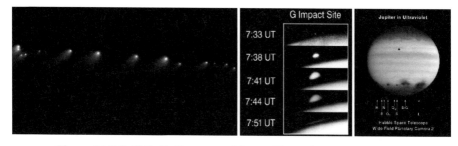

Figure 19.9 (Left) Trail of comets making up Shoemaker–Levy 9. (Centre) Plumes rising up from the impact site. (Right) Scars seen on the surface. Images: STScI, ESA, NASA.

2012, respectively. These latter two were discovered when a search was being made for a possible ring system around Pluto, as such orbiting debris might be a hazard for the New Horizons spacecraft due to arrive at Pluto on 14 July 2015.

The HST has detected many Kuiper Belt objects (KBOs) and, in 2009, its observations enabled the orbit of a small moon, Weywot, around a ~900-km diameter KBO, Quaoar, to be calculated. This has enabled the mass of Quaoar to be found and hence, given its size, its density. This, surprisingly, appears to be ~4.2, which means that it is a rocky body rather than an icy body as had been expected.

Exoplanets in the Galactic Bulge

To show whether the locally measured number of planetary systems orbiting stars is representative of the Galaxy as a whole, the HST observed 180,000 stars towards the Galactic Centre. These observations led to the discovery of 16 candidate planets – a number consistent with the frequency of planets in the solar neighbourhood – indicating that planetary systems are widespread within the Galaxy. Five of these planets orbit their suns in less than one Earth day.

Exoplanet atmospheres

By observing the transit of a planet across the face of a star, the HST has been able to measure the atmospheric composition of two planets. As the planet will block some of the light from the star (typically 1% or 2% in the case of gas giants) the star will dim slightly. However, whilst the planet lies between us and the star, some of the starlight will pass through the planet's atmosphere. By comparing the spectrum of the starlight before and during the transit it is possible to determine the presence and abundance of chemicals and gases that exist in the planet's atmosphere. HST observations of star HD 209458 have shown that its planet's atmosphere contains sodium, oxygen, carbon and hydrogen whilst, even more exciting, star HD 189733's planet shows evidence for carbon dioxide, water vapour and methane!

Gamma-ray bursts

The HST has imaged the host galaxies of many long-duration gamma-ray bursts (GRBs). The observations support the theory that this type of GRB is the result of a massive star collapsing to form a black hole. Gamma-ray bursts will be covered in detail in Chapter 20.

Studies of dark matter and dark energy

The HST has made significant contributions to the study of dark matter by plotting the three-dimensional distribution of dark matter in a slice of the

Universe out to a distance of 6.5 billion light years and, by observing Type Ia supernovae out to cosmological distances, has provided much of the evidence for the presence of dark energy.

Galaxy formation and evolution: the Hubble Deep Fields

One of the key aims of the Hubble Space Telescope was to study distant galaxies at far higher resolution than was possible from the ground. This would enable astronomers to study their evolution as, the more distant they are, the further back in time towards the origin of the Universe we see them. Early studies once the telescope's optics had been corrected indicated that there were substantial differences between the properties of galaxies 'now' (that is, nearby ones) and those that are observed as they were several billion years ago.

The observing time of large telescopes is distributed by 'time allocation' committees, who choose the best observing proposals from astronomers. However, a percentage of a telescope's observing time (up to ~10%) is designated as Director's Discretionary Time. The telescope director is then able to allocate time – often to enable observations of transient phenomena, such as supernovae, to be made. In 1995, once Hubble's corrective optics were shown to be performing well, Robert Williams, the then director of the Space Telescope Science Institute, decided to devote a substantial fraction of his discretionary time to the study of distant galaxies, to produce what became known as the Hubble Deep Field (HDF).

An area of the sky was chosen at a high galactic latitude (that is, well away from the plane of our Milky Way Galaxy) to minimise the dust that obscures our view along the Galaxy's plane and also avoid bright foreground stars and known infrared, ultraviolet and X-ray sources. The chosen field was in the constellation Ursa Major – a northern hemisphere region – so that follow-up observations could be made by optical telescopes such as the two 10-metre Keck Telescopes on Mauna Kea, Hawaii, and the VLA and MERLIN radio instruments. The field of view was just 5.3 square arcminutes in area, just 1/28,000,000 of the sky's total area. In total over 10 days of observations (when the HST made ~150 orbits of the Earth) were made at four wavelengths in the near-infrared, red, blue and near-ultraviolet of the target field, with further observations of the surrounding region to aid follow-up observations by other instruments (Figure 19.10).

A total of 342 individual images were taken and, when cleaned of cosmic-ray hits and satellite tracks and corrected for scattered light, were combined together to yield four monochrome images, one at each wavelength. Three of the images were then combined to give a 'false colour' image that would give an approximation to the actual colours of the galaxies in the image. About 3,000 distinct galaxies could be identified including both spiral and irregular galaxies.

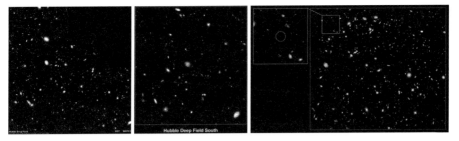

Figure 19.10 (Left) Hubble Deep Field North (1995). (Centre) Hubble Deep Field South (1998). (Right) Hubble Ultra Deep Field (2009) and the VLT image of the most distant galaxy then known. Images: StScI, ESA, NASA.

One of the most significant findings was the discovery of large numbers of galaxies with very high redshifts – many corresponding to distances of about 12 billion light years.

Follow-up observations have been made over a very wide spectral range: infrared emission from 13 of the galaxies visible in the optical images has been observed related to the large quantities of dust associated with star formation; X-ray observations by the Chandra X-ray Observatory revealed six sources in the HDF. The Jodrell Bank MERLIN array detected 16 radio sources in the HDF and was able to determine their precise positions – a great asset when further follow-up observations were made. I feel that this is one of the major pieces of work carried out by MERLIN, and I am proud to have been one of its designers and builders.

In 1998, the HST observed a counterpart region in the southern hemisphere sky. Very similar in appearance, it supported the cosmological principle, which states that at its largest scales the Universe is homogeneous. A wider, but less sensitive, survey has been carried out as part of the Great Observatories Origins Deep Survey; a section of which was then observed for a million seconds (11.3 days) over 400 orbits to create the Hubble Ultra Deep Field, which is the second most sensitive optical deep field image to date. Objects as faint as 30th magnitude were detected, which corresponds to collecting one photon per minute of observing time! (Nearby galaxies would give millions of photons per minute.) The resulting image shows about 10,000 galaxies, one of which, observed by the European Southern Observatory's VLT 8-metre telescope in Chile, has a redshift indicating that it was seen at a time when the Universe was only 600 million years old – the most distant and oldest galaxy in the Universe then observed.

In September 2012, NASA released a further refined version of the Ultra Deep Field dubbed the eXtreme Deep Field (XDF), which is an image of a small part of space in the centre of the Hubble Ultra Deep Field. The total exposure time was

2 million seconds (23 days). The XDF reveals galaxies that span back 13.2 billion years in time, revealing a galaxy theorised to be formed only 450 million years after the Big Bang event.

What have we learned? Galaxies in the distant past were smaller and more irregular in shape that those we see nearby (and hence more recently). This is consistent with the idea that small galaxies collided and accreted gas to gradually build up the larger galaxies we see today. By analysing the rate of star formation in galaxies at different redshifts (and hence age), astronomers have deduced that star formation reached a peak some 8–10 billion years ago and has since declined to about one-tenth of its maximum value. The earliest galaxies were formed when there were only small amounts of heavier elements (beyond the hydrogen and helium created in the Big Bang), so there was almost no dust present. The HST has shown that these galaxies are very blue – just as one might expect. Another important result related to the understanding of dark matter. It was thought that a substantial part might be in the form of red dwarfs and planetary bodies in the galactic halo called massive astrophysical compact halo objects (MACHOs), but the HDF showed that there were not significant numbers of red dwarfs visible (only ~20 foreground stars were observed in the field). The HST observations of galaxy centres have also revealed that the masses of the super-massive black holes that reside there are tightly correlated with the masses of the stellar bulges at the centres of galaxies. This implies that the evolution of galaxies is intimately connected to the black holes within them.

Hubble Heritage Images

There is one further aspect of the HST that has left a legacy to us all. The Hubble Heritage Imaging team have used a small part of the observing time to make some of the most wonderful astronomical images that have ever been seen. These are now used in virtually all astronomy books (including this one) to show us the beauty of the Universe in which we live.

For more on Hubble:

> *Edwin Hubble, The Discoverer of the Big Bang Universe* by Alexander S. Sharov and Igor D. Novikov (Cambridge University Press).
> *Hubble: Imaging Space and Time* by David H. DeVorkin and Robert W. Smith (National Geographic).

20

The violent Universe

This chapter is going to examine the most powerful explosions in the Universe that can be observed today and also study the Big Bang origin of the Universe itself – an explosion of a very different and unique type. The story begins with the serendipitous discovery, which came out of the Cold War, of what are termed 'gamma-ray bursts' or GRBs.

It is an interesting point as to what 'today' means. We see these events now but, as we will see, they arise in galaxies in the distant reaches of the Universe and so we are seeing events that actually happened many billion of years ago.

The discovery of gamma-ray bursts

As nuclear test ban treaties were negotiated in the late 1950s, President Eisenhower's science advisors cautioned him that the USSR might try to secretly carry out nuclear tests in space. It was decided to design and launch a series of satellites that could detect the characteristic double burst of gamma rays (very highly energetic photons) that results from a nuclear blast.

The project was code-named Vela (meaning 'watchers') and the first spacecraft was launched in October 1963, orbiting at an altitude of 120,000 km (74,400 miles). It carried six gamma-ray detectors along with other instruments. The gamma-ray detectors were made using caesium iodide, which scintillates – giving flashes of visible light – when gamma rays pass through it.

The data had to be analysed by hand; in 1969, scientists working with data recorded on 2 July 1967 found in the data a spike, a dip, a second spike and a long, gradual tail off. As the team leader, Ray Klebasabel, said, 'One thing that was immediately apparent was that this was not a response to a clandestine nuclear test.' His team checked for possible solar flares and supernovae and found none that might have caused the mysterious event. The number of recorded events rose rapidly as more sensitive detectors were carried by later generations of Vela satellites.

Later, as pairs of satellites were launched with improved timing capabilities, it became possible to approximately determine the directions from which the gamma-ray pulses originated. The arrival times of the pulses at the satellite pairs Vela 5a and 5b and 6a and 6b could be measured to an accuracy of 1/64 of a second whilst the light travel time between the satellite pairs across their orbital diameters was around 1 second. This enabled the direction of the event relative to the line between each pair of satellites to be determined to about 10 degrees. Given the two pairs of satellites, one could then derive one or two possible directions for the source of the event.

As they suspected, they found that the bursts came from outside the Solar System and also, by their random scatter across the sky, the data hinted that the sources lay not in our Galaxy (in which case one would expect the sources to lie along the plane of the Milky Way) but in the Universe beyond.

Klebasabel published the first results in 1973, detailing 16 confirmed bursts in a paper in the journal *Nature* entitled 'Observations of gamma-ray bursts of cosmic origin'. As a result, a far more sensitive gamma-ray satellite observatory was designed and built. Called the Compton Gamma Ray Observatory, it was launched in 1991 and joined a wide array of Earth satellites and deep space probes that carried much smaller detectors. Over a period of six years it observed nearly 2,000 bursts (Figure 20.1), which showed that they had an isotropic distribution across the sky and so confirmed that they were not associated with our own galaxy.

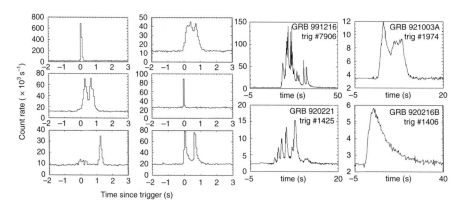

Figure 20.1 Gamma-ray burst profiles: those on the left are typical of the short bursts (less than 2 s) whilst those on the right are typical of long bursts (greater than 2 s). Image: Compton Gamma Ray Observatory, GFSC, NASA.

Figure 20.2 The Swift satellite observing a GRB with an artist's impression of how one might look. Images: GFSC, NASA.

What is the source of the gamma-ray bursts?

For many years after the discovery of GRBs, astronomers searched for a counterpart: an astronomical object whose position agreed with that of a recently observed burst. All such searches were unsuccessful, and where, in a few cases, the position of the GRB was particularly well defined, no bright objects of any nature could be seen. This suggested that the origins of these bursts were either very faint stars or extremely distant galaxies. What was really required were exceedingly fast follow-up observations at other wavebands so that, should a GRB be observed, its source could be immediately identified.

The breakthrough came in February 1997 when the satellite BeppoSAX detected a burst (GRB 970228). Its X-ray camera was immediately pointed towards the direction from which the burst had originated and detected rapidly fading X-ray emission. More significantly still, 20 hours after the burst, the UK's William Herschel Telescope on La Palma was able to identify a fading optical counterpart. Once the GRB had faded, deep imaging was able to identify a faint, distant host galaxy at the location of the GRB.

Because of the very faint luminosity of this galaxy, its exact distance was not measured for several years but, well before, a further breakthrough occurred with the BeppoSAX discovery of GRB 970508 later that year. The position of this event was found within four hours of its discovery, so allowing research teams to begin making observations much sooner than for any previous burst. The spectrum of the object revealed a redshift of $z = 0.835$, placing the burst at a distance of roughly 6 billion light years from Earth, so providing the first accurate determination of the distance to a GRB. This proved that GRBs occur in extremely distant galaxies.

As time is of the essence in making follow-up observations after the detection of a GRB, the locations determined by the current gamma-ray telescopes, such as Swift, are instantly transmitted over the Gamma-Ray Burst Coordinates Network (GRBCN). These positions can then be used to rapidly slew Earth-based telescopes onto the source position in time to observe the afterglow emission at longer wavelengths. The Swift spacecraft (Figure 20.2), which was launched in 2004 and still operational, is equipped with on-board X-ray and optical telescopes that can be rapidly and automatically slewed to observe the afterglow emission following a burst detected by its very sensitive gamma-ray detector.

On the ground, numerous optical telescopes have now been built or modified to incorporate robotic control software that responds immediately to signals sent through the GRBCN. This allows the telescopes to rapidly turn towards a GRB within seconds of receiving the positional data and make follow-up observations whilst the gamma-ray emission is still present.

There was an interesting, though sadly not realised, possibility in 2008. A GRB, 080319B, had an extremely luminous optical counterpart that peaked at a visible magnitude of 5.8. Given a very dark and transparent sky this could have been seen with the unaided eye. Should anyone have been looking in the right direction at this time, the photons that fell on their retinas would have been travelling for 7.5 billion light years, so he or she would have looked back in time more than halfway towards the origin of the Universe!

In 2009, the Swift Gamma-Ray Burst Mission detected GRB 090423 in the constellation Leo. Its afterglow was detected in the infrared and this allowed astronomers to determine its redshift. Having a z of 8.2, this makes GRB 090423 the second most distant object currently known in the Universe. At the time of its discovery it was the earliest object ever detected and its light was emitted when the Universe was only 630 million years old.

(In October 2010, the European Southern Observatory's Very Large Telescope in Chile observed a galaxy in the infrared that has a redshift of 8.55, giving a distance of 13.12 billion light years. Its light was emitted just 600 million years after the origin of the Universe. As the Universe has been expanding since its light was emitted, it is now thought to be at a distance of 30 billion light years!)

So, to summarise what is known: gamma-ray bursts are flashes of gamma rays associated with extremely energetic explosions in distant galaxies and are the most luminous electromagnetic events known to occur in the Universe. Bursts can last from milliseconds to several minutes, although a typical burst lasts a few seconds. The bursts are classified into two types, short – less than

2 seconds in length – and long – greater than 2 seconds. The initial burst is usually followed by a longer-lived 'afterglow' emitted at longer wavelengths (X-ray, ultraviolet, optical, infrared and radio).

How much energy is released?

The measurement of the approximate distance to GRB 970508 in 1997 made it possible to calculate the energy emitted during the event. One knows the energy, say E, falling on the detector, which has a given area – say $0.1\,\text{m}^2$. As the detector is unlikely to be face on to the burst direction its effective area will be less, say $0.05\,\text{m}^2$. So the energy falling on one square metre will be $E \times 20$. If the energy from the burst was emitted isotropically (equally in all directions) then one simply calculates the area in metres, say A, of the surface of a sphere centred on the galaxy at the distance of the Earth and multiplies this by the energy falling on one square metre, giving $E \times 20 \times A$. Though E is very small, A is enormous, and the calculated emitted energy turns out to be roughly equal to the energy that would be released if the total mass of the Sun were instantly converted into electromagnetic energy!

No known process in the Universe can produce this much energy in such a short time and one can thus deduce that the energy must be beamed, almost certainly in a pair of diametrically opposed beams emitted along the rotation axis of the progenitor object. There is no doubt that this must be a black hole. Nuclear fusion can convert just under 1% of the rest mass of an object into energy, but as matter falls into a rotating black hole 30% of the rest mass could be converted into energy – this being the most efficient conversion of mass to energy that is known.

Observations indicate that the beams have an angular width of a few degrees. As a result, the gamma rays emitted by most bursts are expected to miss the Earth and will never be detected. However, when one of the two beams is pointed towards Earth, the focusing of its energy causes the burst to appear much brighter than it would have been were its energy emitted isotropically. This greatly reduces the energy that must be released in the explosion. Taking this into account, typical GRBs appear to convert about 1/2,000 of a solar mass into energy. This *is* theoretically possible and comparable to the energy released in a bright Type Ib supernova.

From the statistics that have been built up over the past 20 years, it appears that GRBs are extremely rare, with only a few occurring per galaxy in a million years. None have been observed within our Milky Way Galaxy, which is not surprising given that there would only be an exceedingly low chance of one occurring in the last 40 years.

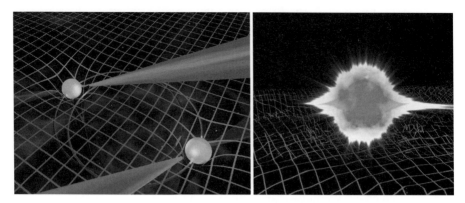

Figure 20.3 Artist's impressions of the 'Double Pulsar' and the distortion of 'space-time' they cause (left) and the GRB as they merge to form a black hole (right). Images: Jodrell Bank Centre for Astrophysics, University of Manchester.

What causes gamma-ray bursts?

It is now believed that the two types of burst, short and long, are the result of two distinct causes. In both cases the burst is emitted as a black hole is formed, but the two scenarios are quite different. Both result from the evolution of giant stars, as was discussed in Chapter 11. Supernovae give rise to neutron stars, some of which may be observed by us as 'pulsars'. Our understanding of the probable cause of the short GRBs was given a fillip by the discovery of a dual pulsar system by astronomers at Jodrell Bank Observatory. Called the 'Double Pulsar', it comprises two pulsars of masses 1.25 and 1.34 solar masses spinning with rotation rates of 2.8 s and 23 ms respectively (Figure 20.3). They orbit each other every 2.4 hours with an orbital major axis just less than the diameter of the Sun. The neutron stars are moving at speeds of 0.01% that of light and it is thus a system in which the effects of general relativity are more apparent than any other known system.

A key prediction of Einstein's General Theory of Relativity is that a pair of co-orbiting objects will emit gravitational waves – ripples of space-time that propagate out into the Universe at the speed of light. As a result, the system will be losing energy. This energy is derived from the angular momentum of the system and its loss causes the two pulsars to gradually come together. At this moment in time, general relativity predicts that the two neutron stars should be spiralling in towards each other at a rate of 7 mm per day. Observations made across the world since then, including those using the Lovell Telescope at Jodrell Bank, have show this to be exactly as predicted.

Eventually these two pulsars will fuse into one. In the case of the Double Pulsar this will not happen for about 84 million years, but there will be many such systems in our own and other galaxies where the combined mass of a pair of neutron stars will exceed the mass limit for neutron stars of about 3 solar masses. It is the final merging of a pair of neutron stars to form a black hole that is believed to be the cause of the short-period GRBs. The explosion is powered by the infall of matter into the new black hole from a surrounding accretion disc.

Some other models that have also been proposed to explain short GRBs include the merger of a neutron star and a black hole, and the collapse of a single neutron star as material falls onto it from a surrounding accretion disc until its mass exceeds that which can be supported by neutron degeneracy pressure and so it collapses into a black hole.

Long-period bursts

Most observed GRBs have a duration of greater than 2 seconds and are classified as 'long gamma-ray bursts'. As these are in the majority and tend to have the brightest afterglows, they have been studied in much greater detail than their short period (less than 2 seconds) counterparts. Virtually every well-studied long GRB has been associated with star-burst galaxies having high rates of star formation and in many cases a very bright supernova as well, thus unambiguously linking long GRBs with the deaths of massive stars. Thus, most observed GRBs are believed to be a narrow beam of intense radiation released during a supernova or hypernova event.

What stars give rise to long-period gamma-ray bursts?

Because of the immense distances of most GRB sources from Earth, identification of the progenitor systems that produce these explosions is a major challenge. The most widely accepted mechanism for the origin of long GRBs is the collapsar model, in which the core of an extremely massive, rapidly rotating star collapses into a black hole in the final stages of its evolution. (The mass of the core is greater than that which can be supported by neutron degeneracy pressure and so the collapse continues until a black hole is formed, as described in Chapter 17.) Matter surrounding the newly formed black hole falls down towards the centre and forms an accretion disc. The infall of this material into the black hole drives a pair of relativistic jets out along the rotational axis, which bore through the outer layers of the star. As they eventually break through the stellar surface the beams of gamma rays radiate into space.

The closest stars to us that are likely to produce long GRBs are likely to be of a type called Wolf–Rayet stars. These are extremely hot and massive stars that

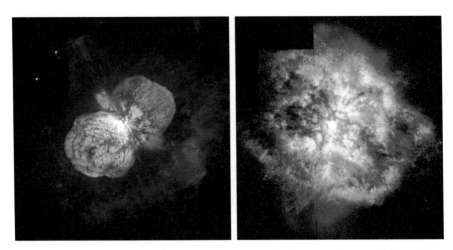

Figure 20.4 Hubble Space Telescope image of the star Eta Carinae (left) and the Wolf–Rayet star WR 124 (right) and their surrounding nebulae. Both stars are candidates for being progenitors of long GRBs. Images: STScI, ESA, NASA.

have shed most or all of their hydrogen due to radiation pressure. The stars Eta Carinae and WR 124, seen in Figure 20.4, are both considered to be possible GRB progenitors.

It should perhaps be pointed out that, as yet, there is still no generally accepted model for how GRBs convert energy into radiation. Any successful model of GRB emission must explain the physical process for generating gamma-ray emission that matches the wide range of light curves, spectra and other characteristics that are observed. A particular challenge is the need to explain the very efficient energy conversions that are inferred from some explosions. It appears that some GRBs may convert as much as half of the explosion energy into gamma rays. Based on the recent observations (in 2008) of the bright optical counterpart of GRB 080319B, whose light curve was correlated with the gamma-ray light curve, it has been suggested that the 'inverse Compton effect' may be the cause. In this model, low-energy photons are scattered by the relativistic electrons that result from the explosion, so augmenting their energy by a large factor and transforming them into gamma rays.

The nature of the afterglows seen at longer wavelengths (from X-ray to radio) that follow GRBs is easier to explain. Any energy released by the explosion that is not radiated away in the burst itself accelerates matter to speeds close to the speed of light. As this matter collides with the surrounding interstellar gas, it creates a shock wave that then propagates forward at relativistic speed into interstellar space. As energetic electrons within the shock wave are accelerated by strong magnetic fields they radiate synchrotron emission (so called because it was first observed being emitted by electrons orbiting at relativistic speeds within a circular synchrotron accelerator) across much of the electromagnetic spectrum, giving rise to the observed afterglows.

Should we be afraid?

At the present time an average of about one GRB is detected per day. As we observe them across most of the observable Universe, so encompassing many billions of galaxies, this suggests that, in a single galaxy, GRBs must be exceedingly rare events. In our own Milky Way Galaxy we might expect one burst every 100,000 to 1,000,000 years. As the radiation is beamed, only a small fraction of these would be beamed towards Earth, so we should not be too worried. However, if there were a GRB close enough to Earth, and beamed towards us, it could have significant effects on the Earth's biosphere. The absorption of radiation in the atmosphere would cause the photodissociation of nitrogen, giving rise to nitric oxide that would destroy ozone. It is thought that a GRB at a distance of ~3,000 light years could destroy up to half of Earth's ozone layer. This would allow more ultraviolet light from the Sun to reach the Earth's surface which, coupled with the direct ultraviolet irradiation from the burst, could have a major impact on the food chain and potentially trigger a mass extinction. It is estimated that one such event might happen every billion years and, though there is no direct evidence, it is possible that the Ordovician–Silurian extinction event, 440 million years ago, could have been the result of such a burst.

There seems to be some evidence that the giant stars that produce GRBs at the end of their lives were far more common in the past when stars were formed largely of hydrogen and helium and there were far fewer heavier elements present. Because the Milky Way now has a good proportion of heavier elements, this effect may reduce the number of such long GRBs that might occur now or in the future. Good! However, the merging of two neutron stars, or a neutron star with a black hole, to give a short duration burst could happen at any time; if one were sufficiently close and beamed towards us, life on Earth could still be at long-term risk.

The Big Bang origin of the Universe

This is rightly said to be the biggest explosion of all time but it was very different from any explosions that we observe today:

It did not happen in space – it created space.
It did not happen in time – it created time.

As St Augustine said, 'The Universe was created *with* time, not *in* time.' (One might argue with these two tenets, which will be further discussed in the final chapter.)

It was everywhere – every part of the Universe that we see today was, at the instant of creation, at the same point.

In the original Big Bang theory, the Universe arose from a 'singularity' of infinite density and zero volume containing all the energy that now exists, much in the form of mass, in our Universe.

Singularities occur in theories when they are no longer capable of describing the physics of extreme conditions. Even if there was no singularity, one real worry is how all the mass that we observe in the Universe (with even more that we do not observe) could have been contained in an almost infinitesimally small region of empty space. The answer is that it wasn't. As we will see, a period of what is called inflation gave rise to a vast amount of energy which, in turn, produced the fundamental particles that make up our Universe. But still, how can all of this come from essentially nothing? Surprisingly, the answer is that the total energy content of the Universe is zero. This sounds stupid. But there are two forms of 'energy': the energy that is related to the matter in the Universe given by Einstein's formula $E = mc^2$ together with its energy of motion (kinetic energy), which we might consider to be 'positive' energy, and also a 'negative' energy in the form of gravitational potential energy. It turns out that in our Universe (as it must) these two forms of energy are equal and opposite giving a zero sum!

Consider a car at the top of a hill; it has a gravitational potential energy of mgh, where m is the mass of the car, g is the acceleration due to gravity and h is the height above the surrounding land. If the car were pushed off the top and rolled down to the bottom, this energy would be released and the car would reach some speed giving it kinetic energy. Einstein's Special Theory of Relativity also states that it will gain some mass. So energy has been converted from one form to another. You can easily see that gravity is associated with negative energy: the car has gained energy of motion (kinetic energy) as it rolled down the hill. But this gain is exactly balanced by a reduction in gravitational energy as it comes closer to the Earth's centre, so the sum of the two energies remains zero.

Another example is provided by comets that come close to the Sun. As they fall in towards the gravitational well of the Sun they lose gravitational potential energy, but as there are no energy losses in space this must be counterbalanced by an increase in kinetic energy – they travel very fast indeed as they round the Sun.

In the sequence of events that made up the Big Bang that will be outlined below, all the matter, antimatter and photons were produced by the energy that was released following a period that we call 'inflation'. All of these particles and their kinetic energy consist of positive energy. This energy is, however, exactly balanced by the negative gravitational energy of everything pulling on everything else. The total energy of the Universe is zero! This idea of a zero energy universe initiated by inflation suggests that all one needed to start off our Universe was just a miniscule volume of energy in which inflation could begin.

The ultimate question is then what produced this pocket of energy and where was it? Some, as we will discuss in the final chapter, think that it might have happened in some pre-existing space and time but, as stated earlier, it could have arisen from nothing and the concepts of space and time were created along with the Universe itself.

Heisenberg's uncertainty principle – a fundamental tenet of quantum theory – provides a natural explanation for how that energy may have come out of nothing. Throughout the Universe, quantum fluctuations cause particles and antiparticles to form spontaneously. Provided they annihilate each other within the time frame determined by the uncertainty principle (the greater the combined mass, the shorter the time), this does not violate the law of conservation of energy. The spontaneous birth and death of these 'virtual particle' pairs are known as 'quantum fluctuations' and are a very well tested part of physics. Indeed, quantum fluctuations must be taken into account when calculating the energy levels of atoms. Unless the effects of virtual particle pairs (such as electrons and positrons) are included, the predicted energy levels disagree with the experimentally measured levels.

Perhaps, before the birth of our Universe, quantum fluctuations were happening. The vast majority may have quickly disappeared, but one lived for a sufficiently long time and had the right conditions for inflation to have been initiated. So the original tiny volume of space inflated by an enormous factor, and our Universe was born. If this hypothesis is true, then the answer to the question as to where it came from is that it came from nothing and its total energy is zero. But, amazingly, this has produced a Universe of incredible structure and complexity and, not least, beauty.

Inflation

The idea of inflation is an integral part of our current understanding of the Big Bang scenario. Apart from its role in creating the matter in the Universe it neatly explains two problems that haunted the standard Big Bang theories.

When Penzias and Wilson discovered the radiation called the Cosmic Microwave Background (to be discussed in detail in Chapter 22) they found that the temperature of the whole of the visible Universe was the same. If we look in one direction the radiation (which tells us the temperature of that part of space) has travelled for 13.6 billion years – the age of the Universe. If we look in the opposite direction we see exactly the same temperature – that radiation has also travelled for 13.6 billion years. In the standard Big Bang models there has not been sufficient time to allow radiation to travel from one of these regions to the other – they cannot 'know' what each other's temperature is, as this information cannot travel faster than the speed of light. So why are they at precisely the same temperature? This is called the 'horizon problem'.

Perhaps an analogy might help. From the crow's nest of a ship two other ships are seen in opposite directions and each is seen to be flying the same flag. Due to the curvature of the Earth the sailors on these two ships cannot see each other, so how can it be that they have identical flags? The obvious answer is that they must once have been so close to each other that flags (travelling at less than the speed of light) could have been given to each ship. They were then at some time in the past 'causally connected'. In the same way the whole of the Universe must have been causally connected at its origin and in a volume of space of smaller diameter than the light travel time across it.

Observations had shown that the space in the Universe was very close (<1%) to being Euclidian (having no curvature and often called 'flat space') so that, in the absence of mass, light would travel in straight lines. The Big Bang theory gives no particular reason why this should be so. Any curvature that the Universe has close to its origin tends to get enhanced as the Universe ages – a slightly positively curved space become more and more so and vice versa. The fact that our observations show us that space has no curvature implies incredibly fine tuning, and there is nothing in the standard Big Bang theory to explain why this should be so. This is called the 'flatness problem'.

These problems were addressed with the idea of inflation, first proposed by Alan Guth and refined by others. In this scenario the whole of the visible Universe was initially contained in a volume of order the size of a proton. Some 10^{-35} s after the origin this volume of space began to expand exponentially and increased in size by a factor of order 10^{50}–10^{60} in a time of ~10^{-32} s – to the size of a sphere of order a metre in size, as shown in Figure 20.5. (Some say a golf ball or a grapefruit.) This massive expansion of space would force the geometry of space to become Euclidian or 'flat', just as the surface of a balloon appears to become flatter and flatter as it expands. (Hence inflation naturally gives a 'flat' universe.) Inflation would also ensure that the whole of the visible Universe would have a uniform temperature, so also addressing the horizon

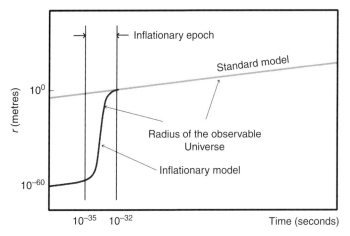

Figure 20.5 The inflationary period of the Universe.

problem. This is a result of the fact that *prior to the inflationary period* the volume of space-time that now forms the visible Universe was sufficiently small that radiation could easily travel across it and so bring it to thermal equilibrium.

A time line of the early Universe

The Planck epoch: 0–10^{-43} s

The pre-inflationary period. A tiny volume of space, perhaps containing a very small amount of matter, at exceedingly high temperatures.

The four forces of the Universe – gravitation, strong nuclear force, weak nuclear force and electromagnetic force – are combined into one. (We currently have no theory that describes this.)

The grand unification epoch: 10^{-43}–10^{-35} s

As the Universe expands and cools from the Planck epoch, gravity begins to separate from the other forces: electromagnetism and the strong and weak nuclear forces. Physics at this scale may be described by a grand unified theory. Eventually, the grand unification is broken as the strong nuclear force separates from the combined weak nuclear force and the electromagnetic (electroweak) force.

The inflationary period: 10^{-35}–10^{-32} s

The strong nuclear force separates from the combined weak nuclear force and electromagnetic force. This initiates the era of inflation when the size of the Universe increases by a factor of order 10^{50}. The final size of the Universe after this time is uncertain, but estimates range from the size of a golf ball, through that of a grapefruit, to up to a metre in diameter. From this point onwards, the Universe expands at a far slower rate – the Hubble expansion – which is initially slowing down due to the gravitational attraction of the matter within it. Quantum fluctuations in the energy density at this time produce the 'seeds' that allow the later structure of the Universe to form.

A 'phase transition' ends the inflationary expansion and releases a vast amount of energy, producing a hot, relativistic plasma of particles and radiation. An almost equal number of particles and antiparticles are initially created but with a very small excess (~1 part in a billion) of matter particles.

The quark epoch: 10^{-12}–10^{-6} s

All the fundamental particles are believed to acquire a mass via the Higgs mechanism. The fundamental interactions of gravitation, electromagnetism, the strong interaction and the weak interaction now take their present forms, but the temperature of the Universe is still too high to allow quarks to bind together to form hadrons (including protons and neutrons). Interestingly, a form of quark–gluon soup has been recreated recently in the Large Hadron Collider at CERN, so enabling physicists to study the conditions that were in existence just a billionth of a second after the origin of the Universe!

The hadron epoch: 10^{-6}–1 s

The quark–gluon plasma that composed the Universe cools sufficiently allowing hadrons, including baryons such as protons and neutrons, to form. At approximately 1 s after the Big Bang neutrinos decouple and begin travelling freely through space.

The lepton epoch: 1–10 s

The majority of particles and antiparticles annihilate each other at the end of the hadron epoch, leaving leptons and antileptons (such as photons) dominating the mass of the Universe. Approximately 10 s after the Big Bang the temperature of the Universe falls to the point at which new lepton/antilepton pairs are no longer created and most leptons and antileptons are eliminated in annihilation reactions, leaving a small residue of leptons.

The era of nucleosynthesis: 3–17 minutes

The up and down quarks form an almost equal number of protons and neutrons. Up quarks have a charge of +2/3 and down quarks a charge of −1/3. A proton is made up of two up quarks and one down quark so has a charge of

(+2/3) + (+2/3) + (−1/3) = +3/3 = +1 and the neutron is made up of one up quark and two down quarks and so has a charge of (+2/3) + (−1/3) + (−1/3) = 0. However, the free neutron is unstable and will decay into a proton, an electron and an antineutrino with a half life of just over 10 minutes. So, over the next few minutes, the number of neutrons falls and the number of protons increases. The only neutrons that can survive are those that have been incorporated into helium nuclei (also called alpha particles) by the process of nuclear fusion. At the same time small amounts of helium 3 (two protons, one neutron), deuterium (one proton, one neutron) and lithium (three protons, four neutrons) are produced. The relative amounts of these elements produced depends on the density of the Universe at this time, and so measurements of their current abundances (taking into account the nuclear synthesis that has taken place in stars since, which has, over time, increased the abundance of helium and reduced that of hydrogen) can tell us about the conditions that must have prevailed in the early Universe. For every proton there is one electron, so overall the Universe is neutral. After about 17 minutes the temperature falls to the point where no further nucleosynthesis can take place.

The photon epoch: 3 minutes to 380,000 years

During this time the energy of the photons is gradually reducing due to the expansion of the Universe, but many still have energies greater than 13.6 electronvolts (eV), the binding energy of the electron in a hydrogen atom. This means that, should a proton capture an electron to form a hydrogen atom, very quickly a photon will come along with sufficient energy to split the electron from its nucleus. Thus no atoms can form and the Universe is composed of photons, electrons, protons and alpha particles (helium nuclei) along with dark matter. The photons scatter off the electrons and, just as light scattering off water droplets forms an opaque fog, so the Universe is opaque to light.

Dark matter begins to clump: 70,000 years

According to the standard theory of cosmology, at this stage, dark matter (discussed in Chapter 21) dominates. As it is not coupled to the bath of radiation that is keeping the normal matter distribution very smooth, the dark matter can begin to concentrate into denser regions under gravity, amplifying the tiny inhomogeneities left by cosmic inflation and so making dense regions denser and rarefied regions more rarefied. As yet, we have no theory that describes its origins earlier in the Big Bang.

A fundamental change came around 380,000 years after the origin. As the Universe continued expanding, the typical photon energy dropped below 13.6 eV and hydrogen and helium atoms were able to form. There were then no free electrons to scatter light and the Universe became transparent. This is thus as far back in time as we can see. The temperature of the Universe was then about 3,000 K, and it was filled with yellow-orange light. Since then the Universe has expanded around 1,000 times and the temperature has dropped by the same ratio to ~3 K (actually 2.73 K). Thus, the radiation is now in the long-infrared/short-wavelength radio part of the radio spectrum and forms what we now call the Cosmic Microwave Background or 'the afterglow of creation', as will be described in Chapter 22. Maps of this radiation across the sky show the imprint of the quantum fluctuations that were formed during the inflationary phase of the Big Bang and provide one piece of evidence that inflation was an integral part of the Big Bang scenario. They essentially show the distribution of dark matter at this time. The concentrations of dark matter acted as gravitational wells that pulled in normal matter to them. Over the next 500 million years or so, these concentrations of dark and normal matter produced sufficiently dense regions of hydrogen and helium that the first stars and galaxies could form.

A note of caution

Though the above scenario appears to fit virtually all of the known facts and is the current 'standard model' of the early evolution of the Universe, there are still one or two puzzles and it may not necessarily be true. There are other theories that, for example, do not regard the origin of our Universe as being the origin of time and, as yet, there is no theory that explains how the dark matter was created.

Suggestions for further reading:

> *Exploding Superstars: Understanding Supernovae and Gamma-Ray Bursts* by Alain Mazure and Stéphane Basa (Springer Praxis).
> *The Biggest Bangs: The Mystery of Gamma-Ray Bursts, The Most Violent Explosions in the Universe* by Jonathan I. Katz (Oxford University Press).
> *The First Three Minutes: A Modern View of the Origin of the Universe* by Steven Weinberg (Basic Books).

21

The invisible Universe: dark matter and dark energy

Perhaps the most amazing aspect of our Universe is that we only see about 1% (in the form of stars and bright nebulae) of its total mass and energy. A further 4% of its mass is in the form of dust and gas, but this leaves about 95% to be accounted for. A major element of this is, we believe, in the form of 'dark matter'. It has been given this name as it does not interact with light and so is invisible but, due to the fact that it does have mass, it does exert a gravitational attraction on normal matter and this is how we have evidence for its existence.

The first question to ask is whether this invisible content is normal (baryonic) matter that just does not emit light, for instance gas, dust or objects such as brown dwarfs, neutron stars or black holes. These latter objects are called MACHOs (massive astronomical compact halo objects) as many would reside in the galactic halos that extend around galaxies.

There are two pieces of evidence that indicate that the total amount of normal matter in the Universe is only ~4% of the total mass/energy content. The first depends on measurements of the relative percentages of hydrogen, helium and lithium and their isotopes that were formed in the Big Bang. These put an upper limit of normal (baryonic) matter at about 4–5%. The second line of evidence is that if a significant amount of mass were in the form of MACHOs then gravitational microlensing studies (as have discovered a number of planets and are described in Chapter 12) would have detected them. Though we know that, for example, pulsars are found in the galactic halo, the total mass of these and other MACHOs cannot explain the missing matter. So we still have to account for ~96% of the total mass/energy content of the Universe. From several lines of observational evidence it is believed that a substantial part of this is in the form of non-baryonic dark matter – usually just called dark matter.

Dark matter in galaxy clusters

The first evidence of a large amount of unseen matter came from observations made by Fritz Zwicky in the 1930s. Zwicky was a Swiss astronomer who spent much of his life working at the Palomar Observatory and became a professor of astronomy at the California Institute of Technology. He made many important contributions to astronomy: for example, in 1934 he and Walter Baade coined the term 'supernova' and hypothesised that they were the transition of normal stars into neutron stars, and were the origin of cosmic rays. In 1937, he predicted the existence of gravitational lenses, as discussed in Chapter 16, 'Proving Einstein right'. Their existence was discovered in a survey made at the Jodrell Bank Observatory just five years after his death in 1974. Had he been alive when this discovery was made I believe that he would have been a very strong contender for a Nobel Prize.

A somewhat acerbic character, throughout much of his later life he felt that his work was underappreciated and he was famous for insults aimed at his colleagues, such as when he was prevented from using the 200-inch Hale Telescope when he is quoted as fuming, 'Those spherical bastards threw me off the 200 goddam-inch telescope! Made up a special rule. No observing after the age of 70! Grrrr, them I could crush!' He expanded on the quote by saying 'A spherical bastard was a bastard any way you looked at it'.

Zwicky was instrumental in obtaining a Schmidt camera for the Palomar Observatory. These are photographic instruments, discussed in Chapter 9, capable of imaging wide fields and are superb survey instruments. Having personally carried, from Germany, the 18-inch diameter correcting lens that had been figured by the designer of this type of telescope, Bernhard Schmidt, he used it to study the Coma cluster of galaxies, 321 million light years distant. He observed that the outer members of the cluster were moving at far higher speeds than were expected. He was able to measure their velocities by using the shift in the spectral lines of the Galaxy.

When the spectra of galaxies were first observed in the early 1900s it was found that their observed spectral lines, such as those of hydrogen and calcium, were shifted from the positions of the lines when observed in the laboratory. In the closest galaxies, the lines were shifted towards the blue end of the spectrum, but for galaxies beyond our Local Group the lines were shifted towards the red. This effect is called a blueshift or redshift and the simple explanation attributes this effect to the speed of approach or recession of the Galaxy, similar to the falling pitch of a receding train whistle, which we know as the Doppler effect.

Suppose a cluster of galaxies were created, none of which were in motion. Gravity would quickly cause them to collapse down into a single giant body. If, on the other hand, the galaxies were initially given very high speeds relative one to another, their kinetic energy would enable them to disperse into the Universe and the cluster would disperse, just as a rocket travelling at a sufficiently high speed could escape the gravitational field of the Earth. The fact that we observe a cluster of galaxies many billions of years after it was created implies that there must be an equilibrium balance between the gravitational pull of the cluster's total mass and the average kinetic energy of its members. This concept is enshrined in what is called the 'virial theorem', so that if the speeds of the cluster members can be found, it is possible to estimate the total mass of the cluster. Zwicky carried out these calculations and showed that the Coma Cluster must contain significantly more mass than could be accounted for by its visible content.

Dark matter in spiral galaxies

In the 1970s a problem related to the dynamics of galaxies came to light. Vera Rubin observed the light from H II regions (ionised clouds of hydrogen such as the Orion Nebula) in a number of spiral galaxies. These H II regions move with the stars and other visible matter in the galaxies but, as they are very bright, are easier to observe than other visible matter. The H II regions emit the deep red hydrogen alpha (H-alpha) spectral line. By measuring the Doppler shift in this spectral line, Rubin was able to plot their velocities around the galactic centre as a function of their distance from it. She had expected that clouds that were more distant from the centre of a galaxy (where much of its mass was expected to be concentrated) would rotate at lower speeds – just as the outer planets travel more slowly around the Sun. This is known as Keplerian motion, with the rotational speed decreasing inversely as the square root of the distance from centre. (This is enshrined in Kepler's third law of planetary motion and can be derived from Newton's Law of Gravitation.)

To her great surprise, Rubin found that the rotational speeds of the clouds did not decrease with increasing distance from the galactic centre and, in some cases, even increased somewhat (Figure 21.1). Not all the mass of a galaxy is located in the centre but the rotational speed would still be expected to decrease with increasing radius beyond the inner regions of the galaxy, although the decrease would not be as rapid as if all the mass were located in the centre. To give a concrete example: the rotation speed of our own Sun around the centre of the Milky Way Galaxy would be expected to be about 160 km/s. It is, in fact, ~220 km/s. The only way these results can be explained is either that the stars in the galaxy are embedded in a large halo of unseen matter – extending well beyond the visible galaxy – or that Newton's Law of Gravitation does not hold

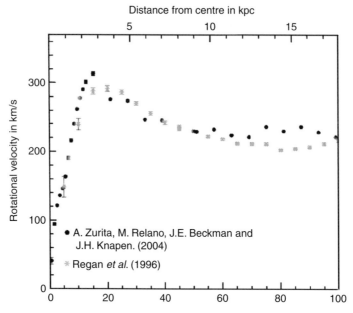

Figure 21.1 The galactic rotation curve for the galaxy NGC 1530.

true for large distances. The unseen matter whose gravitational effects her observations had discovered is further evidence for the existence of dark matter.

A modified form of Newton's law called MOND (MOdified Newtonian Dynamics) was proposed in 1981 by Mordehai Milgrom, who pointed out that Newton's second law when applied to gravitational forces has only been verified when the gravitational acceleration is large and has never been verified where the acceleration is extremely small – as would be the case for stars towards the edge of a galaxy where the gravitational forces are very weak. With a suitable choice of parameters the observed rotation curves of galaxies can be accurately modelled by the MOND theory; however, it has a much harder task explaining other observations that support the existence of dark matter, such as the dynamics of galaxy clusters and gravitational lensing, so MOND will not be considered further.

Weighing a galaxy: there is more mass than we can observe

The observations of the hydrogen line described above can be used to calculate the mass of a galaxy. The plot of Figure 21.2 shows the hydrogen line spectrum of the nearby galaxy M33, which lies at a distance of 2.36×10^{22} m (~2.9 million light years) taken with a small telescope system that I and colleagues

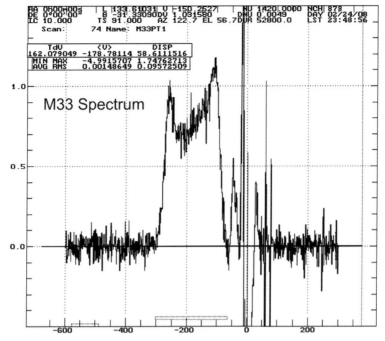

Figure 21.2 The hydrogen line spectrum of M33 in Triangulum taken using a 6.4-metre radio telescope at Jodrell Bank Observatory.

built for use by our undergraduate students at Manchester University. (An image of M33 that I took in 2013 is shown in Figure 14.13 and Plate 14.13.)

The horizontal axis of Figure 21.2 has been converted from frequency to velocity using the Doppler formula. The plot provided us with two pieces of information about the galaxy. Firstly, the spectrum of M33 has a width in frequency due to the fact that it is rotating – one side of the galaxy is coming towards us whilst the other is moving away, and secondly, the centre of the M33 hydrogen emission corresponds to a velocity of −180 km/s. You might well deduce from this latter statement that the galaxy as a whole is moving *away* from us at this speed but, as all but a few galaxies are moving away from us, the sign convention that is used is that galaxies moving away from us are given positive velocities and those moving towards us are given negative velocities. So this indicates that M33 is moving *towards* us at a speed of ~180 km/s. However, our Solar System is moving around the centre of the Galaxy at a speed of ~220 km/s and, having corrected for this, M33 is actually moving towards the Milky Way Galaxy at a speed of ~24 km/s.

The width of the spectral line is ~200 km/s so that the hydrogen at the edge of the galaxy is apparently moving around the centre at a speed of 100 km/s.

However, though the galaxy is presumably circular, its dimensions on a photographic plate are ~71 × 45 arcminutes. This implies that it is inclined to our line of sight at an angle of arcsin(45/71) = ~39 degrees. As a result, the value we measure will be less than the true value due to the projection effect. (If the plane of the galaxy were perpendicular to us, we would not observe any rotational width in the hydrogen line spectrum.) The true rotational velocity of the outer parts of the galaxy about its centre should thus be close to 100/sin(39) km/s = 158 km/s.

Knowing the distance of the galaxy and its angular size we can calculate its radius. M33 is ~71 arcminutes across and lies at a distance of 2.36×10^{22} m and this gives a radius of ~2.4×10^{20} m. If the mass distribution of the galaxy is symmetrical then the gravitational effect on the hydrogen gas at the edge of the galaxy is the same as if all the galaxy's mass were concentrated at its centre. One can thus use an identical method to that used to calculate the mass of the Sun and this gives a total mass for M33 of ~45 billion solar masses.

We have another method of estimating the mass using what is termed the 'mass to light ratio' of stars. This is simply the ratio between the mass of a star or star cluster divided by its luminosity – our Sun has, by definition, a mass of 1 solar mass and a luminosity of 1 so it has a mass to light ratio of 1. One could assume that all the stars in M33 are similar to our Sun in terms of their mass to light ratio. If we then calculate the luminosity of M33 compared to that of our Sun we will directly get an estimate of the mass of M33 in solar masses and this gives a value of ~5 billion solar masses. This is a factor of 10 less than the value derived above. M33 obviously has mass that does not emit light, such as dust and gas, and not all stars will have the same mass to light ratio as our Sun – hot stars are very luminous for their mass compared to our Sun and cooler stars (of which there are many more) less luminous. The average mass to light ratio for stars, gas and dust in our own galaxy is ~1.5 so, assuming a similar mix, this would give M33 a mass of ~8 billion solar masses.

The fact that this is still a factor of ~6 less than that derived dynamically is evidence of the presence of dark matter in the galaxy – there appears to be about five times as much dark matter than normal matter in the makeup of the galaxy.

The early Universe and the formation of galaxies

It was the American physicist George Gamow who first realised that the Big Bang should have resulted in radiation that would still pervade the Universe. This radiation is now called the Cosmic Microwave Background (CMB) and is the subject of the next chapter – 'The afterglow of creation'. Initially in the form of very high energy gamma rays, the radiation became less energetic as the

Universe expanded and cooled, so that by a time some 300 to 400 thousand years after the origin, the peak of the radiation was in the optical part of the spectrum.

Up to that time the typical photon energy was sufficiently high to prevent the formation of hydrogen and helium atoms and thus the Universe was composed of hydrogen and helium nuclei and free electrons – so forming a 'plasma'. The electrons would have scattered photons rather as water droplets scatter light in a fog and thus the Universe would have been opaque. This close interaction between the matter and radiation in the Universe gives rise to a critical consequence: the distribution of the nuclei and electrons (normal matter) would have a uniform density except on the very largest scales as the photons acted rather like a whisk beating up a mix of ingredients.

This fact is important to the argument that follows. When the temperature drops to the point that atoms can form, the matter can begin to clump under gravity to form stars and galaxies. Simulations have shown that, as the initial gas is so uniformly distributed, it would take perhaps 8 to 10 billion years for regions of the gas to become sufficiently dense for this to happen. But we know that galaxies came into existence around 1 billion years after the Big Bang. Something must have aided the process. We believe that this was due to dark matter. As indicated above, dark matter would not have been affected by the radiation so could have begun to gravitationally 'clump' soon after the Big Bang. Thus, when the normal matter became decoupled from the photons, there were 'gravitational wells' in place formed by concentrations of dark matter. The normal matter could then quickly fall into these wells, rapidly increasing its density and thus greatly accelerating the process of galaxy formation.

Gas entrapment

NASA's Chandra X-ray Observatory has revealed that the elliptical galaxy NGC 4555 is embedded in a cloud of gas having a diameter of about 400,000 light years and a temperature of 10 million K (Figure 21.3). At this temperature the molecules of gas will be travelling at very high speeds and the mass of the stars within the galaxy will be far too low to prevent its escape. For the gas to remain in the vicinity of the galaxy the total mass of the system must be about 10 times the combined mass of the stars in the galaxy and about 300 times that of the gas cloud.

Gravitational lensing

Massive objects in space distort the space around them and bend the light that passes near to them. This has the effect of forming what is called a

Figure 21.3 NASA's Chandra X-ray satellite image of hot gas surrounding the galaxy NGC 4555. Image: Chandra, NASA.

Figure 21.4 The cluster Abell 2218 imaged by the Hubble Space Telescope. Image: STScI, ESA, NASA.

gravitational lens. We can observe the lensing effect of single stars (and even planets) and single galaxies. On a much larger scale the mass of a cluster of galaxies can distort the images of more distant objects. The image of the cluster Abell 2218 (Figure 21.4) is a wonderful example showing images of more distant galaxies that have been distorted into arcs. The amount of distortion is a function of the total mass of the intervening cluster, so this gives a way of estimating the total mass of galaxy clusters, confirming the existence of dark matter. Using this technique, astronomers have even shown how the distribution of dark matter has become more 'clumpy' over the last 6 billion years.

How much non-baryonic dark matter is there?

There are several ways of estimating the amount of dark matter. One of the most direct is based on the detailed analysis of the fluctuations in the CMB. The percentage of dark matter has an observable effect, and the best fit to current observations corresponds to dark matter making up ~27% of the total mass/energy content of the Universe. Other observations support this result. Only ~5% is made up of normal matter, leaving two further questions: what is dark matter and what provides the remaining ~68% of the total mass/energy content?

What is dark matter?

The honest answer is that we do not really know. The standard model of particle physics does not predict its existence and so extensions to the standard theory (which have yet to be proven) have to be used to predict what it might be and suggest how it might be detected.

Dark matter can be split into two possible components: hot dark matter would be made up of very light particles moving at close to the speed of light (hence hot) whilst cold dark matter would comprise relatively massive particles moving more slowly. Simulations that try to model the evolution of structure in the Universe – the distribution of the clusters and superclusters of galaxies – require that most of the dark matter is 'cold', but astronomers do believe that there is a small component of hot dark matter in the form of neutrinos. There are vast numbers of neutrinos in the Universe but they were long thought to have no mass. However, recent observations attempting to solve the solar neutrino problem, discussed in Chapter 2, show that neutrinos can oscillate between three types: electron, tau and muon. This implies that they must have some mass but current estimates put this at less than 1 millionth of the mass of the electron. As a result they would only make a small contribution to the total amount of dark matter – agreeing with the simulations.

A further confirmation of the fact that hot dark matter is not dominant is that, if it were, the small-scale fluctuations that we see in the WMAP data would have been 'smoothed' out and the observed CMB structure as described in Chapter 22, 'The afterglow of creation', would show far less detailed structure.

Axions

One possible candidate for cold dark matter is a light neutral axion whose existence was predicted by the Peccei–Quinn theory in 1977. There would be of order 10 trillion in every cubic centimetre. If axions exist, they could theoretically change into photons (and vice versa) in the presence of a strong magnetic field. One possible test would be to attempt to pass light through a wall. A beam of light is passed through a magnetic field cavity adjacent to a light barrier. A photon might rarely convert into an axion, which could easily pass through the wall where it would pass into a second cavity where (again with an incredibly low probability) it might convert back into a photon.

Another experiment at Lawrence Livermore Laboratory is searching for microwave photons within a tuned cavity that might result from an axion decay, whilst in Italy polarised light is being passed back and forth millions of times through a 5-tesla field. If axions exist, photons could interact with the field and become axions, causing a very small anomalous rotation of the plane of polarisation. The most recent results do indicate the existence of axions with a mass of about three times that of the electron, but this has to be confirmed and there may well be other causes for the observed effect on the light.

WIMPS

An extension to the standard model of particle physics called 'supersymmetry' suggests that WIMPS (weakly interacting massive particles) might be a major constituent of cold dark matter. A leading candidate is the neutralino – the lightest neutral supersymmetric particle. Billions of WIMPS could be passing through us each second and there are a number of ways by which their existence might be detected either directly or indirectly. The latter will be covered first.

Indirect detection experiments

Indirect detection experiments search for the products of WIMP annihilation or decay. If WIMPs are their own antiparticle then two WIMPs could annihilate to produce gamma rays or particle–antiparticle pairs and, if they are unstable, WIMPs could decay into particles of normal matter. The results of these processes could be detected through an excess of gamma rays, antiprotons or positrons emanating from regions where it is expected dark matter might be concentrated, such as the centre of our Galaxy. However, other processes, as yet not fully understood, can give rise to these decay products so the detection of such a signal is not conclusive evidence for dark matter.

The Fermi Gamma-ray Space Telescope is searching for gamma rays from dark matter annihilation and decay. It was reported in April 2012 that an analysis of data from its Large Area Telescope instrument has produced strong evidence of a 130 GeV line in the gamma radiation coming from the centre of the Milky Way and that WIMP annihilation seemed to be the most probable explanation.

An instrument package called PAMELA was flown on a Russian Earth orbiting satellite. On 5 November 2008, it was reported that it had detected an excess of high-energy positrons coming from the centre of our Galaxy. This excess could be the result of an interaction between two dark matter particles and so, as the authors of the *Nature* paper say, 'may constitute the first indirect evidence of dark-matter particle annihilations'. There would be expected to be a major concentration of dark matter towards the centre of the Galaxy, although they add that there could yet be other explanations, such as the presence of a nearby pulsar.

The International Space Station is carrying an instrument called the Alpha Magnetic Spectrometer, which is designed to directly measure the fraction of cosmic rays that are positrons. The first results, published in April 2013, indicated an excess of high-energy cosmic rays that could potentially be due to the annihilation of dark matter.

As WIMPs pass through the Sun or the Earth, they may scatter off atoms and lose energy, giving rise to an accumulation at their centres, so increasing the chance that two will collide and annihilate. This could produce a flux of high-energy neutrinos originating from the centre of the Sun or the Earth. If such a signal were found, its detection would constitute a powerful indirect proof of WIMP dark matter and it is being searched for by high-energy neutrino telescopes such as AMANDA, IceCube and ANTARES.

Direct detection experiments

Very occasionally, it is thought that a WIMP will interact with the nucleus of an atom making it recoil – rather like the impact of a moving billiard ball with a stationary one. In principle, but with very great difficulty, these interactions could be detected.

Though a million WIMPS might pass through every square centimetre of the Earth each second, they will very rarely interact with a nucleus of a heavy atom. It is estimated that within a 10-kilogram detector only one interaction might occur, on average, each day. To make matters worse we are being bombarded with cosmic rays that, being made of normal matter, interact very easily. Any

WIMP interactions would be totally swamped! One way to greatly reduce the number of cosmic rays entering a detector is to locate it deep underground – such as at the bottom of the Boulby Potash Mine in North Yorkshire, UK, at a depth of 1,100 m. At this depth, the rock layers will have stopped all but one in a million cosmic rays. In contrast, only about three in a billion WIMPs would have interacted with nuclei in the rock above the detector.

To make matters worse, natural radioactivity in the rocks surrounding the experimental apparatus increases the 'noise' that can mask the WIMP interactions, so the detectors are surrounded by radiation shields of high purity lead, copper wax or polythene and may be immersed within a tank of water. The chosen detectors may also emit alpha or beta particles so care must be taken over the materials from which they are made. Photomultiplier tubes (to detect scintillation) cause a particular problem and 'light guides' are used to transfer the light from the crystal, such as sodium iodide, in which the interaction takes place to the shielded photomultiplier tube.

There are two main techniques used by current experiments to detect WIMP interactions. One aims to detect the heat produced when a particle hits an atom in a crystal absorber such as germanium. Cryogenic detectors, operating at temperatures below 100 mK, are used to detect such events. The second technique use a tank of liquid such as xenon or argon surrounded by photodetectors to detect the flash of light produced when a particle interacts in the liquid. Importantly, both of these techniques are capable of distinguishing between the events produced when background particles scatter off electrons and those when dark matter particles scatter off nuclei.

Possible success?

One possible way to show the presence of WIMP interactions in the presence of those caused by local radioactivity is due to the fact that, in June, the motion of the Earth around the Sun (29.6 km/s) is in the same direction as that of the Sun in its orbit around the Galactic Centre (232 km/s). So the Earth would sweep up more WIMPS than in December when the motions are opposed. The difference is ~7%, so one might expect to detect more WIMPs in June than in December. As the number of interactions from local radioactivity should remain constant, this gives a possible means of making a detection. In the DArk MAtter (DAMA) experiment at the Gran Sasso National Laboratory, 1,400 m underground in Italy, observations have been made of scintillations within 100 kg of pure sodium iodide crystals. The results of seven annual cycles have given what is regarded as a possible detection but, again, there may be other explanations.

In 2011, researchers using the CRESST (Cryogenic Rare Event Search with Superconducting Thermometers) experiment in Italy, which is also located in the Gran Sasso National Laboratory, presented evidence of 67 collisions occurring in detector crystals from sub-atomic particles and calculated that there is a less than 1 in 10,000 chance that all were caused by known sources of interference or contamination. The detectors used in CRESST have an advantage in that three types of nuclei – calcium, tungsten and oxygen – are bonded together in calcium tungstate. Each of these nuclei has a different mass so that if, for example, the recoil comes from a tungsten nucleus – the heaviest of the three – it implies that the WIMP itself is heavy. Their results suggested that many of these collisions were caused by WIMPs, and/or other unknown particles.

The LUX experiment

The hopes outlined above have been somewhat dashed by the first results, published in October 2013, from the LUX (Large Underground Xenon) experiment located in the Davis Laboratory of the Sanford Underground Research Facility (SURF) located deep in the Homestake Mine in South Dakota, USA. This laboratory is where, as described in Chapter 2, Ray Davis discovered the solar neutrino problem and is at a depth of 1,500 m so that the flux of cosmic rays is reduced by a factor of a million compared to the surface.

In the hope of capturing WIMP interactions, LUX employs a target of 368 kg of liquefied ultra-pure xenon, cooled to −100 °C, at the heart of a 318,000-litre tank of water to provide additional shielding. Xenon is a scintillator that produces light proportional to the amount of energy released in any particle interactions within it. The light is then detected by an array of light detectors (photomultiplier tubes) sensitive to a single photon.

Because the xenon is very pure, it produces very little intrinsic background radiation itself and, as it is three times as dense as water, it prevents the majority of radiation originating from outside the detector from reaching its centre. This produces a 'very quiet region' in the middle of the target volume in which to search for dark matter interactions.

Particle interactions inside the LUX detector produce both photons and electrons. The photons, moving at the speed of light in xenon (~0.64 the speed of light), are immediately detected by the photomultiplier tubes. This photon signal is called S1. By analysing the photons detected by the array of 61 photodetectors above and below the xenon target the x–y position of the interaction within the target can be found. An electric field that pervades the liquid xenon causes the electrons to drift upwards (at about 1 km/s) towards the liquid surface. When they reach the surface and enter the gas above, they produce photons (the S2 signal) by electroluminescence, exactly as light is produced in neon signs. The lower down in the xenon tank the interaction takes place, the longer it will take for the electrons to drift up to the surface and hence the greater the difference in time between the S1 and S2 signals, so allowing the depth (the z co-ordinate) at which the interaction occurred to be found. Thus LUX can determine at which point within the target interactions happen.

There is a fundamental problem of distinguishing any WIMP detections from the numerous interactions caused by the remaining neutrons that have penetrated the shielding surrounding the xenon target. Neutrons have a very high chance of interacting with several xenon nuclei as they lose their kinetic energy and so produce multiple events within a short time frame. On the other hand, any WIMPS passing through the target that do interact will have an almost nonexistent chance of interacting a second time so will produce only a single event. (Of course, since WIMPs are so weakly interacting, most will pass through the detector unnoticed.) In this way LUX is able to distinguish any WIMP interactions that might occur. Less easy to distinguish from WIMP interactions are those resulting from electron interactions, which will also only produce a single event. However, these electron recoil events do have a recognised signature so can be separated out.

I have included a full account of the LUX detector as I feel that it is a beautiful example of experimental technique.

The first results

The first results from LUX, based on the first 85-day observing run, were reported at the end of October 2013. In this time, LUX produced 160 single events that passed the data analysis selection criteria, all of which were consistent with the electron recoil background, and so it is not thought that any WIMPs had been detected. This is the most sensitive dark matter direct detection result yet made and rules out the hints of positive detections made by other experiments. For example, a far smaller ultra-cold silicon detector had produced three possible WIMP detections but, if these were real, LUX's far larger detector should have detected more than 1,600 events – one every 80 minutes – but none were detected.

What of the future?

LUX will be undertaking longer runs to hopefully track down these enigmatic objects but waiting in the wings is a far larger detector called LUX-ZEPLIN (LZ). The ZEPLIN group, whose experiments were conducted in the Boulby Mine in North Yorkshire, UK, pioneered many of the techniques now in use in LUX and, following their final experiment with ZEPLIN-III in 2012, joined in the LUX collaboration, so LZ will be based on the techniques developed for both the LUX and the ZEPLIN experiments, hence the name. It will, however, incorporate further capabilities beyond those already demonstrated in the LUX and ZEPLIN experiments. Compared to the 0.49 m diameter of LUX, LZ will be 1.2 m across and will contain ~21 times as much xenon so greatly increasing its sensitivity. A principal advance in LZ will be the fact that the xenon tank will be surrounded by a clear-acrylic tank filled with a liquid scintillator, which will substantially enhance the ability to eliminate events caused by background particles. If LUX does not make the first detection of dark matter particles, then there is every hope that LZ will.

However, dark matter particles may be unlike anything in the standard particle models, and if so, 'passive' detectors such as LUX may never detect them. But there is another approach that is, in effect, an 'active' detector, where a very high energy accelerator such as the Large Hadron Collider is used to create WIMPS and identify them from their decay products.

How much normal and dark matter is there and what else must there be?

Normal and dark matter can between them account for some 32% of the total mass–energy of the Universe. It appears that the majority, some 68%, must be something else. It is thought to be a form of energy – dark energy – latent within and totally uniform in space. In fact this could be exactly what was invoked by Einstein to make his 'static' universe – the cosmological constant or Lambda (Λ) term. A positive Λ term can be interpreted as a fixed positive energy density that pervades all space and is unchanging with time. Its net effect would be repulsive. There are, however, other options and a range of other models are being explored where the energy is time dependent. These are given names such as 'quintessence', meaning fifth force. As the total amount of this energy and its repulsive effects are proportional to the volume of space, the effects of dark energy should become more obvious as the Universe ages and its size increases.

In the Big Bang models of the Universe, the initial expansion slows with time as gravity reins back the expansion, and the expansion rate would never increase. However, if there is a component in the Universe whose effect is repulsive and increasing with the volume of space, the scale size of the Universe will vary in a quite different way with time. Initially, when the volume of the Universe is small, gravity will dominate over dark energy and the initial expansion rate of the Universe will slow – just as in the standard Big Bang models, but there will come a point when the repulsive effects of the dark energy will first equal and then overcome gravity and the Universe will begin to expand at an ever increasing rate. If this is the case, distant galaxies will be further away from us than would have been the case in the standard Big Bang models.

Evidence for dark energy

In the 1990s it became possible to measure the distance to very distant galaxies. We can estimate the distance to remote objects if we have what is called a standard candle – an object of known brightness – some of which have been observed at known distances nearby. Hubble used such a technique to measure the distances of galaxies using Cepheid variable stars of known peak brightness. These had first been observed in the Small Magellanic Cloud (SMC) at a known distance from us. Suppose one of these stars is observed in a distant galaxy and appears 1/10,000 as bright as a similar Cepheid in the SMC. Assuming no extinction by dust it would, from the inverse square law, be at 100 times the distance of the SMC.

But, though Cepheid variable stars are some of the brightest stars known, there is a limit to distances that can be measured by using them. Something brighter is required. For a short time, supernovae are the brightest objects in the Universe and there is one variant, called a Type Ia supernova, that is believed to have a well-known peak brightness. It might be useful to consider an analogy. Imagine a ball of plutonium of less than critical mass. If one then gradually added additional plutonium uniformly onto its surface it would, at some point in time, exceed the critical mass and explode. The power of this explosion should be the same each time the experiment is carried out, as a sphere of plutonium has a well-defined critical mass. As you will see, this is rather similar to what occurs when a Type Ia supernova occurs.

A Type Ia supernova occurs in a binary system. The more massive star of the pair will evolve to its final state first and its core may become a white dwarf about the size of the Earth. Later, its companion will become a red giant and its size will dramatically increase. Its outer layers may then be attracted onto the surface of the white dwarf whose mass will thus increase (Figure 21.5). At some critical point, when its mass nears the Chandrasekhar limit of roughly 1.44 times the mass of the Sun, the outer layers will ignite and, in the resulting thermonuclear explosion, the entire white dwarf star will be consumed. As all such supernovae will explode when they reach the same total mass, it is expected that they will all have similar peak brightness (about 5 billion times brighter than the Sun) and should thus make excellent standard candles.

Because Type Ia supernovae are so bright, it is possible to see them at very large distances. The brightest Cepheid variable stars can be seen at distances out to about 10–20 Mpc (~32 to 64 million light years). Type Ia supernovae are about

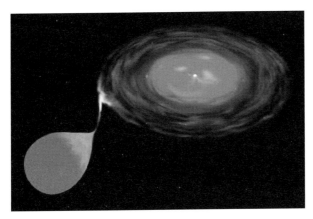

Figure 21.5 Material accreting onto a white dwarf. Image: JPL, NASA, Wikimedia Commons.

a quarter of a million times brighter than Cepheid variables and so can be seen about 500 times further away, corresponding to a distance of around 1,000 Mpc – a significant fraction of the radius of the known Universe. However, supernovae are rare, with perhaps only one occurring in a typical spiral galaxy every 300 years. Observations are now examining thousands of distant galaxies on a regular basis and sophisticated computer programs look for supernovae events. Once initially detected, observations continue to look for the characteristic light curve of a Type Ia supernova that results from the radioactive decay of nickel-56; first to cobalt-56 and then to iron-56.

Hubble (and later others) plotted the apparent expansion velocity of galaxies against their distance and produced a linear plot. This plot would not be expected to continue linearly out to very great distances due to changes in the expansion rate of the Universe over time. It was expected that the curve would fall below the linear line for great distances as it was thought that the expansion rate of our Universe was slowing down. Observations of distant Type Ia supernovae have recently enabled far greater distances to be measured which, together with the corresponding redshifts, have enabled the Hubble plot to be extended to the point where the plot would no longer be linear. As expected, the plot is no longer linear but the curve falls above the linear extrapolation, not below. This implies that the expansion of the Universe is speeding up, not slowing down as expected, and is thus evidence that dark energy exists.

This discovery was based on observations of Type Ia supernovae by the High-z Supernova Search Team in 1998, followed in 1999 by those made by the Supernova Cosmology Project. In 2011, the Nobel Prize in Physics was awarded for this work. The prize, 'for the discovery of the accelerating expansion of the

Universe through observations of distant supernovae', was divided, one half awarded to Saul Perlmutter who led the Supernova Cosmology Project, the other half jointly to Brian P. Schmidt and Adam G. Riess who led the High-z Supernova Search Team.

The nature of dark energy

Dark energy is known to be very homogeneous, not very dense (about 10^{-29} g/cm^3) and appears not to interact through any of the fundamental forces other than gravity. This makes it very hard to detect in the laboratory.

The simplest explanation for dark energy is that a volume of space has some intrinsic, fundamental energy as hypothesised by Einstein with his cosmological constant, Λ (Lambda). Einstein's Special Theory of Relativity relates energy and mass by the relation $E=mc^2$ and so this energy will have a gravitational effect. It is often called a vacuum energy because it is the energy density of empty vacuum. In fact, most theories of particle physics predict vacuum fluctuations that would give the vacuum exactly this sort of energy. One can perhaps get some feeling of why a pure vacuum can contain energy by realising that it is not actually empty! Heisenberg's uncertainty principle allows particles to continuously come into existence and (quickly) go out of existence again. A pure vacuum is seething with these virtual particles.

According then to Heisenberg's uncertainty principle there is an uncertainty in the amount of energy that can exist. This small uncertainty in energy, ΔE, allows a non-zero energy to exist for short intervals of time, ΔT, where $\Delta E \times \Delta T$ is of order $h/2\pi$, with h being the Planck constant, which is equal to 6.626×10^{-34} m^2 kg s^{-1}.

Because of the equivalence between matter and energy, these small energy fluctuations can produce particles of matter (a particle and its antiparticle must be produced simultaneously) that come into existence for a short time and then disappear. As an example, consider the proton and the antiproton, which have masses of 1.7×10^{-24} g. If a virtual pair were to be created, their equivalent energy would be (from $E=mc^2$) 3×10^{-3} ergs and thus they could only exist for a time of order 3×10^{-25} s.

A number of experiments have been able to detect this vacuum energy. One of these is the Casimir experiment in which, in principle, two metal plates are placed very close together in a vacuum. In practice it is easier to use one flat plate and one plate that is part of a sphere of very large radius. One way to think of this is that the virtual particles have associated wavelengths: the wave–particle duality. Virtual particles whose wavelengths are longer than the separation of the plates cannot exist between them so there are more virtual particles on their outer sides and this imbalance gives the effect of an attractive force between the plates.

An interesting analogy is when two ships sail alongside each other to transfer stores or fuel in the open sea in conditions with little wind but a significant swell. Between them, only waves whose wavelength is smaller than the separation of the hulls can exist but on the outside all wavelengths can be present. This inequality gives rise to a force that tends to push the two ships together, thus requiring that the ships actively steer away from each other.

The cosmological constant is the simplest solution to the problem of cosmic acceleration, with just one number successfully explaining a variety of observations, and it has become an essential feature in the current standard model of cosmology. Called the 'Lambda-CDM' model, as it incorporates both cold dark matter and the cosmological constant, it can be used to predict the future of the Universe, as will be discussed in the final chapter – 'To infinity and beyond'.

Two more books:

> *Heart of Darkness: Unraveling the Mysteries of the Invisible Universe* by Jeremiah P. Ostriker and Simon Mitton (Princeton University Press).
> *The Cosmic Cocktail: Three Parts Dark Matter* by Katherine Freese (Princeton University Press).

22

The afterglow of creation

Predictions of a 'hot' Big Bang and hence the presence of radiation within the Universe

Two American scientists, George Gamow and Richard Dicke, independently predicted that very high temperatures must have existed at the time of the Big Bang – both for somewhat the wrong reasons. Gamow wanted the temperature to be very high so that all the elements found in nature could be synthesised in the early stages of the Universe. We now know that apart from nitrogen (in the CNO process) the elements are created during the later stages in the life of stars and, in the case of massive stars, during the end of their life in a supernova explosion.

The other prediction was made by Richard Dicke, a physicist at Princeton University. He, like Fred Hoyle, did not like the idea of a singular start to the Universe (what was there before?) and came up with the idea of an oscillating universe in which the Universe expands up to a maximum size and then collapses again to a singularity (called the Big Crunch) of the maximum density possible before expanding again. Since Gamow's failure to explain the formation of the elements heavier than helium, Hoyle and others had shown how the heavier elements had been formed in stars. Thus, to start afresh with a new expanding universe, all the heavier elements had to be destroyed. Dicke realised that extreme heat – temperatures of at least a billion degrees – would do the job nicely as the heavy elements would crash together and split up into their constituent protons, neutrons and electrons.

An undeniable consequence of both these predictions is that such a hot phase in the life of the early Universe would have been filled with very high energy photons, initially in the form of gamma rays. As the Universe expanded the radiation would become less energetic and would now be in the far infrared and very short wave radio parts of the spectrum. But this means that if one could place a thermometer in the space between the galaxies it would not read absolute zero but a value of a few degrees above absolute zero. This radiation is now normally called the 'Cosmic Microwave Background' radiation (CMB) but has had alternative names such as 'the relict radiation' and the name of this chapter – 'The afterglow of creation'.

In 1948, Gamow produced an important paper with his student Ralph Alpher, which was published as 'The origin of chemical elements' in the journal *Physical Review*. (Gamow had the name of Hans Bethe listed as one of the authors on the article even though he had played no part in its preparation. The paper became known as the Alpher–Bethe–Gamow paper to make a pun on the first three letters of the Greek alphabet, alpha, beta and gamma!)

The paper showed how the present levels of hydrogen and helium in the Universe (which are thought to make up ~98% of all matter) could be largely explained by reactions that occurred during the 'Big Bang'. This lent theoretical support to the Big Bang theory, although it did not explain the presence of elements heavier than helium. In this paper, no estimate of the strength of the present-day residual CMB was made, but shortly afterwards his students Ralph Alpher and Robert Herman predicted that the afterglow of the Big Bang would have cooled down after billions of years, and would now fill the Universe with a radiation whose effective temperature was five degrees above absolute zero. The names of Alpher and Herman have been largely ignored since then, which is somewhat unfair. Their prediction was essentially forgotten about for 20 years and even they thought, incorrectly, that the technology then available would be unable to detect such weak radiation.

Ironically, the effective temperature of the Universe had been measured by this time, but was published in a chemistry journal, and so this did not become common knowledge for many years after. The radical CN can exist in space due to the photodissociation of HCN in dense molecular clouds. It was first discovered in 1941 by A. McKellar using a Coudé spectrograph mounted on the 100-inch Hooker telescope on Mount Wilson. The radical has some excited rotational vibration levels and McKellar observed one that corresponded to an effective temperature of 2.3 degrees above absolute zero. This implied that the CN lay in a bath of radiation at this effective temperature – the first observation of the CMB!

Richard Dicke had been a pioneer in the field of radio astronomy and he realised that this radiation, now largely in the microwave part of the spectrum, should be detectable. He recruited two young physicists, David Wilkinson and Peter Roll, to build a horn receiver and detector to look for

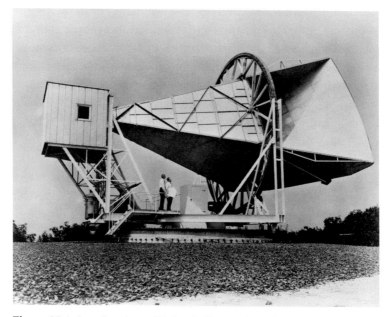

Figure 22.1 Arno Penzias and Robert Wilson at the Holmdel antenna with which they discovered the Cosmic Microwave Background. Image: Great Images in NASA, Wikimedia Commons.

what they called 'the primeval fireball'. They began work in the spring of 1964.

The serendipitous discovery of the Cosmic Microwave Background

Our story now moves to the Bell Telephone Laboratories where two radio astronomers, Arno Penzias and Robert Wilson, had been given use of the telescope (Figure 22.1) and receiver that had been used for the very first passive satellite communication experiments using a large aluminium covered balloon called 'Echo'. It had been designed to minimise any extraneous noise that might enter the horn-shaped telescope, and the receiver was one of the best in the world at that time. They tested it thoroughly and found that there was more background noise produced by the system than they expected. They wondered if it might have been caused by pigeons nesting within the horn – being at ~290 K they would radiate radio noise – and bought a 'Haveaheart' pigeon trap (now in the Smithsonian Air and Space Museum in Washington) to catch the pigeons. They used the internal mail system to send them 40 miles away to the Bell Telephone Whippany site where they were released but, as pigeons do, they returned and had to be 'removed' by a local pigeon expert. During their time within the horn antenna, the pigeons had covered much of the interior with what, in their letter to the journal *Science*, was called 'a white dielectric substance' – we might call it 'guano'. This was cleaned out as well, but having removed both the pigeons and the guano there was no substantial difference. The excess noise remained the same wherever they pointed the telescope – it came equally from all parts of the sky.

Another radio astronomer, Bernie Burke, when told of the problem suggested that they contact Robert Dicke at Princeton University. As described above, Dicke had independently theorised that the Universe should be filled with radiation resulting from the Big Bang and his students were building a horn antenna on top of the physics department in order to detect it. Learning of the observations made by Penzias and Wilson, Dicke immediately realised that his group had been 'scooped', and told them that the excess noise was not caused within their horn antenna or receiver but that their observations agreed exactly with predictions that the Universe would be filled with radiation left over from the Big Bang. Dicke was soon able to confirm their result, and it was perhaps a little unfair that he did not share in the Nobel Prize that was awarded to Penzias and Wilson.

The cause of the Cosmic Microwave Background: what we now believe

We believe that the Universe began in a burst of inflation that expanded a volume of space smaller than a proton by a factor of perhaps 10^{60} up to the size of order a metre in diameter in a time of less than 10^{-23} s. This released a massive amount of energy. Half of the gravitational potential energy that arose from this inflationary period was converted into kinetic energy from which arose an almost identical number of particles and antiparticles, but with a very small excess of matter particles (about one part in several billion). All the antiparticles annihilated with their respective particles giving rise to a bath of radiation (the CMB) within which remained a residual number of particles.

The free electrons scattered the light – rather as water droplets scatter light – and so, as in a fog, the Universe was opaque. As the Universe expanded and cooled there finally came a time, ~380,000 years after the origin, when the typical photon energy became low enough to allow atoms to form. There were then no free electrons left to scatter radiation so the Universe became transparent. This is thus as far back in time as we are able to see. At this time the Universe had a temperature of ~3,000 K. As the radiation and matter were then in thermal equilibrium, the radiation would have a very well defined power spectrum – known as the black body spectrum – with a peak of energy in the yellow part of the visible spectrum. Since that time, the Universe has expanded by about 1,000 times. The wavelengths of the photons that made up the CMB will also have expanded by 1,000 times and so will now be in the far infrared and short wavelength radio part of the spectrum – but will still have a black body spectrum. The effective black body temperature of this radiation will have fallen by just the same factor and would thus now be ~3 K.

This prediction agrees well with the average temperature, now measured, of the CMB of 2.725 K. However, it was not until 1992, nearly 30 years later, that measurements made by the COBE spacecraft were able to show that the CMB had the precise black body spectrum that would result from the Big Bang scenario. Since then, it has been very difficult to refute the fact that there was a 'hot' Big Bang.

The Cosmic Background Explorer

The Cosmic Background Explorer (COBE) was a satellite whose goal was to investigate the CMB radiation of the Universe. COBE was originally planned to be launched on Space Shuttle mission STS-82-B in 1988 from Vandenberg Air Force Base, but the Challenger explosion delayed this plan and eventually, having redesigned it to drastically reduce its weight, COBE was placed into Sun-synchronous orbit on 18 November 1989 aboard a Delta rocket.

Its first significant result was not long in coming, using the Far-InfraRed Absolute Spectrophotometer (FIRAS) – a spectrophotometer used to measure the spectrum of the CMB – and with just 9 minutes worth of data COBE was able to plot a spectrum (Figure 22.2) where the error bars on the data points lay within the theoretical black body curve that should be seen if the CMB really is the radiation from a time when the radiation and matter were in thermal equilibrium.

The second significant result came from the Differential Microwave Radiometer (DMR), a microwave instrument that would map variations (or anisotropies) in the CMB. The ground-based results, with their limited accuracy, showed a uniform CMB temperature, but COBE's greater sensitivity soon showed that the temperature was not constant across the sky; it was slightly hotter in the direction of the constellation Leo, and slightly cooler in the opposite direction towards Aquarius. The hottest region is +3.5 mK above the average and the coolest −3.5 mK below the average.

Such an observation was predicted due to the fact that our Solar System is moving through space, orbiting the centre of our Galaxy, and so our motion should mean that photons arriving from the direction of motion are boosted in energy, and hence appear 'hotter', whilst those from behind lose energy and

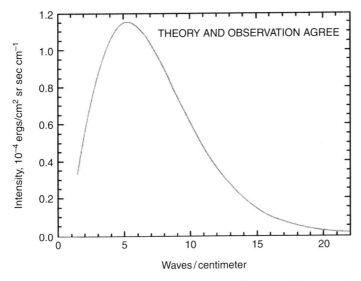

Figure 22.2 The spectrum of the CMB measured by the COBE spacecraft. Image: GFSC, NASA.

thus appear cooler (the Doppler effect). But, to everyone's surprise, the direction of motion determined from these observations was not aligned with the orbital motion of the Solar System around the Galaxy, but was in nearly the opposite direction. This means that our Galaxy and the whole Local Group of galaxies are moving with a speed of more than 2 million miles per hour (600 km/s or ~1/500 the speed of light) with respect to the Universe at large. What could cause this? It appears to be a result of the clustering and superclustering of galaxies in our local (within 100 million light years) neighbourhood so that there is a net gravitational pull towards the direction of Leo.

Incidentally, the COBE DMR observations clearly show changes in the observed velocity of 30 km/s – a 5% effect – due to the motion of the Earth around the Sun – good evidence that Galileo was right!

The face of God

The most significant observation made by COBE – and initially its most controversial one – was when NASA and Berkeley announced that COBE had detected the expected fluctuations in the CMB. Against the advice of some of the team, it was decided to accompany the press release with a, now famous, image. This image *did* show tiny fluctuations in temperature – but these *did not* show the fluctuations in the CMB. Information accompanying the image did try to make this clear but was subtle and widely ignored by the press. The majority of the fluctuations were caused by noise and only about 10% of the signal was attributed to the CMB. The point was that the fluctuations were just a little greater than would have been expected by noise and thus, statistically, indicated the presence of underlying fluctuations.

Suppose you randomly shook sand over a metre-square flat wooden board. It would then be covered by piles of sand and you could measure the highest peaks that occurred – let's say that they were ~5 cm high. Now sprinkle the sand randomly over a metre-square board that has a rough surface having surface irregularities with peaks of, say, 1 cm. You might now expect that the highest peaks would be a little higher, perhaps 6 cm rather than 5 cm, and you could deduce that the board below was not smooth and even estimate its roughness. This is exactly how the COBE team deduced the presence of the background fluctuations. When asked by someone in the audience at the Press Conference how important this result was, George Smoot said, 'Well, if you are a religious person, it's like seeing the face of God.' This naturally gave rise to intense media interest!

John Mather and George Smoot received the Nobel Prize in Physics in 2006 for this work, which the Nobel Prize Committee said had begun the period of 'precision' cosmology.

The cause of the fluctuations in the CMB

Why are these small variations present? To answer this we need to understand a little about 'dark matter' as fully discussed in Chapter 21. Though not yet directly detected, its presence has been inferred from a wide variety of observations.

As described above, for ~380,000 years following the Big Bang, the matter and radiation were interacting as the energy of the photons was sufficient to ionise the atoms giving rise to a plasma of nuclei and free electrons. This gives two results:

(1) The radiation and matter are in thermal equilibrium and the radiation will thus have a black body spectrum as was proven by the COBE observations.
(2) The plasma of nuclei and electrons will be very homogeneous as the photons act rather like a whisk beating up a mix of ingredients.

It is worth repeating an argument from the last chapter on dark matter relating to the second of these points. When the temperature drops to the point that atoms can form, the matter can begin to clump under gravity to form stars and galaxies. Simulations have shown that, as the initial gas is so uniformly distributed, it would take perhaps 8 to 10 billion years for regions of the gas to become sufficiently dense for this to happen. But we know that galaxies came into existence around 1 billion years after the Big Bang. Something must have aided the process. We believe that this was non-baryonic dark matter. As this would not have been coupled to the radiation, it could have begun to gravitationally 'clump' immediately after the Big Bang. Thus, when the normal matter became decoupled from the photons, there were 'gravitational wells' in place formed by concentrations of dark matter. The normal matter could then quickly fall into these wells, rapidly increasing its density and thus greatly accelerating the process of galaxy formation.

The concentrations of dark matter that existed at the time the CMB originated have an observable effect due to the fact that if radiation has to 'climb out' of a gravitational potential well it will suffer a type of redshift called the 'gravitational redshift'. So the photons of the CMB that left regions where the dark matter had clumped would have had longer wavelengths than those that left regions with less dark matter. This causes the effective black body temperature of photons coming from denser regions of dark matter to be lower than those from sparser regions – thus giving rise to the temperature fluctuations that are observed. As such observations can directly tell us about the Universe as it was just 380,000 or so years after its origin it is not surprising that they are so valuable to cosmologists!

Wilkinson Microwave Anisotropy Probe and Planck

Observations by the COBE spacecraft first showed that the CMB did not have a totally uniform temperature and, since then, observations from the Wilkinson Microwave Anisotropy Probe and Planck spacecraft, balloons and high mountain tops have been able to make maps of these so-called 'ripples' in the CMB – temperature fluctuations in the observed temperature of typically 60 microkelvin.

The Wilkinson Microwave Anisotropy Probe (WMAP) was originally known as the Microwave Anisotropy Probe (MAP) but was renamed after the death of one of the pioneers of CMB observations, David Wilkinson, who sadly died of cancer whilst its data were being analysed. The WMAP spacecraft was launched on 30 June 2001 and flew to the Sun–Earth L2 Lagrangian point, arriving there on 1 October 2001. This lies at a distance of 1.5 million km from the Earth directly away from the Sun. In general, as one moves further away from the Sun, a planet (or space probe) will orbit more slowly, but if it lies on the Sun–Earth line the additional gravitational pull from the Earth will add to that from the Sun and the satellite will orbit more rapidly. So, at just the right distance, a satellite can be made to orbit the Sun in one Earth year and so move round the Sun in consort to the Earth. This position is called L2. Here, as the spacecraft is observing the half of the sky away from the Sun, it can produce a map of the sky in six months. WMAP completed its first all-sky survey in April 2002.

The fundamental problem in producing an all-sky map is the contamination by foreground emission from our own galaxy and other, more distant, radio sources. To overcome this, WMAP observed in five frequencies that allow these to be determined and subtracted from the map. Synchrotron radiation, from electrons spiralling round the magnetic field of the Galaxy, dominates the low frequencies whilst emissions from dust dominate the higher frequencies. These emissions contribute different amounts to the five frequencies, thus permitting their identification and subtraction, leaving a map without any evidence of the fact that it was made within our Galaxy! (I actually find this quite amazing.)

The three-year WMAP data alone showed that the Universe must contain dark matter and that the age of the Universe is 13.7 billion years. The five-year data included new evidence for the cosmic neutrino background, showed that it took over half a billion years for the first stars to re-ionise the Universe, and provided new constraints on cosmic inflation.

At a time ~400,000 years after the Big Bang, WMAP showed that 10% of the Universe was made up of neutrinos, 12% of atoms, 15% of photons and 63% of dark matter. The contribution of dark energy at the time was negligible.

The seven-year WMAP data were released on 26 January 2010. According to these data the Universe is 13.75 ± 0.11 billion years old and they confirmed some asymmetries in the data that will be discussed below. The WMAP spacecraft continued to take data in perfect working order until September 2010. Figure 22.3 shows the fluctuations as observed by WMAP of the CMB. (Colour versions can be found by searching for 'WMAP CMB map'.)

Observations from balloons and the ground

Observations of the CMB are greatly hampered by the presence of water vapour and so balloon flights such as Boomerang and Maxima and ground-based

Figure 22.3 All-sky map of the CMB ripples produced by WMAP in 2008. Image: WMAP Team, GFSC, NASA.

antenna arrays such as the Very Small Array and the Cosmic Background Imager located in high, dry locations have been used to observe relatively small regions of the sky, but with higher resolution than that initially achieved by the all-sky satellite imagers COBE and WMAP. They have thus greatly aided the extension of the power spectrum, discussed below, to smaller angular scales.

The Planck mission

Planck was a space observatory built by the European Space Agency to complement and improve upon observations of the CMB made by NASA's WMAP. Planck was launched into orbit on 14 May 2009 and flown to the same L2 region as WMAP, where it was injected into its final 'figure of 8' orbit on 3 July, by which time the high-frequency receiver systems had reached the operational temperature of just 1/10 of a degree above absolute zero. It had the ability to make observations at smaller angular scales than WMAP and had significantly higher sensitivity. In addition, it made observations at nine wavelengths rather than the five made by WMAP, which should help remove the 'foreground' contamination of the CMB data. Two of the receiver systems were designed and built at the Jodrell Bank Observatory and are the most sensitive receivers ever made at their operating wavelengths.

Though the main objective of the mission was to observe the total intensity and polarisation of the primordial CMB, its data will create a catalogue of galaxy clusters using the Sunyaev–Zel'dovich effect and observe the effects of gravitational lensing on the CMB. In order to remove the effects of the Milky Way from the raw images it made detailed observations of the local interstellar medium, the galactic synchrotron emission and magnetic field.

Planck started its first all-sky survey on 13 August 2009, followed by the second on 14 February 2010, with 100% sky coverage achieved by mid June 2010. The first all-sky maps of the CMB were published in the spring of 2013. (Note: The Planck CMB map is so detailed that it is very difficult to reproduce in monochrome. Colour versions can be easily found by searching for 'Planck CMB map'.)

The Planck results

What have the observations of the CMB told us about the Universe?

1 The curvature of space and hence the density of the Universe

The photons that make up the CMB have travelled across space for billions of years and will thus have been affected by the curvature of space.

330 A Journey through the Universe

This curvature is a function of the amount of matter (and energy) in space and hence its density. If the density is higher than some critical amount, the space will be positively curved (as, in two dimensions, on the surface of a sphere). If the density is critical, Euclidian geometry holds true and space is said to be 'flat'. If the density is less than this the space will be negatively curved (as, in two dimensions, on the surface of a saddle).

Theory tells us the expected angular scale of the fluctuations of the CMB. If we observed it through flat space then we would see exactly this angular scale. If, however, we observed it through positively curved space – which acts rather like a convex lens – we would observe a larger pattern than predicted, and if we observed it through negatively curved space – which acts rather like a concave lens – we would observe a smaller pattern than predicted. It is possible to simulate the expected pattern of fluctuations if space were negatively curved, positively curved or flat and these can be compared with observations.

The upper part of Figure 22.4 shows the fluctuations observed by the Boomerang balloon experiment. Below are the simulations of what would

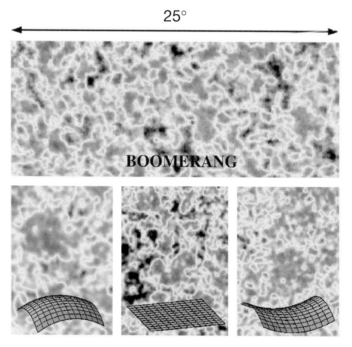

Figure 22.4 Boomerang observations of the CMB with simulations below of what would be observed if space were positively curved (left), flat (centre) and negatively curved (right). Image: NSBF, NASA.

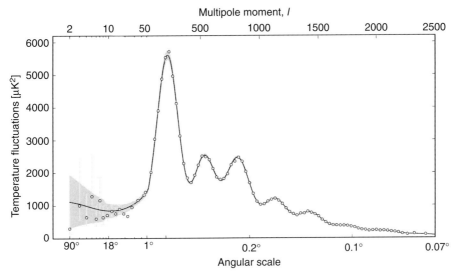

Figure 22.5 The power in the CMB fluctuations as a function of angular scale derived from the Planck data. Image: ESA and the Planck Collaboration.

be expected if space were either flat or positively or negatively curved. It is fairly obvious that the 'flat' pattern agrees most closely with that observed.

These observations can be put on a more quantitative basis by analysing the relative amounts of structure seen at different angular scales. Looking at the 'blobs' in the Boomerang data it appears that there is a significant amount of structure seen on scale sizes of order 1 degree, and this is borne out by the position of the first peak in the Planck angular scale plot shown in Figure 22.5.

In fact, the peak in the plot is at ~0.8 degrees, which is exactly what we would expect if space were 'flat', and the WMAP and Planck results show that space is 'flat' to an accuracy of 1%. This important result leads to a value of the density of the Universe of 9.9×10^{-30} g/cm^3, which is equivalent to only 5.9 protons per cubic metre. This is in terms of mass, but we now know that the major part is in the form of energy.

2 The relative amounts of the various components in the Universe

In the sections below, the values provided by analysis of the Planck data are given first as these are perhaps expected to be more accurate than those given by WMAP and other experiments. But these are also given in brackets after each result and I find it a little surprising that the values have changed so significantly.

Normal matter 4.9% (4.5%) This implies that ~95% of the energy density in the Universe is in a form that has never been directly detected in the laboratory!

Dark matter 26.8% (22.7%) The major part of this is 'cold dark matter' (CDM; 'cold' equating to relatively slow moving particles). Dark matter is likely to be composed of one or more species of sub-atomic particles that interact very weakly with ordinary matter. As we have seen, dark matter plays a very important part in the formation of the galaxies and has significant gravitational effects. Particle physicists have many plausible candidates and experiments are perhaps on the verge of their detection as described the previous chapter. This 27% includes a small proportion of 'hot dark matter' (HDM; 'hot' equating to light particles such as neutrinos moving at speeds close to that of light). Neutrinos do make a small contribution to the dark matter total but too great a component of HDM would have prevented the early clumping of gas in the Universe, delaying the emergence of the first stars and galaxies. However, WMAP and Planck do see evidence that a sea of cosmic neutrinos does exist in the numbers that are expected from other lines of reasoning.

Dark energy 68.3% (72.8%) In the 1990s, observations of supernovae were used to trace the expansion history of the Universe, showing that the expansion appeared to be speeding up, rather than slowing down. If 68% of the energy density in the Universe is in the form of dark energy, which has a gravitationally repulsive effect, it is just the right amount to explain both the flatness of the Universe and the observed accelerated expansion.

3 The time and redshift of reionisation

The 'reionisation redshift' relates to the time when the first stars formed and the ultraviolet light produced by them was able to split electrons off hydrogen atoms giving ionised hydrogen. It is termed 'reionisation' as when the Universe was hot – up to about 380,000 years after its origin – the hydrogen had been previously ionised. At that time, when the atoms formed and cooled, the Universe entered the so-called 'Dark Ages', which lasted until the Universe was again filled with light from the first stars. Data from WMAP showed that this began at a time starting just 400 million years after the Big Bang and lasted for ~500 million years.

4 The age of the Universe

The data from Planck give an age of 13.8 billion years, slightly greater than the WMAP result of 13.75 billion years. Combining the Planck data with those of WMAP and other experimental values, such as that obtained from gravitational lensing, gives a best fit to the age of 13.79 billion years.

5 Hubble's constant

Planck gave a value for Hubble's constant of 67 which, in combination with other data, provides values that extend up to ~68 ± 1. This is in reasonable agreement with, but having greater precision than, the final result from the Hubble Space Telescope Key Project of ~72. I am pleasantly pleased that this result agrees very well with that obtained from observations of the Double Quasar discovered by astronomers at Jodrell Bank Observatory in the early 1980s, which gave a value of 67 ± 13. At the time most other results were either significantly higher or lower, and it is good to know that our result (admittedly with large error bars) was very close to the currently accepted value.

Planck's legacy

On 19 October 2013, the instruments and cooling systems on the Planck spacecraft were switched off, marking the end of the observational part of the Planck mission after spending more than four years mapping the CMB. A day later a piece of software was uploaded to prevent the systems ever being switched on so that the transmitter that had sent the spacecraft's data back to Earth could never cause interference to future space probes. Prior to this, the spacecraft had been manoeuvred out of the L2 position and will spend eternity silently orbiting the Sun.

But vast amounts of data still needing to be analysed and the numerous maps and catalogues will provide a legacy for both today's and future generations of cosmologists. The peaks and troughs in the CMB spectrum contain highly detailed information about the basic cosmological parameters that have been outlined above. WMAP established the basic details, then Planck not only confirmed this picture but was able to make it far more precise.

Build your own universe

There is an excellent piece of software produced by the WMAP team, which can be found by searching for 'WMAP CMB power spectrum analyzer'. This shows how the CMB spectrum would appear with varying values of the amounts of normal and dark matter and the relative amount of dark energy. It is important to set the value of Hubble's constant to ~70, the 'reionisation redshift' to 11 and the 'spectral index' to 0.95 first. The spectral index is a measure of the relative amounts of fluctuations on different scales in the very early Universe.

As they say, you can build your own universe and see what WMAP and Planck would have observed!

Inflation and gravitational waves

One prediction of the theories relating to the Big Bang origin of the Universe and the inflation that vastly increased its size is that the tiny quantum fluctuations present prior to the inflationary phase would have become far greater in extent and would have launched gravitational waves travelling through space. As these waves distorted space as they passed through, they would have had the effect of slightly polarising the light, so giving rise to a 'curling' pattern in the polarised orientation of light in the ancient Universe. If this is true, then this imprint should still be visible in the microwave radiation that makes up the CMB.

As the amount of polarisation is very small its detection is exceedingly difficult, but finally, in March 2014, the scientists using an experiment called BICEP2 located at the South Pole (where it is so cold that water vapour is frozen out of the atmosphere and microwave radiation can pass through unattenuated) announced that they had detected the predicted curling pattern of polarisation in the cosmic microwave radiation.

If this result were to be confirmed then it is proof of an inflationary phase in the early Universe, gives a further indirect detection of gravitational waves and, perhaps most importantly, that gravity has to be quantised. However, after these results were announced, other groups suggested that the effect of dust in our Galaxy – which can produce very similar polarisation results – may have been underestimated. And in June 2014, when the results of BICEP2 were formally published, the group indicated that they were less confident of their result. Data from the Planck spacecraft released since their initial announcement had indicated that, in general, the polarisation effects of nearby dust were greater than had been assumed, though the Planck data relating to the area of sky specifically observed by BICEP2 was yet to be published – this being due later that year. In their published paper in *Physics Research Letters* the group states: '[Our] models are not sufficiently constrained by external public data to exclude the possibility of dust emission bright enough to explain the entire excess signal.' The results from the Planck spacecraft, one of whose objectives was to measure the polarisation present in the CMB, were thus awaited eagerly so that the BICEP2 results could be confirmed or, perhaps, shown to have been over optimistic.

Does it all add up?

The CMB is the most precise portrait of the young Universe that we have and can be used to scrutinise the models for the origin and evolution of the cosmos. The agreement with the standard model – called the ΛCDM model as it incorporates both cold dark matter and the Λ constant from Einstein's General Theory of Relativity – is pretty good, but Planck has confirmed that some anomalous features in the CMB that were seen by WMAP are more serious than previously thought. This suggests that something fundamental may be missing from the standard model and some aspects may need revision. In particular, the Universe may not be totally uniform (or isotropic) at the largest scales.

The all-sky CMB map shows that, whilst the observations on small and intermediate angular scales agree extremely well with the model predictions, the fluctuations detected on large angular scales on the sky – between 6 and 90 degrees – are about 10% weaker than would have been expected. There are two other anomalies: there appears to be a substantial asymmetry in the CMB signal observed in the two opposite hemispheres of the sky, with one of the two hemispheres appearing to have a significantly stronger signal than the other; and a feature first seen in the WMAP data and confirmed by Planck is the presence of a so-called 'cold spot', where one of the low-temperature spots in the CMB extends over a patch of the sky that is much larger than expected.

Cosmologists are thus facing an interesting dilemma: it appears that the standard model of cosmology is still the best way to describe the CMB data (although we do not yet really understand dark matter, dark energy and inflation) but the anomalies seen by Planck highlight that the model should be extended and, perhaps, might need to be radically modified.

There is an intriguing theory that might explain the cold spot: could this be the result of the interaction of 'our' Universe with a second that formed along with ours during the initial inflationary period when, perhaps, two expanding 'bubbles' of space, rather than one, formed together? Could this be the first evidence of another universe beyond our own?

Two good books:

>   *Afterglow of Creation* by Marcus Chown (Faber and Faber).
>   *Wrinkles in Time* by George Smoot (Avon Books).

23

To infinity and beyond: a view of the cosmos

Modern cosmology arose from Einstein's General Theory of Relativity, which is essentially a theory of gravity. As gravity was the only force of infinite range that could act on neutral matter, Einstein realised that the Universe as a whole must obey its laws. He was led to believe that the Universe is 'static', or unchanging with time, and this caused him a real problem as gravity, being an attractive force, would naturally cause stationary objects in space to collapse down to one point. To overcome this he had to introduce a term into his equation that he called the cosmological constant, Lambda or Λ. This represents a form of antigravity that has the interesting property that its effects become greater with distance. So, with one force decreasing and the second increasing with distance it was possible to produce a static solution. He later realised that this was an unstable situation, and that a static universe was not possible, calling this 'the greatest blunder of his life'. He could have predicted that the Universe must be either expanding or contracting. However, as we will see, perhaps he was not quite as wrong as he thought.

Big Bang models of the Universe

A Russian meteorologist, A. A. Friedmann, solved Einstein's equations to produce a set of models in which the Universe expanded from a point, or singularity. These were given the name 'Big Bang' models by Fred Hoyle – some say that this was meant to be a disparaging term as Hoyle was an advocate of another theory, the Steady State theory, to be discussed below, and he did not like them. In all of these models, the initially fast rate of expansion is slowed by the attractive gravitational force between the matter of the Universe. If the density of matter within the Universe exceeded a critical amount, it would be

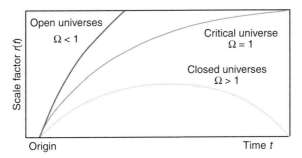

Figure 23.1 The Friedmann models of the Universe.

sufficient to cause the expansion to cease and then the Universe would collapse down to a 'Big Crunch' (these are called closed universes). If the actual density were less than the critical density, the Universe would expand for ever (called open universes). In the critical case that is the boundary between the open and closed universes, the rate of expansion would fall to zero after infinite time (called the 'flat' or 'critical' universe).

A useful analogy is that of firing a rocket from the Earth: if the speed of the rocket is less than 11,186 km/s, the Earth's escape velocity, the rocket will eventually come to a halt and fall back to Earth (equivalent to closed universes), if it equals 11,186 km/s it will just escape the Earth with its speed reducing with time (equivalent to the critical universe), whilst if it exceeds 11,186 km/s it will leave the Earth more quickly (equivalent to open universes).

The models, shown in Figure 23.1, are distinguished by a constant, Ω (Omega), that is defined as the ratio of the actual density to the critical density. In closed universes Ω is greater than 1, space has positive curvature, the angles within a triangle add up to more than 180 degrees and two initially parallel light rays would converge. In open universes Ω is less than 1, space has negative curvature, the angles within a triangle add up to less than 180 degrees and two initially parallel light rays would diverge. In the critical case, Ω is equal to 1, space is said to be 'flat' (I much prefer the term 'Euclidian' rather than flat), the angles within a triangle add up to 180 degrees and two initially parallel light rays would remain parallel. It should be pointed out that this refers to a universe on the very large scale and, in the region of a massive object, such as a star or galaxy, the space becomes positively curved.

The expansion of the Universe and a problem with age

As has been described in Chapter 19, Edwin Hubble showed that the Universe was expanding and was even able to deduce its age, which was derived to be ~2 billion years. In fact, in all the Friedmann models, the real age must be less than this as the Universe would have been expanding faster in the past and, in the case of the 'flat' universe, the actual age would be 2/3 that of the Hubble age or ~1.3 billion years. This result obviously became a problem as the age of the Solar System was determined (~4.5 billion years) and calculations relating to the evolution of stars made by Hoyle and others indicated that some stars must be much older than that, ~10 to 12 billion years old. During the blackouts of World War II, Walter Baade used the 100-inch telescope to study the stars in the Andromeda Galaxy and discovered that there were, in fact, two types of Cepheid variable. As a result, Hubble's constant reduced to ~250 (km/s)/Mpc. There still remained many problems in estimating distances, but gradually the observations have been refined and, as a result, the estimate of Hubble's constant has reduced in value to about 70 (km/s)/Mpc.

It is unlikely that the true value of Hubble's constant will differ greatly from this. But the Hubble age of ~14 billion years that is derived corresponds to the age of a 'flat' universe of only ~9.3 billion years. From observations of globular clusters, which contain some of the oldest stars in the Universe, and of the white dwarf remnants of stars, we suspect that the Universe must be older than 12 billion years so, if we believe the current value of Hubble's constant, there is still an age problem with the Big Bang models. This is really important and is rarely ever pointed out. The standard Big Bang models cannot be correct!

The cosmological redshift

In the previous chapters, the blueshifts and redshifts seen in galaxies were regarded as being due to the Doppler effect, and this would be perfectly correct when considering the blueshifts shown by the galaxies in the Local Group. However, in the cases of galaxies beyond our Local Group there is a far better way of thinking about the cause of the redshifts that we see. As Hubble showed, the Universe is expanding so it would have been smaller in the past. In addition, it is not right to think of the galaxies (beyond the movements of those in our Local Group or within other groups or clusters of galaxies) moving through space but, rather, that they are being carried apart by the expansion of space. To repeat the nice analogy, think of baking a currant bun. The dough is packed with currants and then baked. When taken out of the oven the bun will (hopefully) be bigger and thus the currants will be further apart. They will not have moved *through* the dough, but will have been carried apart by the *expansion* of the dough.

When a photon was emitted in a distant galaxy corresponding to a specific spectral line, the Universe would have been smaller. In the time it has taken that photon to reach us whilst travelling through space, the Universe has expanded and this expansion has stretched, by exactly the same ratio, the wavelength of the photon. This increases the wavelength, so giving rise to a redshift that we call the 'cosmological redshift'. A simple analogy is that of drawing a sine wave (representing the wavelength of a photon) onto a slightly blown up balloon. If the balloon is then blown up further, the length between the peaks of the sine wave (its wavelength) will increase.

The Steady State model of the Universe

Because of this 'age' problem many astronomers did not give much credence to the Big Bang models, and in 1948 Herman Bondi, Thomas Gold and Fred Hoyle, who disliked the idea of an instantaneous origin of the Universe, proposed an alternative theory called the 'Steady State' theory. All cosmological theories embrace what is called the 'cosmological principle'; that is, on the large scale at any given time, the view of the Universe from any location within it will be the same. (This has been nicely proven by the Hubble Space Telescope in that the two Hubble Deep Fields, one in the northern sky and one in the south, have identical characteristics.) Bondi, Gold and Hoyle extended this principle to give what they called the 'perfect cosmological principle' where the words 'at any given time' were replaced by 'for all time'. Their universe was unchanging on the large scale. That did not mean that it was not expanding. At the heart of their theory was the idea that, as the galaxies moved further apart due to the expansion of the Universe, new matter, in the form of hydrogen, was created in the space between them, which eventually formed new galaxies to keep the observed density of galaxies constant. This universe had no beginning and will have no end and is, as the theory's name implies, in a 'steady state'. As new matter is continuously being created, it is also called the theory of 'continuous creation'.

Big Bang or Steady State?

In the early 1960s observational tests were made to decide between the two theories. Suppose one could measure the galaxy density close to us – to give the number of galaxies per cubic megaparsec. As these galaxies are close to us we see them essentially at the present time. If we could then measure the density of galaxies in the far Universe, we would be measuring it at some time in the past. In the Steady State model these results should be the same, but in the

Big Bang model the density should have been higher in the past. Martin Ryle at Cambridge attempted such measurements by counting radio sources. Though there were problems with his initial data, these results did finally indicate a greater density of radio sources in the past, so disproving the Steady State theory. The deathblow to the Steady State theory came in 1963 when radiation, believed to have come from the Big Bang, was discovered, as described in the previous chapter.

Inflation and the formation of the primeval elements

In Chapter 20, the concept of inflation and the formation of the primeval elements were described in some detail. Here follows a brief summary of the main points.

Observations had shown that the Universe was very close to being 'flat', with $\Omega \sim 1$, and the Big Bang theory gives no particular reason why this should be so. Any curvature that the Universe has close to its origin tends to get enhanced as the Universe ages – a slightly positively curved space become more and more so and vice versa. In fact, had Ω not been in the range 0.999999999999999 to 1.000000000000001 one second after its origin the Universe could not be as it is now. This is incredibly fine tuning, and there is nothing in the standard Big Bang theory to explain why this should be so. This is called the 'flatness' problem.

There was a second 'horizon' problem in that, in the Friedmann models, there has not been sufficient time since the origin of the Universe for all regions of space to be in thermal equilibrium – all at the same temperature of ~ 2.7 K.

The idea of 'inflation', first proposed by Alan Guth, solves these problems by postulating an initial, very rapid, phase of expansion which would force space to become 'flat' and also, as all the matter in the Universe was in thermal equilibrium prior to the inflationary period, solves the 'horizon' problem.

Half of the gravitational potential energy that arose from this inflationary period was converted into kinetic energy from which arose an almost identical number of particles and antiparticles, but with a very small excess of matter particles (about one part in several billion). All the antiparticles annihilated with their respective particles leaving a relatively small number of particles in a bath of radiation. The bulk of this 'baryonic matter' was in the form of quarks that, at about one second after the origin, grouped into threes to form protons and neutrons. (Two up quarks and one down quark form a proton, and one up quark and two down quarks a neutron. The up quark has +2/3 charge and the down quark −1/3 charge, so the proton has a charge of +1 and the neutron 0 charge.) An

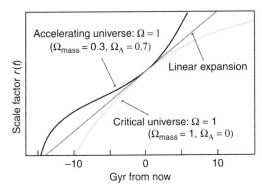

Figure 23.2 A plot showing the scale size of the Universe with time.

almost equal number of protons and neutrons were produced but, as free neutrons are unstable, with a half life of 10.3 minutes, the only ones to remain were those that were incorporated into helium nuclei comprising two protons and two neutrons. So, after a few minutes, the normal (baryonic) matter in the Universe was very largely composed of hydrogen nuclei (protons), helium nuclei (alpha particles) and electrons – with one electron for each proton. (We now believe that several times more 'dark matter' was also created but, at present, we do not know how.)

The makeup of the Universe

The observations of the Cosmic Microwave Background (as described in the previous chapter), Type Ia supernovae, Hubble's constant and the distribution of galaxies in space all now give a consistent model of the Universe. From the latest observations of the Cosmic Microwave Background made by the Planck spacecraft, it appears that normal matter accounts for just ~5% and dark matter ~27%, with the remaining ~68% of the total mass–energy content of the Universe being in the form of dark energy. Figure 23.2 shows how we believe the scale size of the Universe has changed with time in the past and how it will expand ever faster in the future. You will see that the actual age of the Universe is similar to the Hubble age (linear expansion from zero size) with a value now believed to be 13.8 billion years.

A Universe fit for intelligent life

The very fact that you are reading this book tells us that our Universe has just the right properties for intelligent life to have evolved. But why should this be so? As eloquently described in the book *Just Six Numbers* by Professor Martin Rees (now Lord Rees), there are a number of parameters that have a major influence on how universes can evolve and how stars produce elements that are needed for life. One of these has already been covered in this chapter: the constant Omega. If Ω had been higher the Universe would have rapidly collapsed without allowing life a chance to evolve; if it had been smaller, galaxies and stars would not have formed. In addition, if the cosmological constant, Λ (which is surprisingly small), had been larger, it would have prevented stars and galaxies forming.

You have also read in Chapter 22 how the galaxies formed as a result of fluctuations in the density of the primeval Universe – the so-called 'ripples' that are observed in the Cosmic Microwave Background. The parameter that defines the amplitude of the ripples has a value of $\sim 10^{-5}$. If this parameter were smaller the condensations of dark matter that took place soon after the Big Bang (and were crucial to the formation of the galaxies) would have been both smaller and more spread out, resulting in rather diffuse galaxy structures in which star formation would be very inefficient and planetary systems could not have formed. If the parameter had been less than 10^{-6}, galaxies would not have formed at all! But if this parameter were greater than 10^{-5} the scale of the 'ripples' would be greater and giant structures, far greater in scale than galaxies, would form and then collapse into super-massive black holes – a violent universe with no place for life!

One parameter of our Universe is so well known that it is barely given a moment's thought – the number of spatial dimensions, three. But if this were either two (no complex structures) or four (forces fall as the inverse cube and atoms could not form), life could not exist.

Einstein's famous equation, $E = mc^2$, relates the amount of energy that can be extracted from a given amount of mass, so the value of c is obviously fundamentally important. In practice only a small part of the energy bound up in matter can be released, as in the conversion of hydrogen to helium. This process releases 0.7% of the mass of the four protons that form helium – a percentage closely linked to the strength of the strong nuclear force. This parameter, 0.007, has been called 'nuclear efficiency'. However, if this value were too small, say 0.006, the sequence of reactions that build up helium could not take place. In the first of these reactions, two protons form a deuterium nucleus but, given a value of 0.006 for the nuclear efficiency, deuterium would be unstable so preventing the further reactions that give rise to helium – stars would be inert. On the other hand, if this parameter were 0.008, meaning that nuclear forces were stronger relative to electrostatic forces, the electrostatic repulsion of two protons would be overcome and they could bind together so no hydrogen would have remained to fuel the stars. A critical reaction in the evolution of stars is the formation of carbon in the triple alpha process. As described earlier, Fred Hoyle played a key role in the understanding of this reaction and pointed out that even a change of a small percentage from the observed value of 0.007 would have severe consequences on the amount of carbon that would be formed in stars – with obvious consequences for life as we understand it.

A 'multiverse'

So how can it be that all the parameters described above are finely tuned so that we can exist? There are two possible reasons. The first is that our Universe was 'designed' by its creator specifically so that it could contain intelligent beings, a view taken by some scientist-theologians. A second view is that there are many universes each with different properties; the term 'multiverse' has been applied to this view. We have no knowledge of what lies in the cosmos beyond the horizon of 'our' visible Universe. Different regions could have different properties; these regions could be regarded as different universes within the overall cosmos. Our part of the cosmos is, like baby bear's porridge, just right.

String theory: another approach to a 'multiverse'

Theoretical physicists have a fundamental problem. Einstein's General Theory of Relativity which relates to 'gravity' is a classical theory, whereas the other forces are described by quantum mechanics. A 'theory of everything' has yet to be found that can bring together all the fundamental forces. One approach that is being actively pursued is that of 'string theory'. The early string theories envisioned a universe of ten dimensions, not four, making up a ten-dimensional space-time. The additional six beyond our three of space and one of time are compacted down into tiny regions of space of order 10^{-35} m in size and called strings. These are the fundamental building blocks of matter. Different 'particles' and their properties depend on the way these strings are vibrating – rather like the way a string of a violin can be excited into different modes of vibration to give harmonically related sounds. As these strings move they warp the space-time surrounding them in precisely the way predicted by general relativity. So string theory unifies the quantum theory of particles and general relativity.

In recent years five string theories have been developed, each with differing properties. In one there can be open strings (a strand with two ends) as well as closed strings where the ends meet to form a ring. The remaining four only have closed rings. More recently, Ed Witten and Paul Townsend have produced an eleven-dimensional 'M-theory', which brings together the five competing string theories into a coherent whole. This eleventh dimension (and it not impossible that there could be more) gives a further way of thinking about a multiverse.

We can think of a simple analogy: take three slices of bread and suspend them (perhaps with a knitting needle) with each slice separated by, say, three centimetres. On the same side of each of these slices place some ants. The ants could survive, at least for a while, by eating the bread of what is effectively a two-dimensional universe. To them the existence of other colonies of ants on adjacent slices would be unobservable. If they could think, they would believe that they existed in a single two-dimensional universe. But *we* can see that all of these exist within a cosmos that actually has a third dimension.

In just the same way, rather than being individual regions of one large spatially linked cosmos, it could be that other 'universes' exist in their own space-time – hidden from ours within a further dimension.

Beyond our imagination

In the description of the final talk that I gave as Gresham Professor of Astronomy, I ended with: '... or even that our Universe could just be one small part of a multiverse that extends beyond our imagination!' I was implying that the totality of the cosmos and the laws that govern its existence might well be beyond the ability of our human minds to grasp. (Certainly mine!)

I was interested to read comments made later that year by Lord Rees, Astronomer Royal, past president of the Royal Society and previous holder of the Gresham Chair of Astronomy:

- Some of the greatest mysteries of the Universe may never be resolved because they are beyond human comprehension. Rees suggests that the inherent intellectual limitations of humanity mean we may never resolve questions such as the existence of parallel universes, the cause of the Big Bang, or the nature of our own consciousness.
- He even compares humanity to fish, which swim through the oceans without any idea of the properties of the water in which they spend their lives.
- 'A 'true' fundamental theory of the Universe may exist but could be just too hard for human brains to grasp.

Perhaps I am not alone!

The future of the Universe

The accelerating expansion of the Universe that is now accepted has a very interesting consequence. It used to be thought that, with a slowing rate of expansion, as the Universe became older we would see an increasing number of galaxies (as the distance we could see becomes greater). In a universe whose expansion is accelerating the exact opposite will be true – yes, we will be able to see further out into space, but there will be increasingly less and less for us to see as the expansion carries galaxies beyond our horizon.

On the large scale, the space between the galaxy clusters will be expanding – carrying them apart ever faster – but it is believed that clusters like our own Local Group will remain gravitationally bound and, in fact, its members will merge into one single galaxy largely made up from our own Milky Way Galaxy and the Andromeda Galaxy. If one looks forwards in time to ~100 billion years, any observers in existence within this 'galaxy' would see a totally empty Universe! The expansion of space will have carried all other galaxies beyond our horizon – the edge of the visible Universe.

It would be virtually impossible for such observers to learn about the evolution of the Universe for a number of reasons, including the fact that the peak of the energy spectrum of the Cosmic Microwave Background (CMB) will have redshifted down to ~1 m and would be virtually impossible to detect.

From theoretical studies of stellar evolution and how the relative abundances of the elements change with time (for example, the amount of hydrogen is reducing and that of helium increasing as a result of nucleosynthesis in stars) it might well be possible to estimate the age of the galaxy, but it would be not be possible to infer that its origin involved a Big Bang.

We happen to live at the only time in the history of the Universe when the magnitude of dark energy and dark matter are comparable and also when the CMB is easily observable, so enabling us to infer the existence of dark energy and the way in which the Universe has evolved since the Big Bang, and to postulate its future runaway expansion.

Any observers present when the Universe was young would not have been able to infer the presence of dark energy as, at that time, it would have had virtually no effect on the expansion rate. Those in the far future will not be able to tell that they live in an expanding universe at all, and not be able to infer the existence of dark energy either. As the longest lived stars come to the end of their lives, the evidence that lies at the heart of our current understanding of the origin and evolution of the Universe will have disappeared.

Now is about the best time in the life of the Universe to unravel its mysteries. Thank you for joining with me in this quest.

More in-depth reading:

Just Six Numbers by Martin Rees (Phoenix).
The Elegant Universe by Brian Greene (Vintage).
Cosmology by Steven Weinberg (Oxford University Press).

Index

2003 UB$_{313}$, 26
3C 273, 125
75-metre Mark I radio telescope, 247

active galaxies, 246
   NGC 4261, 248
Adams, John Couch, 60, 77
Airy, George, 60
albedo, 33
   Earth, 33
   Enceladus, 63
   Mars, 33
   Mercury and Moon, 63
   Venus, 33, 63
Alpher, Ralph, 321
Alpher–Bethe–Gamow paper, 321
anthropic principle, 148
apehelion, 28
Apian, Peter, 94
astronomical unit, 10
Aurora Australis, 21, 22
Aurora Borealis, 21, 22
auroral corona, 23

Baade, Walter, 302, 338
Backus, Peter, 191
Banks, John, 11
Barringer, Daniel, 101

Bayer, Johann, 212
Bell Telephone Laboratories, 322
Bell, Jocelyn, 159, 161
Belville, Ruth, 264
Bessel, Friedrich, 121, 151
Bethe, Hans, 146, 321
Bevis, John, 154
Big Bang theory, 294
   closed universes, 337
   flat or critical universe, 337
   flatness problem, 296
   Friedmann models, 338
   Heisenberg's uncertainty principle, 295
   horizon problem, 296
   inflation, 295
   initial singularity, 294
   Omega, 337
   zero energy content, 294
binary pulsar, 233
Birr Castle, 122
black holes, 239
   A0620–00, 246
   angular momentum, 241
   Ariel IV, 245
   Cygnus X-1, 245
   evaporation, 253
   event horizon, 239, 241

   gravitational microlensing, 243
   Hawking radiation, 252
   Hawking temperature, 253
   intermediate mass, 240
   Kerr black holes, 242
   Monoceros X-1, 245
   neutron degeneracy pressure, 240
   no-hair theorem, 241
   Omega Centauri, 240
   Penrose process, 242
   Schwarzschild radius, 240, 241
   singularity, 240
   size of emitting region, 249
   super-massive, 240
   what are they?, 239
   X-ray binary systems, 244
Bode, Johann Elert, 55
Bondi, Herman, 339
Brahe, Tycho, 4, 91, 155
British eclipse expedition, 229
Brown, Mike, 28, 67
Burke, Bernie, 323
Burney, Venetia, 62
Butler, Paul, 169

Campbell, Bruce, 169
Campbell, William, 229

Index

celestial sphere, 3
Cepheid variables, 123
Ceres, 26
Challis, James, 60
Chandrasekhar,
    Subramanyan, 150
Cherenkov radiation, 17, 156
Chrétien, Henri, 123
Christy, James, 64, 89
Clark, Alvan, 121, 151
Clarke, Arthur C., 113
climate change, 32
comets, 91
  1P/Halley, 92
  antitail, 95
  Bayeaux Tapestry, 92
  Borrelly, 95
  coma, 94, 95
  comet of 1577, 91
  Deep Impact Mission, 98
  dust tail, 94
  gas or ion tail, 94
  Giotto image, 95
  Giotto space probe, 95
  great comets, 95
  Hale–Bopp, 91
  Halley's, 92
  Holmes, 95
  icy conglomerate theory, 94
  ion tail, 95
  Kuiper Belt, 93
  long-period, 92
  nuclei, 92, 93
  Oort Cloud, 93
  Shoemaker–Levy 9, 65
  short-period, 93
  Stardust mission, 97
  Tempel 1, 98
  water on Earth, 99
  Wild 2, 97
constant of gravitation, 9
Cook, Captain James, 11
Copernican model of the
    Solar System, 2

coronal mass ejections, 22
Cosmic Microwave
    Background, 252,
    265, 296, 306, 321,
    345
  black body spectrum, 324
  Boomerang balloon, 328,
    330
  cause, 323
  cause of fluctuations, 326
  COBE spacecraft, 324
    CMB fluctuations, 325
    differential microwave
      radiometer, 324
    far-infrared absolute
      spectrophotometer,
      324
    mapping anisotropies,
      324
  Cosmic Background
    Imager, 329
  Maxima balloon, 328
  Planck mission, 329
  Planck results, 329
    age of Universe, 332
    components making up
      Universe, 331
    curvature of space, 329
    Hubble's constant, 333
    reionisation redshift, 332
  Planck spacecraft, 327
  serendipitous discovery,
    322
  Sunyaev–Zel'dovich effect,
    329
  Very Small Array, 329
  WMAP spacecraft, 327
    CMB Power Spectrum
      Analyzer, 333
    results, 328
cosmological constant, 336
cosmology, 336
  beyond our ability to fully
    understand, 344

  Big Bang models, 336
    closed universes, 337
    flat or critical universe,
      337
    Friedmann models, 338
    Omega, 337
  Big Bang or Steady State, 339
  consistent model, 341
  continuous creation, 339
  cosmological principle, 339
  cosmological redshift, 338
  flatness problem, 340
  Freidmann models, 340
  future of the Universe, 345
  horizon problem, 340
  Hubble's constant, 338
  inflation, 340
  multiverse, 343
  perfect cosmological
    principle, 339
  Steady State theory, 336, 339
  string theory, 343
    M-theory, 344
  Universe fit for life, 341
Crab Nebula, 9
Currie, Thayne, 166

d'Arrest, Heinrich, 60
dark energy, 315
  Casimir experiment, 318
  evidence, 316
  High-z Supernova Search
    Team, 317
  Hubble plot, 317
  Lambda (Λ) term, 315
  nature of, 318
  standard model of
    cosmology, 319
  supernova cosmology
    project, 317
  Type Ia supernova, 316
dark matter, 301
  Abell Cluster 2218, 308
  amount, 309

cold dark matter, 309
  Alpha Magnetic
    Spectrometer, 311
  AMANDA, IceCube and
    ANTARES, 311
  axions, 310
  CRESST experiment, 312
  cryogenic detectors, 312
  DAMA experiment, 312
  direct detection, 311
  indirect detection
    experiments, 310
  LUX experiment, 313
  LUX first results, 314
  LUX-ZEPLIN, 314
  neutralino, 310
  PAMELA, 311
  percentage mass, 315
  weakly interacting
    massive particles
    (WIMPs), 310
gas entrapment, 307
gravitational lensing, 307
hot dark matter, 309
  neutrinos, 309
in galaxy M33, 304
in spiral galaxies, 303
MACHO, 301
mass to light ratio, 306
role in galaxy formation,
  306
what is it?, 309
Davis, Ray, 17, 313
de Grasse Tyson, Neil, 27
de Lacaille, Nicolas Louis,
  220, 221
deferents, 1
di Bondone, Giotto, 92
Dicke, Richard, 320, 321, 323
Dixon, Jeremiah, 10
Dolland, John, 120
Double Pulsar, 235
Drake, Frank, 181, 184
Duhalde, Oscar, 220

Dusky Sound, 11
dwarf planets, 28, 68
  Ceres, 69
  Eris, 68
  Haumea, 69
  Makemake, 69
  Pluto, 69

Earth, 29, 42
  Cambrian era, 43
  central core, 44
  Chicxulub crater, 43
  equatorial bulge, 43
  mass extinctions, 43
  photosynthesis, 42
  plate tectonics, 43
  secondary atmosphere, 42
Earth–Moon system, 42
eclipses, 23
  annular, 24
  eclipse track, 24
  lunar, 23
  ring of fire, 24
  saros, 25
  solar, 24
Eddington, Arthur, 228
Einstein rings, 243
Einstein, Albert, 12, 17, 265,
  315
Einstein's General Theory of
  Relativity, 84, 226,
  233
Einstein's Special Theory of
  Relativity, 156, 233
epicycles, 1
Eris, 29, 30
Essen, Louis, 261
European Southern
  Observatory, 130
European Space Agency, 275,
  329
Ewen, Harold, 198
exoplanets, 164
  51 Pegasai, 169

astrometry, 173
COROT, 175
detection in infrared, 164
detection in visible light, 165
Formalhaut b, 166
Gaia spacecraft, 174
Gliese, 174
gravitational microlensing,
  171
HD 209458b, 175
Kepler, 175
Kepler Space Observatory,
  176
OGLE-2005-BLG-390Lb, 172
OGLE-TR-56b, 175
planetary transits, 174
radial velocity method of
  detection, 167
visual detection, 164

Fermi Gamma-ray Space
  Telescope, 311
Flagstaff Observatory, 51
Flamsteed, John, 60
Fowler, William, 148
Friedmann, A. A, 336

galaxies, see also Milky Way
  active galactic nucleus, 224
  Andromeda Galaxy, 204
  Coma Cluster, 206
  Coma Supercluster, 207
  cosmic web, 208
  distances, 207
  elliptical, 203
  groups of, 204
  Hercules Cluster, 206
  Hubble sequence, 203
  Hubble Ultra Deep Field, 208
  irregular, 204
  Large Magellanic Cloud,
    204, 219
    supernova 1987A, 220
    distance, 220

galaxies (cont.)
  Tarantula Nebula, 219
  Local Group, 204
  M33 in Triangulum, 205
  Magellanic Clouds, 217
  Omega Centauri, 212
    intermediate-mass black hole, 213
  realm of, 205
  Sgr A*, 211
  Small Magellanic Cloud, 204, 219
  spiral, 204
  superclusters, 207
  super-massive black holes, 211
  Virgo Cluster, 206
  Virgo Supercluster, 207
  Whirlpool Galaxy, 204
  white nebulae, 203
Galilei, Galileo, 1, 59, 118, 259
Galle, Johann, 60
gamma-ray bursts, 253, 285
  afterglow, 289
  black hole formation, 291
  causes, 287, 290
  Compton Gamma Ray Observatory, 286
  discovery, 285
  discovery of GRB 970508, 287
  energy, 289
  Gamma-Ray Burst Coordinates Network, 288
  GRB 090423, 288
  GRB 970228, 287
  long, 289, 291
  short, 288
  Swift spacecraft, 288
  Vela spacecraft, 285
Gamow, George, 306, 320
General Theory of Relativity, 240, 336

gravitational time dilation, 266
Giotto di Bondone, 92
Global Positioning System network, 233
global warming, 32
Gold, Thomas, 161, 339
Goodricke, John, 270
Gran Sasso National Laboratory, 312
Grand Unified Theory, 155
gravitational lenses, 230
  Double Quasar, 231
gravitational wave detectors, 237
gravitational waves, 226, 235
gravitons, 237
Green, Charles, 11
Guccione, Giuseppe, 180
Guth, Alan, 296, 340

Hale, George Ellery, 124, 269
Hall, George, 111
Halley, Edmond, 10, 91
Harriot, Thomas, 118
Harvard College Observatory, 270
Haumea, 28
Haveaheart pigeon trap, 322
Hawking, Stephen, 252, 254
Hayden Planetarium, 26
Heisenberg's uncertainty principle, 252
Helmholtz, Hermann von, 12
Herman, Robert, 321
Herschel, Caroline, 58
Herschel, John William, 212, 223
Herschel, William, 57, 74, 89, 122, 151
Hewish, Antony, 159
Homestake Mine, 17
Horowitz, Paul, 182, 187

Hoyle, Fred, 147, 160, 320, 336, 339, 343
Hubble age, 274
Hubble sequence, 274
Hubble Space Telescope
  Hubble Heritage Images, 284
  Wide Field Camera 3, 278
Hubble Space Telescope science
  Cepheid distance scale, 279
  dark matter and dark energy, 281
  exoplanet atmospheres, 281
  exoplanets, 281
  eXtreme Deep Field, 283
  galaxy formation and evolution, 282
  gamma ray bursts, 281
  Hubble's constant, 273, 279
  Hubble Deep Field, 282
  Hubble Ultra Deep Field, 283
  Pluto, Nix and Hydra, 280
  Shoemaker–Levy 9, 280
  supernova 1987A, 278
Hubble, Edwin, 121, 123, 203, 266, 267, 337
  an attorney, 268
  at University of Chicago, 267
  cometary nebula, 268
  distance to Andromeda Galaxy, 272
  expanding universe, 273
  galaxy classification, 274
  infantry, 269
  Legion of Merit, 270
  Mount Wilson Observatory, 269
  Rhodes scholarship, 267
  study for a PhD, 268
Hubble's constant, 266, 273
Hubble's law, 272
Hulse, Joseph, 233

Huygens, Christiaan, 83, 85, 260
hydrogen Balmer series, 137

ice dwarfs, 66
  Ixion, 66
  Varuna, 66
impacts, 100
  Asclepius, 107
  asteroid 2004 FH, 107
  asteroid 2008 TC3, 107
  Barringer Crater, 101
  Chelyabinsk meteor, 108
  Chicxulub asteroid, 103
  Chicxulub crater, 103
  Comet Encke, 106
  crater formation, 101
  eyewitness report, 105
  Great Daylight Fireball, 107
  Hubble Space Telescope, 109
  impact craters, 100
  instruments on Galileo spacecraft, 110
  Meteor Crater, 101
  Nördlinger Ries crater, 102
  Shoemaker–Levy 9, 109
  Steinheim crater, 102
  Tunguska event, 104
  Tycho, 101
International Astronomical Union, 27
inverse square law, 8
Island of Hven, 4

Jodrell Bank MERLIN array, 283
Jodrell Bank Observatory, 11, 96, 125, 235
Jupiter
  atmosphere, 81
  clouds, 80
  core, 80
  Galilean moons, 82
    Europa, 83
    Io, 83
  Great Red Spot, 81
  mass, 80
  ring system, 82

Kamioka Underground Observatory, 155
Kepler, Johannes, 5, 119, 155, 176
Kepler's third law of planetary motion, 6
Klebasabel, Ray, 285
Kuiper Belt objects, 66, 281
Kulik, Leonid, 106

L2 Lagrangian point, 327
Laplace, Pierre-Simon, 239
Large Hadron Collider, 237, 253
Las Campanas Observatory, 220
Laser Interferometer Gravitational Wave Observatory, 236
Lassell, William, 61
laws of planetary motion, 6
Le Verrier, Urbain, 60, 77
Leavitt, Henrietta, 270
Lemonnier, Pierre, 60
Levy, David, 109
Lindblad, B., 201
Lippershey, Hans, 118
Lowell Observatory, 121, 271
Lowell, Percival, 51, 57, 62, 121

Magellan spacecraft, 30
Magellan, Ferdinand, 219
Makemake, 28
Marcy, Geoffrey, 169, 187
Mars, 29, 49
  atmosphere, 49
  axial tilt, 49
  Beagle II lander, 51
  canali, 50
  closest approaches, 29
  Curiosity rover, 53
  Gusev Crater, 52
  Mariner 9, 53
  Mariner spacecraft, 51
  Mars Global Surveyor, 51
  Mars Reconnaissance Orbiter, 52
  Martian soil analysis, 54
  Meridiani Planum, 52
  Olympus Mons, 51
  permafrost, 53
  Phobos and Deimos, 54
  polar ice caps, 49
  south pole, 49
  Spirit and Opportunity, 52
  surface temperature, 49
  Viking landers, 51
Mason, Charles, 10
Mather, John, 326
Mauna Kea, Hawaii, 129
Maxwell, James Clerk, 85
Mayor, Michel, 169
McKellar, A., 321
McMillan, Robert, 66
Méchain, Pierre, 10
Mercury, 29, 30, 31, 38
  BepiColombo, 39
  ice near poles, 38
  Magnetospheric Orbiter, 39
  Mariner 10, 38
  observed by radar, 38
  Planetary Orbiter, 39
  precession, 227
meridian, 4
MERLIN array, 162
Messier, Charles, 9, 205
meteor showers, 95
  bolide, 96
  Leonids, 97
  meteor storm, 97
  Orionids, 96
  Perseids, 97
  radiant, 96

352 Index

Milgrom, Mordehai, 304
Milky Way, 192
  21-cm line, 198
  central bar, 211
  Cepheid variables, 197
  constellation Carina, 215
    Eta Carinae, 215
    Eta Carinae Nebula, 215
    Southern Pleiades, 215
  constellation Centaurus, 211
    Centaurus A, 223
  constellation Crux, 214
    Alpha Crucis, 214
    Coal Sack, 215
    Jewel Box, 214
  constellation Tucanae
    47 Tucanae, 221
      millisecond pulsars, 222
      search for planets, 221
  constellation Vela, 217
    Vela pulsar, 217
    Vela supernova remnant, 217
  dark nebula, 195
  disc, 193
  Eagle Nebula, 195
  Galactic Centre, 192, 210
  galactic rotation curve, 198
  globular clusters, 193
    47 Tucanae, 194
    M13 in Hercules, 194
  Horsehead Nebula, 195
  hydrogen line, 198
  hydrogen line profiles, 200
  interstellar medium, 195
  neutral hydrogen (H I), 198
  obscuration by dust, 192
  Omega Centauri, 194
  open clusters, 192
    Hyades and Pleiades clusters, 193
    Perseus Double Cluster, 193
  Orion Nebula, 195
  Orion Spur, 202
  Perseus Arm, 201
  RR Lyrae stars, 196
  Scutum-Centaurus Arm, 201
  Sgr A*, 202
  size, shape and structure, 196
  spiral arms, 198
  spiral structure, 201
  spiral structure puzzle, 200
  super-massive black hole, 202
  visible constituent, 192
Millikan, Robert, 267
minor planets, 55
  Celestial Police, 55
  Ceres, 55
  main asteroid belt, 55
  Titus–Bode law, 55
modified Newtonian dynamics, 304
Moon
  craters
    Copernicus, 47
    Tycho, 47
  diameter, 44
  distance, 44
  gravitational pull, 44
  highlands, 45
  libration, 44
  exploration, 48
    Apollo missions, 48
    Jade Rabbit, 49
    Luna 16, 20, and 24, 48
    Lunar Orbiters, 48
  maria, 45
    Mare Crisium, 46
    Oceanus Procellarum, 45
  Moon illusion, 44
  orbit, 47
  regolith, 47
  tidal force, 47
Moore Hall, Chester, 120

Morrison, Philip, 180
Mount Wilson Observatory, 123, 269

near-Earth objects, 112
  1036 Ganymed, 112
  433 Eros, 112
  asteroids, 112
  Don Quijote, 116
  extinct comets, 112
  Large Synoptic Survey Telescope, 113
  LINEAR programme, 112
  NEAT, 113
  Palermo scale, 114
  Pan-STARRS, 114
  Spaceguard, 113
  Spacewatch, 113
  The Discovery Channel Telescope, 114
  Torino scale, 114
Near-Earth Objects, 66
Neptune, 29, 77
  atmosphere, 77
  cloud-top temperature, 77
  diameter, 77
  discovery, 77
  Great Black Spot, 77
  ice giant, 77
  mantle, 77
  mass, 77
  ring system, 78
  rotation period, 77
  seen by Galileo, 77
  Triton, 78
neutron-degenerate matter, 241
Newton, Isaac, 7, 91, 122, 143
Newton's Law of Gravitation, 226

occultations of Pluto, 64
Öpik–Oort cloud, 93
orbital inclination, 30

Palitzsch, Johann Georg, 91
Palomar Observatory, 124
Palomar Sky Survey, 126
Paranal Observatory, 130, 211
Paris Observatory, 60
Parkes Telescope in Australia, 235
parsec, 134
Parsons, William, 122
Peach, Damian, 80
Penfield, Glen, 103
Penrose, Roger, 242
Penzias, Arno, 296, 322
perihelion, 28
Perlmutter, Saul, 318
Petersen, Dan, 111
Phobos, 30
Piazzi, Giuseppe, 55
Pickersgill Harbour, 11
Pigafetta, Antonio, 219
Planet X, hunt for, 57
　discovery of Uranus, 58
　Galileo recorded Neptune, 59
　Uranus, 57
planetary atmospheres, 33
　Eris, 35
　evolution of Earth's, 36
　　subduction, 36
　　volcanic activity, 36
　Jupiter, 35
　law of equipartition of energy, 34
　Mars, 36
　Maxwell–Boltzmann distribution, 34
　Mercury, 35
　methane, 33
　Moon, 35
　nitrogen, 34
　nitrous oxide, 33
　oxygen, 34
　Pluto, 35

　secondary atmospheres, 35
　solar nebula, 33
　Titan, 35
　Triton, 35
　Venus, 36
　volcanic eruptions, 35
planetary densities, 30
planetary rotation periods, 31
planetary temperatures, 31
　carbon dioxide, 32
　greenhouse effect, 32
　methane, 32
　water vapour, 32
Pluto, 26, 29, 30, 31
　Charon, 64, 89
　Kerberos and Styx, 65
　New Horizons, 65
　Nix and Hydra, 65, 89
Pounds, Ken, 245
preons, 241
primeval fireball, 322
Princeton University, 323
proton–proton cycle, 13
Ptolemaic model of the Solar System, 1
Purcell, Edward, 198

quantum gravity, 240
quantum mechanics, 237
quantum theory, 240
quantum tunnelling, 13
Quaoar, 28
quark-degenerate matter, 241
quarks, 241
quasar, 247
　3C 273, 247
quasi-stellar objects, 125, 247
Queloz, Didier, 169

radio galaxies, 246
Rees, Martin, 342, 344
relict radiation, 321
Riess, Adam G., 318
Ritchey, George Willis, 123

rocky planets, 38
Rømer, Ole Christensen, 82
Roll, Peter, 321
Rosse, Third Earl of, 10, 122, 154, 204
Rubin, Vera, 303
Ryle, Martin, 340

Sagan, Carl, 78, 80
Samuel Oschin Telescope, 66
Sanford Underground Research Facility, 313
Saturn, 30
　moons
　　Enceladus, 89
　　Huygens Probe, 87
　　lakes on Titan, 88
　　Mimas, 87
　　Titan, 85, 87
　rings, 85
　　A Ring, 86
　　Bright Ring, 86
　　Cassini Division, 85
　　Crepe Ring, 86
　　Encke Division, 86
　　F Ring, 86
　　Roche limit, 86
Schiaparelli, Giovanni, 50
Schmidt, Bernhard, 125, 302
Schmidt, Brian P., 318
Schmidt, Maarten, 125, 247
Scotti, James V., 109
search for life beyond the Earth, 179
search for other worlds, 164
Sedna, 28
SETI, the Search for Extra-Terrestrial Intelligence, 180
　300-metre Arecibo dish, 72
　76-metre Lovell Radio Telescope, 72, 183
　Allen Telescope Array, 189

SETI (cont.)
  Arecibo dish, 182
  Drake equation, 184, 186
  evidence of other
      civilisations, 188
  Jodrell Bank, 183
  Mark I radio telescope, 180
  optical SETI, 186
  Project Ozma, 181
  Project Phoenix, 72, 182,
      183, 186
  Project SERENDIP, 182
  projects META and BETA,
      182
  SETI Institute, 183
  SETI@home, 183
  Square Kilometre Array,
      191
  Wow!, 181
Shapiro delay, 85, 232, 235
  Cassini spacecraft, 232
  Viking spacecraft, 232
Shapiro, Irwin A., 232
Shapley, Harlow, 197, 210,
    271
Shelton, Ian, 220
Shoemaker, Carolyn, 109
Shoemaker, Eugene M., 100,
    108
Slipher, Vesto, 62, 121, 271
Smoot, George, 326
solar maximum, 22
Space Telescope Science
    Institute, 275
Special Theory of Relativity,
    13, 226, 265
  time dilation, 265
Spitzer Space Telescope, 201
Spitzer, Lyman, 274
St John's College, Cambridge,
    60
stars, 133
  0.5 to ~8 solar masses, 146
  aging, 145

Alpha Centauri AB, 211
apparent brightness, 133
black dwarf, 146, 152
black holes, 162
carbon–nitrogen–oxygen
    cycle, 146
Cepheid variables, 270, 279
  luminosity, 270
  period–luminosity
      relation, 270
  Small Magellanic Cloud,
      270
  standard candle, 271
Chandrasekhar limit, 150
classification, 137
colour, 136
Crab Nebula, 154, 158, 162
Delta Cepheus, 270
discovery of pulsars, 159
distances, 133
  GAIA, 135
  Hipparcos, 135
Eta Carinae, 216, 292
  hypernova, 217
Hertzsprung–Russell
    diagram, 139
high-mass, 152
intrinsic brightness, 133
lifetimes, 143
low-mass, 145
luminosity, 133
masses, 142
neutron stars, 157, 235
optical interferometry, 141
planetary nebula, 149
proper motion, 135
Proxima Centauri, 212
Sirius B, 151
size, 141
  direct measurement, 141
spectra, 137
spectroscopic parallax, 138
stellar parallax, 133
supernova 1987A, 155

surface temperature, 136
triple alpha process (3α),
    147
Type Ia supernova, 280
Type II supernova, 154
variable, 149
white dwarfs, 145, 149
Wolf–Rayet stars, 291
Stjerneborg, 4
string theory, 237
Sun, 12
  atmosphere
    chromosphere, 18
    photosphere, 18
    solar corona, 18
  convective zone, 15
  magnetic field, 19
  Maunder minimum, 20
  prominences, 21
  proton–proton cycle, 14, 16
  radiative zone, 15
  rotation period, 19
  solar neutrino problem, 16,
      17
  solar wind, 21
  sunspot cycle, 20
  sunspot pairs, 20
  sunspots, 19, 20
    maximum, 20
    number, 20
    penumbra, 19
    plage, 19, 20
    umbra, 19
super-massive black holes,
    231, 246
  galaxy M84, 249
  mass, 249
  Milky Way Galaxy, 250
  Sgr A*, 246

Tautenburg Observatory, 126
Taylor, Russell, 233
telescopes
  18.5-inch refractor, 121

Index

achromatic doublet, 120
active optics, 127
adaptive optics, 127
Airy pattern, 120
Alfred-Jensch-Telescope, 126
Cassegrain, 123
eight-power, 118
Faulkes Telescopes, 131
Gemini North, 129
Gemini South, 129
Hale Telescope, 200-inch, 124
Hooker Telescope, 100-inch, 123, 269
Hubble Space Telescope, 121, 275
 COSTAR, 277
 flawed mirror, 276
 guidance system, 275
 servicing missions, 275, 277
James Webb Space Telescope, 277
Keck Telescopes, 10-metre, 129
Large Binocular Telescope, 129
Leviathan 72-inch reflecting telescope, 122
Multiple Mirror Telescope, 129
Newtonian reflecting, 122
optical interferometers, 130
Palomar 200-inch, 274
Ritchey–Chrétien, 123
robotic, 131
Samuel Oschin Schmidt Telescope, 126
Schmidt camera, 125
segmented mirrors, 129
simple, 120
Spitzer Space Telescope, 274
spun cast mirrors, 128
Very Large Telescope, 130
Yerkes Observatory Telescope, 120
time, 255
 atomic time, 257
  caesium beam frequency standards, 257
 clocks
  atomic, 261
   caesium fountain, 262
   caesium beam, 261
   hydrogen maser, 261
   NIST-F1, 262
   rubidium, 262
  Helio-chronometer, 258
  invar pendulum, 260
  long case or grandfather, 260
  pendulum, 259
  quartz, 260
  quartz time standards, 261
  Riefler pendulum, 260
  Shortt free pendulum, 260
  sundials, 258
  water clocks, 258
 cosmic time, 265
 Equation of Time, 256
 Global Positioning System, 257
 Greenwich Mean Time, 255
 leap second, 257
 Local Solar Time, 255
 marine chronometers, 263
 pulsars, 264
 sidereal day, 257
 sidereal time, 257
 time transfer, 263
  Belville family, 264
  Greenwich Time Lady, 264
  radio-controlled clocks, 264
  time ball, 263
 Universal Time, 256
time line of the early Universe, 297
 a note of caution, 300
 dark matter begins to clump, 299
 era of nucleosynthesis, 298
 grand unification epoch, 297
 hadron epoch, 298
 inflationary period, 297
 lepton epoch, 298
 photon epoch, 299
 Planck epoch, 297
 quark epoch, 298
Titius, Johann Daniel, 55
Tombaugh, Clyde, 62
Townes, Charles, 186
Townsend, Paul, 344
trans-Neptunian objects, 66
 K31021A and K31021B, 67
 K31021C, 67
 Orcus, 67
 Sedna, 67
 Xena, 68
Turner, Herbert Hall, 267

Universal Law of Gravitation, 9
Uraniborg, 4
Uranometria, 212
Uranus, 31, 75
 axial tilt, 75
 distance from the Sun, 75
 mass, 75
 moon Miranda, 76
 radiation belts, 76
 rings, 75, 76
 rotational period, 75

Venus, 29, 31, 40
 Aphrodite Terra, 41
 atmosphere, 42
 farra, 42

Venus (cont.)
  high albedo, 40
  internal structure, 42
  Ishtar Terra, 41
  Magellan spacecraft, 41
  Mariner 2 spacecraft, 40
  radar observations, 41
  temperature, 40
  transits, 40
  Veneras 7, 9 and 10, 40
violent universe, 285
VLBA array, 162
von Weizsäcker, Carl, 146
voyages to the outer planets, 70
  asteroid belt, 70
  Cassini–Huygens spacecraft, 84
  Galileo spacecraft, 79
  gravitational 'slingshot', 73
  gravity assist, 73
  heliopause, 73
  New Horizons spacecraft, 281
  New Horizons mission, 89
  Pioneer 10, 70
  Pioneer 10 plaque, 71
  Pioneer 11, 72
  The Grand Tour, 73
  Voyager 2, 74
    Neptune, 77
    Uranus, 75
  Voyagers 1 and 2, 73
    message, 78

Wales, William, 11
Walker, G. A. H., 169
Werthimer, Dan, 187
Wesley, Anthony, 111
Wheeler, John, 239
Whipple, Fred L., 94
Whirlpool Galaxy, 122
Wilkinson, David, 321, 327
Williams, Robert, 282
Wilson, Robert, 296, 322
Witten, Ed, 344
Woolsthorpe Manor, 7

Yang, S., 169
Yerkes, Charles Tyson, 268

Zwicky, Fritz, 126, 302
  Coma Cluster, 302
  gravitational lenses, 302
  Schmidt camera, 302
  supernova, 302
  virial theorem, 303